LONG-WAVE RUNUP MODELS

LONG-WAVE RUNUP MODELS

Friday Harbor, USA   12 - 17 September 1995

Editors

Harry Yeh
University of Washington

Philip Liu
Cornell University

Costas Synolakis
University of Southern California

World Scientific
Singapore • New Jersey • London • Hong Kong

Published by

World Scientific Publishing Co. Pte. Ltd.

P O Box 128, Farrer Road, Singapore 912805

USA office: Suite 1B, 1060 Main Street, River Edge, NJ 07661

UK office: 57 Shelton Street, Covent Garden, London WC2H 9HE

ISBN 981-02-2909-7

Printed in Singapore.

Preface

The Second International Workshop on Long-Wave Runup Models was held at Friday Harbor, San Juan Island, Washington, USA, on September 12–16, 1995. (For the record, the first workshop was held at Two Harbors, Catalina Island, California, USA, in 1990 and the summary of the first workshop was reported by Liu, *et al.*[1]) The objectives of the workshop were to review the research progress in tsunami runup and to identify future research directions in tsunami hazard mitigation. Additional objectives were comparisons of numerical, analytical, and physical prediction models with existing laboratory and field data. Just as the organizers of the other two tsunami related workshops held in 1995, i.e., the Tsunami-Sediment Deposit Workshop (organized by J. Bourgeois and B. Atwater), and the Tsunami Measurement Workshop (organized by J. Lander and H. Yeh), we were motivated by the recent six tsunami events, i.e., the 1992 Nicaragua, the 1992 Flores, Indonesia, the 1993 Okushiri, Japan, the 1994 East Java, Indonesia, the 1994 Shikotan, Russia, the 1994 Mindoro, Philippines, and the 1994 Skagway, Alaska tsunamis.

A total of 55 scientists and students participated in the workshop: one from Australia, two from Brazil, one from Canada, two from England, one from Indonesia, eight from Japan, one from Russia, one from Thailand, and thirty-eight from the United States. Student participation was encouraged, and 21 of the 55 participants were students.

In order to foster close interactions among a wide range of experts, a secluded workshop location was chosen: Friday Harbor is located approximately 120 km north of Seattle, Washington. Unlike the 1990 Catalina workshop, however, the format of this workshop was specifically designed to focus more on discussions than on formal presentations. To accomplish this goal, four benchmark problems were selected before the workshop so that predictive models could be compared, both qualitatively and quantitatively evaluated and discussed among the participants during the workshop. The workshops organizers provided interested participants with initial and topographic data for the four propagation problems and requested predictions at specific locations in a specified format. The actual laboratory or physical measurements were only presented during the workshop in the same format, allowing comparisons of predictions with measurements. The four benchmark problems were the following:

1) the prediction of an edge-wave packet propagation along a uniformly sloping beach,
2) the interaction and runup of incident solitary waves with a conical island,
3) the runup of solitary waves on a vertical wall, and
4) the tsunami runup around Okushiri Island, Japan.

The first problem models a tsunami propagating along a shore from its nearshore generation region, e.g. a scenario for landslide generated tsunamis. The second problem was

---

[1] Liu, P. L.-F., Synolakis, C. E. & Yeh, H. H. 1991, Report on the international workshop on long-wave run-up. *J. Fluid Mech.*, **229**, 675–688.

motivated by the tsunami disaster on Babi Island during the 1992 Flores tsunamis.[2] These first two problems were intended to be used for validating three-dimensional models and/or depth-integrated two-dimensional models. The third problem was for two-dimensional models in a vertical plane and/or one-dimensional models. Detailed laboratory data for these problems were generated by the workshop organizers, but only limited and necessary information were given to the participants prior to the workshop. As we will discuss below, this information provided was only sufficient for initializing calculations.

The fourth problem was motivated by the challenge of hindcasting a real tsunami catastrophe, and it was selected because of the availability of extensive field measurements on Okushiri Island and of detailed high resolution bathymetry and topography data. Also, this tsunami produced extreme runup of 30 m and massive overland flow velocities of 10–18 m/sec, and we felt that the simulation of such an extreme event is important to test the ability of current state-of-the-art models.

Nine months prior to the workshop, the bathymetry and topography data, and either the wave-generator motion or the assumed initial water-surface displacement of the benchmark-problem were given to the participants. By the fiftieth day prior to the workshop, each participant was asked to submit his/her written discussion of the solution of one or more problems. Thirty days prior to the workshop, the written discussions were distributed to everyone, allowing more focused discussions during the workshop. During the workshop, the detailed laboratory data were presented in the same format, allowing comparisons of predictions with the data for the first time. In addition to the discussions on the benchmark problems and the models, and several review sessions, special evening sessions were held to discuss field observations of recent tsunamis: Nicaragua, Flores, Okushiri, East Java, Shitokan, Mindoro, and Skagway. During the workshop, seven discussion themes were organized as follows, laboratory, analytical, finite-difference, finite-element, vertical-plane models, boundary-integral-element models, and marker-and-cell models.

All the articles presented in this proceedings were revised based on the in-depth discussions made during the workshop. There are six review articles and sixteen discussion papers collected in this proceedings. Detailed benchmark problem descriptions and data are presented in Appendices

All local arrangements were made by Ms. Elaine Donovan, Julie Golding and many others in the Department of Civil Engineering at the University of Washington. The idea to use benchmark problems for the workshop discussions was originally made by Dr. Clifford Astill. The workshop was sponsored by the US National Science Foundation.

Harry Yeh
Philip Liu
Costas Synolakis

---

[2]   Yeh, H., Liu, P., Briggs, M., & Synolakis, C. 1994. Propagation and Amplification of Tsunamis at Coastal Boundaries. *Nature* **372**, 353–355.

Contents

Contributors and Participants

Arnason, Halldor

University of Washington, Department of Civil Engineering, Seattle, WA 98195-2700

Astill, Clifford J.

National Science Foundation; 4201 Wilson Blvd., Room 545, Arlington, VA 22230

Atwater, Brian

U.S. Geological Survey, University of Washington, Seattle, WA 98195-1310

Baptista, Antonio M.

Oregon Graduate Institute of Science & Technology, P.O. Box 91000, Portland, OR 97291-1000

Barnes, Tim

University of Bristol, School of Mathematics, University Walk, Bristol, BS8 1TW, England

Beck, Bradley C.

Oregon Graduate Institute of Science & Technology, P.O. Box 91000, Portland, OR 97291-1000

Bourgeois, Jody

University of Washington, Geological Sciences, Seattle, WA 98195-1310

Briggs, Mike

Waterways Experiment Station, CERC, 3909 Halls Ferry Road, Vicksburg, MS 39180-619

Carrier, George

Harvard University, 311 Pierce Hall, Cambridge, MA 02138

Chacaltana, Julio Tomas

Institute of Material Science, Laboratory of Mechanical Fluids PEM, Coppe UFRJ Caica Postal 68503, Rio de Janeiro, Brazil

Chen, Yongze

Scripps Institute for Oceanography, 9500 Gilman Drive, La Jolla, CA 92093-0209

DaSilva, Fernando Teles

Institute of Material Science, Laboratory of Mechanical Fluids PEM, Coppe UFRJ Caica Postal 68503, Rio de Janeiro, Brazil

Fadda, Dani

Southern Methodist University, Mechanical Engineering Department, Dallas, TX 75275-0337

Fujima, Koji

National Defense Academy, Department of Civil Engineering, 1-10-20 Hashirimizu, Yokosuka, Kanagawa-ken 239, Japan

Gardarsson, Sigurdur

University of Washington, Department of Civil Engineering, Seattle, WA 98195-2700

Gelfenbaum, Guy

USGS Center for Coastal Geology, 600 4th Street South, St. Petersburg, FL 33701

Grilli, Stephan

University of Rhode Island, Department of Ocean Engineering, Narragansett, RI 02882-1197

Hoeg, Hans

CalTech, Keck Laboratory of Hydraulic and Water Resources, Pasadena, CA 91125

Lee, HoJun

Disaster Control Research Center, Tohoku University, Sendai 980-77, Japan

Imamura, Fumihiko

Disaster Control Research Center, Tohoku University, Sendai 980-77, Japan

Jaffe, Bruce

US Geological Survey, 345 Middlefield Rd., Menlo Park, CA 94025

Kajardi, Entin

University of Delaware, Ocean Engineering Laboratory, Newark, DE 19716

Kanoglu, Utku

1194 W. 30th Street, Apt 5, Los Angeles, CA 90007

Kennedy, Andrew

Monash University, Department of Mechanical Engineering, Clayton, Victoria, Australia 3168

Kobayashi, Nobu

University of Delaware, Ocean Engineering Laboratory, Newark, DE 19716

Kondo, Hirokazu

Chuo University, Department of Civil Engineering, Kasuga 1-13-27 Bunkkyo-ku, Tokyo 112, Japan

Liu, Philip L.-F.

Cornell University, School of Civil & Environmental Engineering, Ithaca, NY 14853-3502

Mazova, Raissa

N. Novgorod Technical University, 24 Minin Str., Nizhny Novgorod 603600, Russia

Meyer, Richard

University of Wisconsin, Center for the Mathematical Sciences, 1308 West Dayton Street, Madison, WI 53715-1149

Moore, Andrew

University of Washington, Department of Civil Engineering, Seattle, WA 98195-2700

Myers, EdOregon

Graduate Institute of Science & Technology, P.O. Box 91000, Portland, OR 97291-1000

Noguchi, Kenji

Public Works Research Institute, Ministry of Construction, 1 Asahi, Tsukuba City, Ibaraki pref. 305, Japan

Opishinski, Thomas

University of Rhode Island, Department of Ocean Engineering, Narragansett, RI 02882-1197

Peregrine, D. Howell

University of Bristol, School of Mathematics, University Walk, Bristol, BS8 1TW, England

Petroff, Catherine

University of Washington, Department of Civil Engineering, Seattle, WA 98195-2700

Pourtaheri, Hassan

US Army Corps of Engineers, P.O. Box 60267, New Orleans, LA70160

Prasetya, G.S.

Coastal Engineering Laboratory, LPTP BPP Teknologi Gedung II, LT.22. Jl. M. H. Thamarin No. 8, Jakarta Pusat 10340, Indonesia

Raad, Peter E.

Southern Methodist University, Mechanical Engineering Department, Dallas, TX 75275-0337

Rabinovich, Alexander

Institute of Ocean Sciences, P.O. Box 6000, Sidney BC V8L 4B2, Canada

Raichlen, Fred

CalTech, Keck Laboratory of Hydraulic and Water Resources, Pasadena, CA 91125

Sabau, Adrian

Southern Methodist University, Mechanical Engineering Department, Dallas, TX 75275-0337

Satake, Kenji

Seismotectonics Section, Geological Survey of Japan, Tsukuba 305, Japan

Synolakis, Costas   University of Southern California, Civil Engineering
                             Department, Los Angeles, CA 90089-2531

Tadepalli, Srinivas   44C Escondido Village, Stanford, CA 94305

Takagi, Toshimitsu   Chuo University, Department of Civil Engineering,
                             Kasuga 1-13-27 Bunkkyo-ku, Tokyo 112, Japan

Takahashi, Tomoyuki   Disaster Control Research Center, Tohoku University,
                             Sendai 980-77, Japan

Tega, Yukiko   University of Delaware, Ocean Engineering Laboratory,
                             Newark, DE 19716

Titov, Vasily   University of Southern California, Civil Engineering Dept.,
                             Los Angeles, CA 90089

Troshina, Elena   Institute of Marine Science, University of Alaska,
                             Fairbanks, AK 99775-1080

Walters, Roy A.   U.S. Geological Survey, MS 412, Denver Federal Center,
                             Bldg. 53, Rm H2716, Box 25046, Lakewood,
                             CO 80225-0046

Wakahara, Toshihiro   SHIMIZU Corp., 4-17, Etchujima 3-chome, Koto-ku,
                             Tokyo 135, Japan

Watson, Gary   United States Geological Survey, 1201 Pacific Avenue,
                             Suite 600, Tacoma, WA 98402

Watts, Phillip   CalTech, Keck Laboratory of Hydraulic and Water
                             Resources, Pasadena, CA 91125

Yeh, Harry   University of Washington, Department of Civil
                             Engineering, Seattle, WA 98195-2700

The participants of the workshop.

Review Articles

On the Interaction of Water Waves with Islands

R. E. Meyer

Abstract. The nature of surface waves about islands and submerged banks of geophysically realistic seabed topography is described. For reasons, and with qualifications, discussed in Section I, this is done on the basis of linear theory for ideal water. Attention is focused mainly on plain waves incident from the distant ocean upon axisymmetric islands, because simple analytical answers available for them illuminate also the more general case. Particular consideration is given to the monochromatically periodic case (Sections II-V) and to the cases of incident waves which are a single 'solitary' wave (Section VII) or a semi-infinite periodic wave train (Section VIII).

I Introduction

The historical [1] and contemporary observations of damage caused by tsunamis indicate very strongly that it must be difficult to account for them physically without strongly three-dimensional water motion and possibly, also some resonance effects. This evidence, and ship hydrodynamics, have prompted attempts of mathematical oceanographers to face up to the daunting theoretical difficulties of three-dimensional waves with free surface, which possess mechanisms differing strikingly from any familiar ones. Naturally, they started with the linearized form of the 'Classical Equations' of 'Ideal Water' meaning (here and in all that follows) those governing the irrotational motion of inviscid, incompressible fluid without surface tension or interaction with bottom sediment and without drop formation or air bubbles, etc.

The Classical Equations are nonlinear and electronics has made them accessible to supercomputing with quantitative accuracy beyond the scope of linear theory. The degree of nonlinearity accessible quantitatively is limited physically, however, by the restriction to Ideal Water: dissipation, wave breaking and many other affects are not covered. In the realm of 'Moderate Nonlinearity' actually opened up, linear theory turned out to have some unexpected successes, e.g., nonlinear amplification by shoaling and the resulting runup on beaches are dominated by the old Green's Law [2]-[3] predicted already by linear theory [4]. All the same, direct numerical work on the Classical Equations has led to gratifying improvement of quantitative agreement between theory and experiment in wave tanks. Oceanographic reality, however, turns out to cause further disappointments.

The main reason is that the present shape of the seabed almost everywhere has been formed by sedimentation [5], which implies slopes ε of the order of 10^{-3} (10^{-4} on continental shelves and 10^{-2} only exceptionally); in the laboratory, by contrast, slopes much smaller than 10^{-1} tend to be impractical. As a result, oceanographical wave-travel distances exceed those in a laboratory by orders of magnitude and so, therefore, do the cumulative effects of dissipation, which may be unimportant in even large wave tanks [6], but are well documented [7]-[11] to play a major quantitative role over geophysically realistic seabed slopes. In sum, Ideal-Water can simulate waves of Moderate Nonlinearity in the laboratory with good accuracy, but cannot be of similar quantitative help for such waves in the field, on which the following will focus.

While the Classical Equations are quantitatively insufficient, they are nevertheless very necessary for the elucidation of the basic physical mechanisms of surface waves, and while the lessons they teach us have only qualitative significance in geophysics, those lessons are an absolute prerequisite for progress towards quantitative realism. In turn, the predictions of those Equations for Moderate Nonlinearity differ mainly in quantitative respects from the predictions of the linear approximation to the Classical Equations. For the qualitative lessons of primary relevance to realistic oceanography, Moderate Nonlinearity offers only a small advantage over the linear theory and, in some respects [3], [12], [13], this advantage can be gained by a mere re- interpretation of the 'linear' predictions. Linear theory, on the other hand, offers a major advantage for the elucidation of basic wave mechanisms in three dimensions by a coverage of their dependence on the physical parameters, which is very expensive to obtain by direct numerical approaches. For instance, linear analysis will be seen in Sections IV to VIII to reveal a severe frequency-sensitivity of a basic mechanism which defeats even a highly sophisticated, direct numerical approach [14] not taking full account of the best analytical predictions (Section V).

Moreover, linear theory shows a way to take advantage of oceanographic realism by use of the exact dispersion relation to replace, at no expense in analytical or computational difficulty or labor, the common recourse to shallow-water equations. This is particularly relevant in the present context because the waves excited near islands turn out (Section IV) not to be longwaves, regardless of how long the waves incident from the distant ocean may be. A further benefit of the linear theory are the indications it gives of economical and efficient approaches to computation by clarifying the relative importance of different objectives and the relative value of accuracy in the successive steps on the shortest way towards them (Section VI). In short, while the Classical Equations of Ideal Water cannot give quantitatively realistic answers in the geophysical context, the following will concentrate on their linear theory on account of the quite remarkable qualitative insights which it gives into the unexpected character of the physical mechanisms underlying three-dimensional water waves over natural topographies.

The analytical shortcut through the linear theory of waves is via their spectrum. It is not proposed here to explain the mathematical analysis, to which the bibliography below offers ample access, but to focus attention on the utterly unexpected physical mechanisms of three-dimensional water waves which it has illuminated. The most efficient approach will be to concentrate first on the nature of the wave patterns (Sections II-IV) and on simple, oceanographically realistic algorithms for the spectra (Section V), before applying the relevant lessons from them to elucidate the response to single waves (Section VII) and wave trains (Section VIII) from the distant ocean.

Interest in the spectra was also kindled historically by the intriguing question of surface-wave resonance. The possibility of resonance is obvious in basins, lakes and wave tanks, where the waves must travel to and fro. That is impossible in an open ocean, where the waves must come from infinity and be scattered away to infinity; such waves are not resonant, whence resonance on unbounded domains is impossible. The fallacy of this long-known mathematical truth was discovered by Ursell [15] and, to give an experimentally accessible example, he pioneered [16] the theory of edgewaves; surprising cousins of them will be relevant.

It is now known, even from the fully nonlinear theory of the Classical Equations on beaches [17], that water-wave resonance on an unbounded domain is possible only when the motion is three- dimensional. The pioneering analyses were made simultaneously by Longuet-Higgins [18] and Shen [19]. They used different approaches, and the next Section outlines Shen's, which proved in the end to be more realistic.

II   Semiclassical Mechanics

J. B. Keller suggested [19] that the small slope of realistic seabed topography offers a small parameter ε which makes surface waves accessible by his method of 'geometrical optics', later published in [20]. For an outline of the method in relation to water waves, the review articles [21], [22] may be helpful. Keller's method has three notable advantages. The first is that it can be phrased, and carried all through, for arbitrary seabed topography. It is also irrelevant whether the waves are long or short or in between; that helps to understand the violent resonance of storm waves around Browns Bank which was captured on film. The method applies equally to lakes and wave tanks and to islands and submerged reefs and banks in the open ocean.

The second advantage is that Keller's method is based conceptually on ray-tracing, although that plays a minor quantitative role. One aims only to understand the most important structural properties of the wave patterns. Naturally, this simplifies further for one-dimensional topographies, where it extends edgewave theory, and for axisymmetric topographies on which the following focuses primarily because they offer the simplest case revealing clearly the surprising and quite fundamental differences distinguishing the three-dimensional mechanism of water waves from the two-dimensional mechanisms which have preoccupied fluid dynamics for two centuries. For axisymmetric topography, ray-tracing is easily disposed of analytically, but the ray concept reveals readily which properties found for axisymmetry are common to all shapes of islands or submerged banks. The main question addressed in this way is whether all waves are scattered off to infinity or whether it is possible that patterns exist which are "trapped" in the sense that their waves travel only over a finite area of the unbounded ocean surface?

For axisymmetric topography it follows readily [19] from the dispersion relation for surface waves that such trapped patterns always exist for islands and also for many submerged reefs, and that they occupy ring-domains around the island shore or reef crown;

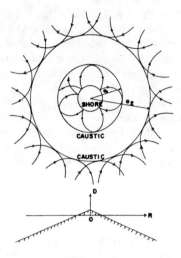

Figure 1

and this generalizes to arbitrary island and reef shapes. Fig. 1 (from [19]) shows such a ring pattern schematically; the curves (other than the 'caustic' circles bounding the rings) represent rays qualitatively. Fig. 2 (from [18]) shows such a pattern over a submerged cylinder; this figure represents the crests of a "normal" or "free" wave mode.

Since a trapped pattern contains only a finite kinetic energy, it raises the possibility of resonance, but while such trapping is a necessary condition, it is not sufficient: there must also be enough phase coherence, which appears in semiclassical mechanics [20] as a uniqueness requirement in phase space. One such condition is obvious in axisymmetry where natural wave modes must depend on the angular coordinate θ and frequency ω according to $\exp i(n\theta - \omega t)$ with integer n; here $n > 0$ for waves traveling anticlockwise around the ring, if $\omega > 0$, and $n < 0$ for their mirror images traveling clockwise. The uniqueness requirement in the radial direction is more sophisticated [19]. The immediate result of semiclassical Mechanics that trapped wave patterns always come in such clockwise-anticlockwise pairs is of practical importance: The waves interfere additively (on linear theory) and where the respective crests of a pair collide, the local doubling of crest height may threaten damage. Plausibly, such pair collisions may occur first on the 'lee-side' of an island or bank, i.e., the side opposite to the one facing the incident waves, and it will be seen (Sections VII, VIII) that the maximal crest doubling always occurs at the very lee. This is, in fact, the process by which trapped waves forced abandonment of a radar tower on Browns Bank and caused tsunami damage on Babi Island [23].

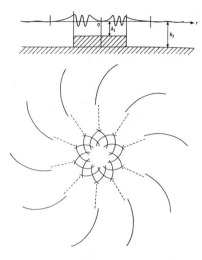

Figure 2

The third advantage of Keller's method is that the outcome of this qualitative analysis covers a large variety of phenomena by a remarkably simple quantitative algorithm requiring a minimum of computation: for the discrete eigenfrequencies of an axisymmetrical island, it reads [19]

$$k(r) \tanh[k(r)h(r)] = \varepsilon^{1/2}\omega, \tag{1}$$

$$ak(a) = \varepsilon^{1/2}n/\omega, \tag{2}$$

$$\omega \int_R^a [\{rk(r)\}^2 - \varepsilon(n/\omega)^2]^{1/2}\frac{dr}{r} = \left(n_1 + \frac{1}{2}\right)\pi\varepsilon^{1/2}. \tag{3}$$

Eq. (1) is the familiar dispersion relation between the wave number k and water depth h; R denotes the shore radius and (2) determines the outermost radius a of the ring pattern; and (3) then determines the eigenfrequencies because it turns out to be satisfied by, and only by, a countably infinite set of eigenvalues ω^2/ε dependent on the counting parameters $n(= 0, \pm1, \pm2, \ldots)$ and $n_1(= 0, 1, 2, \ldots)$. As before, n counts the zeros of the waves around the island at fixed radius r, while n_1 counts the zeros in a manner related to the radial direction. Caution: The notation is suggestive, but its symbols refer to physically intrinsic units dependent on the wave pattern; a scaling based on the island radius (Sections V--VIII) leads to eigenfrequencies $|\omega| = O(\varepsilon^{-1/2})$. Reference to [19] (or [21], [22]) is also needed for restrictions on the values of n and n_1 stemming from the topography and the continuous spectrum of scattered waves. This contrasts with the wave tank, which has no continuous spectrum and much larger slopes.

In sum, semiclassical mechanics furnishes a huge amount of information, but for our subject, it turns out to be only preliminary information.

III   Radiation Damping

There is a decisive flaw in the preceding Section. As seen in Figs. 1, 2, the wave ring around the island shore or the crown of a submerged reef is only part of the wave pattern. There is also an outer wave field which extends indefinitely into the open ocean, and semiclassical mechanics shows this outer field to be always a part of the normal or 'free' wave modes of islands and submerged banks and reefs, regardless of their shape.

To see why that feature of the patterns is of great practical importance, one needs to recall the distinction between resonance and quasiresonance. The familiar type of resonance occurs when energy conservation implies that self-sustainable oscillations (i.e., 'normal' or 'free' modes or 'eigenfunctions') have finite energy. The linear theory governing small waves to first order in amplitude is the natural model for detecting 'free' modes, and their finite energy implies that their frequency is real, as it must be on physical grounds. It also implies that their energy flux cannot extend effectively to infinite distances of an unbounded domain and therefore, that the linear model **cannot** include a mechanism for exciting such waves by input from the far field. Edgewaves on a straight coast are an example: they are described [16] by linear theory, have finite energy per unit coast length, and possess no linear excitation mechanism. Ursell [16] used a special wavemaker to exhibit them, but their excitation in nature depends on mechanisms, such as subharmonic interaction, which are small of higher order.

The free modes of an island do extend to infinity and their waves in the far field carry energy away to the ocean. Hence, though another part of those free modes is trapped, their energy cannot be trapped! The free modes must decay in time and the ω in $\exp(-i\omega t)$ must be complex; $Re\ \omega$ is then the frequency and $-Im\ \omega$ is the damping rate.

However, if a first-order model devoid of dissipation, such as used here, has radiation damping, then it must also have the reverse mechanism of radiation excitation and that must also be of **first order**. Such "quasiresonance" may therefore be much more violent than the familiar resonance?

Semiclassical mechanics, as known 30 years ago and outlined in Section II, could not touch the issue of radiation damping and could predict only real eigenvalues ω^2. Shen could only refer [19] to a conjecture of theoretical physics that the quasiresonances of smallest radiation damping would be most effective, and plausibly, those might be associated with the largest spatial gaps between the 'trapped' inner wave ring and the waves of the far field. Those gaps occur for the smallest ratios n_1/n and then (1)-(3) reduce [19] to

$$\omega^2 \sim (2n_1 + 1)n\varepsilon_s g/R \qquad (4)$$

as $(n_1 + \frac{1}{2})/n \to 0$, where ε_s is the beach slope and R, the island radius, because the 'trapped' wave ring is then narrow; g is gravity. It's as simple as that?

IV   Radiation Excitation

But, it is not. Longuet-Higgins noted [18] rightaway that radiation damping is of prime importance, that (4) is far too crude, and that reliable estimates of damping and excitation rates are indispensable for ascertaining the degree of quasiresonance of individual eigenvalues.

Unfortunately, mathematically "non-selfadjoint" problems, of which not every eigenvalue is real, pose analytical difficulties greater by orders of magnitude than those encountered in "self- adjoint" problems (which include familiar resonance). Some of the best-known analytical and numerical methods fail for non-selfadjoint problems. Longuet-Higgins therefore sought a trade- off by applying [18] a longwave approximation to the oceanographically unrealistic topography of a submerged circular cylinder on a flat ocean bed (Fig. 2).

This problem is accessible by Fourier-Bessel expansions, and he performed the rare feat of computing accurately all the eigenvalues of a non-selfadjoint problem. He showed there are no real ones in his model, all have $Im \, \omega < 0$, almost all have substantial $|Im \, \omega|$ implying rapid decay of the free modes, but for rather large $|n|$, a very few modes have notably small $|Im \, \omega|$ and can retain energy long enough to favor observation.

He then applied [18] this to the response of the model to a plane-wave train of real frequency σ and unit amplitude incident from infinity. The wave train generates free modes and scattered waves and, above the cylinder (Fig. 2) the surface elevation is

$$\zeta = \sum_{-\infty}^{\infty} A_n J_n(k_c r) \exp i(n\theta - \sigma t)$$

in terms of Bessel functions J_n; k_c is the wave number above the cylinder, and the depth ratio (Fig. 2) will be denoted by $\delta^2 < 1$. On matching the solution to the Fourier-Bessel expansion of the incident wave train, A_n is found given by

$$\delta A_n \Delta(\sigma) = 2i^{n+1}/\pi,$$

where Δ is the characteristic function of which the roots are the eigenvalues. Clearly, the amplitude $|A_n|$ is large when σ is close to an eigenvalue ω. An impression is offered by

Fig. 3 (from [18]) of $|A_n|$ vs. σ (for $\delta^2 = 1/16$); the amplification for $|n| = 6$ is seen to reach nearly 600, that for $|n| = 9$ exceeds 10^4, and so on, because an eigenvalue of very small $|Im\ \omega|$ exists for such $|n|$.

Another striking feature of great practical importance is the narrowness of the largest spikes in Fig. 3. The "response coefficient" max $|A_n|$ is [18] of order $1/(\delta|Im\ \omega|)$, but the half-width of its frequency band is only of order $|Im\ \omega|$ (which defeats (4) and makes the success of even highly sophisticated [14] direct numerical approaches expensive).

In sum, quasiresonance can be extremely violent, but is then also extremely frequency-sensitive!

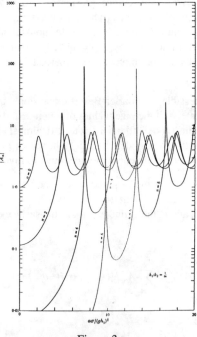

Figure 3

The analysis in [18] was extended in [24] to the longwave approximation for an island formed by a further cylinder atop the one in Fig. 2. The theory [18], [19] also prompted a laboratory test [25] of the response for the submerged cylinder which turned into one of the most painstaking and rigorous water-wave experiment yet performed. A significant effect measured was the substantial change of the incident waves as they travel over the flat tank bed from the wavemaker toward the cylinder. At the same time (and several continents away), a theoretical study [6] of the error sources in the model of [18] was undertaken, and the corrections for them were computed. As indicated by [26], replacement [6] of the longwave limit by the full linear theory shifted the quasiresonant frequency bands substantially, as seen in Fig. 4 (from [6]) (even though the dispersion relation predicted small wave numbers), but other corrections [6], including that for viscosity [18], [6], proved rather insignificant. The major disagreement - see Fig. 5 (from [25]) - between the resonance effects found in the tank and the quasiresonances predicted in the open ocean remains unexplained. (A possible source of such a discrepancy may be the difficulty of simulating in a tank the far field and radiation of the free modes; the experiment was arranged to fit the linearized Classical Equations closely, but the domain distinction made theory and experiment differ significantly in the applicable form of energy conservation.)

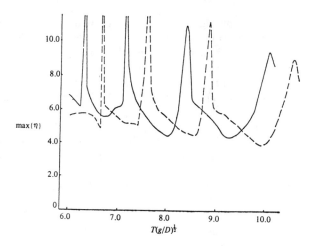

Figure 4

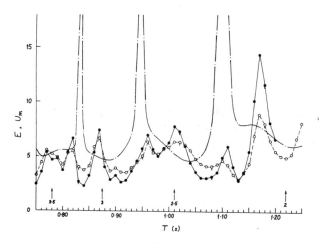

Figure 5

V   Oceanographic Quasiresonance

The frequency shift and sensitivity (Fig. 4) found in [6] indicate an imperative need for an accurate analysis modeling realistic topography on the full linear theory. That challenge was addressed in [26] and especially, in [27], on the basis of a small-slope approximation which places no restrictions on wavelength or topography for geologically realistic seabeds. Fourier transform in θ reduces the problem to an ordinary differential equation in the radial coordinate r which accounts with sophisticated accuracy for the vertical structure of the motion.

To carry the analysis to simple algorithms for the relevant results, Lozano added [27] the assumption that the depth $h(r)$ is analytic at real r, firstly because depth-sounding cannot distinguish the true depth from such an approximation, and secondly because this makes the differential equation accessible to rigorous Mathematical Asymptotics in the slope parameter ε for at least those free modes which decay slowly enough to be of interest.

The mathematics yields the asymptotic expansion to all orders of the solution in powers of $\varepsilon^{1/2}$ and the corresponding characteristic function $\Delta_e(\omega; n, \varepsilon, \ldots)$, but the analytic structure implies that all roots of Δ_e must be real. (Not surprisingly, the first approximation to them is given by (1)-(3); its relative error is only $O(\varepsilon^2)$.) This forced Lozano to study the exact characteristic function Δ dependent on the exact solutions, which are only marginally supercomputable, but no matter, they can be denoted by symbols. By exploiting a subtle, structural symmetry related to energy conservation, he split Δ into a sum $\Delta = \Delta_e + \Delta_t$, where (by contrast to Δ_e) Δ_t depends transcendentally on ε.

It follows [27] that, as long as $|\Delta_t|$ is not too large, Δ has one, simple root ω_k near each root of Δ_e, and that the first approximation to the rate of radiation damping is then

$$- \operatorname{Im} \omega_k \simeq P_k \exp(-4\pi d_k)/|\operatorname{Re} \omega_k|, \tag{5}$$

where d_k is a physically intrinsic measure of the width of the gap between the trapped-wave ring and the far-field waves (Fig. 1). Also, the prefactor $P_k > 0$ and P_k^{-1} and d_k are given [27], [22] by WKB integrals similar to that in (3) (in which dependence on the solutions is already transformed into dependence on the main physical parameters and the topography); while P_k is neither small nor large, d_k is large of order ε^{-1} at a few of the eigenvalues corresponding to each value of $|n|$ that exceeds substantially the smallest $|n|$ for which Δ_e has a root. (E.g., for a conical island of realistic slope, that requires [19] $|n| \geq 3$.) Those are the eigenvalues of small radiation damping.

Comparison of the energy flux levels in the trapped and far-field wave regions of a standing wave at real frequency σ shows [27] the amplitude amplification to be

$$\rho(\sigma) = |\Delta(\varepsilon\sigma^2; n, \varepsilon, \ldots)|^{-1}.$$

This varies from an exponentially small minimum between successive $Re\ \omega_k$ for the same n to an exponentially large maximum

$$\rho_{max} \simeq \exp(2\pi d_k)[1 + O(\varepsilon)] \tag{6}$$

at $\sigma = Re\ \omega_k$, which is the 'response coefficient'. However, the half-width of the frequency band of large ρ near $Re\ \omega_k$ is [18], [27] only about $48^{1/2}|Im\ \omega_k|/|Re\ \omega_k|$. In fact, these "spectral concentrations" look much like δ- function approximations: a high, narrow spike with roughly unit area under it, which will be relevant in Section VII.

Caution: Mathematical Asymptotics uses 'exponential' primarily to denote a type of transcendental, by contrast to algebraic, structure; in this sense, the term applies to the radiation damping of most of the eigenvalues. However, the quantitative impression suggested by the term may be exaggerated, only a very small minority of those eigenvalues generate quantitatively sensational spectral concentrations.

It should also be noted that the free modes so excited have their energy concentrated in narrow rings around the island and have their maximal crest height at the island shore (Semiclassical Mechanics has a simple algorithm for tracking energy flux along rays [20], [22]) and that they come as symmetrical, clockwise and anticlockwise traveling, mode pairs which interfere by addition: at some points of the shore, a mode pair will cause no run-up, but at some other points, it will cause twice as much as indicated by ρ_{max}.

VI Flaws and Corrections

For brevity, many complications have been ignored which occur in less common situations. For instance, patterns with two disjoint trapped-wave rings, or patterns with none at all, can be generated by some seabed topographies, and [19] needs to be consulted. Similarly, [27] needs scrutiny for what may not be covered.

A glaring flaw of all the work mentioned so far (except [18]) is the assumption of perfect shore reflection. That has been remedied in [28], [22] by modifying the results of [19], [27] for the effect of a given reflection coefficient, but the practical use of this correction is spoiled by present ignorance of the coefficient's value in many circumstances [11].

Viscous effects have been considered by many authors. They may be minor in the laboratory [6] and there are oceanographical issues [18], [24] on which they do not have much influence. As mentioned in Section I, however, there are others of major importance [7]-[11] to water waves near islands, for which a realistic, quantitative description is impossible without better estimates of viscous dissipation than are available so far. This is a main reason why theory focusing on the geophysical context, as in this account, cannot yet have more than qualitative scope.

Perfect axisymmetry is geophysically unrealistic, and perturbation of it has been found [18], [24] to cause frequency splitting between the clockwise and anticlockwise free modes with the same $|n|$, which leads to beats. Frequency shifts and splits, and beats, are also caused by Coriolis effects [18], [24]. Of course, beats can be observed only for mode pairs of sufficiently long 'life- time'.

Most real islands are far from axially symmetrical. That may call for ray-tracing, which has been known for 50 years to be an efficient computational approach to surface waves. However, [18], [19] and [27] have changed its application drastically [22]. The main objective must be a credible estimate of the effective, intrinsic gap width d_k in (5); it requires only a rough approximation to the integrand in the WKB-integral for d_k [19]. Better ray-tracing to approximate the analogous integrands [19] for $Re\ \omega_k$ makes sense only for the very few modes with large d_k. In this way, the analytical advances have improved the focus, and reduced the expense, of computing by an order of magnitude.

An apparent flaw, which may already have tried the patience of the Reader, is the intuitive implausibility of 'monochromatic' waves of a single frequency in regard to a description of many processes of practical interest, and of tsunami effects in particular. The reason why so much preoccupation with the monochromatic case in Sections II--V is actually a shortcut is that this case concerns a fundamental solution of the wave equation in the sense that its Fourier transform is a delta function. Without a clear grasp of the nature of the wave patterns and their spectrum, and of the sophisticated and decisive, unfamiliar physical structure introduced by radiation damping and radiation excitation, an intelligible discussion of practical applications gets very difficult. The information in Sections II, IV, V, on the other hand, makes a plethora of applications readily accessible. Among them, the most instructive, qualitative information is offered by the examples of a short incident wave (Section VII) and of a semi-infinite, incident wave train (Section VIII).

VII   Response to a Solitary Wave

The epitome of a short wave train is a single hump or 'solitary wave', but contemporary literature uses this term in several senses. The discussion can be clarified greatly by restricting attention first of all to the strict solitary wave [29], [30], [31], because it can then be based on a physically consistent far field of a wave traveling over an ocean of uniform depth with invariant shape, amplitude and speed. Such a strict solitary wave is a one-parameter family $S(t - x/c)$ of single surface elevations consisting of a 'body' flanked by infinitely long, exponentially weak 'tails'. If the amplitude (i.e., maximum of surface elevation) be denoted by a^2, then the width of the body is roughly $1/a$ and the speed c is close to $(gh_\infty)^{1/2}$, where g denotes gravity and h_∞, the ocean depth. It should be noted that $S(t - x/c)$ is a family of exact solutions [30] of the Classical Equations, provided the amplitude is strictly limited, and these solutions have a paradoxical property favoring the application of linear theory: on the one hand, S does not satisfy a linear differential equation (however small the amplitude a^2 is), but on the other hand, substantial nonlinear modifications of S during approach to, and interaction with, an island require distances of wave-travel which are surprisingly large multiples [32] of the 'body-width' $1/a$.

The Fourier transform

$$A(\sigma) = \int_{-\infty}^{\infty} S(\tau)e^{i\sigma\tau}d\tau$$

of S with respect to

$$\tau = t - x/c$$

looks similar to another 'solitary wave' in that it is characterized by a 'body' flanked by infinitely long, exponentially weak 'tails'. However, the 'amplitude' and 'body-width' of

$A(\sigma)$ are both of order a, so that a solitary wave S of relatively strong amplitude a^2, and correspondingly narrow body, has a transform A which is both fairly strong and wide. Conversely, a 'weak' (and therefore wide) solitary wave has a weak and narrow transform.

One can profit from the fundamental solution explored in Sections II-V by writing the far field $\zeta_\infty = S(\tau)$ as the inverse of its transform,

$$\zeta_\infty(\tau) = \frac{1}{2\pi} \int_{-\infty}^{\infty} A(\sigma)e^{-i\tau\sigma}d\sigma = \sum_{-\infty}^{\infty} \zeta_{n\infty}e^{in\theta},$$

$$\zeta_{n\infty}(\tau) = \frac{i^n}{2\pi} \int_{-\infty}^{\infty} A(\sigma)J_n(\sigma r/c)e^{-i\sigma t}d\sigma.$$

It then follows on the linear theory that the solution with this far field is

$$\zeta = \sum_{-\infty}^{\infty} \zeta_n e^{in\theta},$$

$$\zeta_n = \frac{1}{2\pi} \int_{-\infty}^{\infty} A(\sigma)\Psi_n(r,\sigma)e^{-i\sigma t}d\sigma \qquad (7)$$

where $\Sigma\Psi_n e^{in\theta}$ denotes the surface elevation for which the far-field is the monochromatic wave train of frequency σ; [27] gives useful, simple approximations to $\Psi_n = w_n(r;\sigma)/\Delta_n(\varepsilon\sigma^2)$, where w_n denotes the normalized eigenfunction and Δ, the characteristic function, as in Section V. It follows right away that the solitary wave excites all the free modes explored in Sections II-V, in fact, their excitation starts long before the body of the wave approaches the island, because the solitary wave has an infinitely long precursor 'tail'. However, the exponentially weak start of the interaction is of minor interest, the realistic concern is with the response to the body of the solitary wave.

To understand the nature of that response, it helps to look briefly at (7) with $\Delta(\sigma)$ replaced by unity (corresponding to a delta-function far field [18]). Then (7) integrates over the spectral concentrations of Ψ_n (Section V), each of which looks like an approximation to a delta-function with roughly unit area under it, and every spectral concentration therefore captures a roughly equal fraction of the incident energy flux: none captures much.

If the response involves any substantial amplification or frequency-sensitivity, that must therefore depend almost entirely on the nature of the incident wave and its Fourier transform $A(\sigma)$. Moreover, the contribution to (7) of the exponentially weak 'tails' of $A(\sigma)$ is rather negligible, the integration needs to be extended only over the frequency interval I_σ containing the transform's 'body'. Since S is even in τ, $A(\sigma)$ is even in σ and $I(\sigma)$ is centered on $\sigma = 0$. This makes it relevant to note that free modes of significantly long life are unlikely for $n < 3$ because there are certainly [19] none for conical islands of geophysically realistic slope. If $I(\sigma)$ is narrow, because the solitary wave $S(\tau)$ is broad (and therefore also weak), then the response is likely to differ little from a brief transient of wave scatter, because eigenvalues of long life have[27] frequency $\sigma = O(\varepsilon^{-1/2})$. Such initial transients are best analyzed by unsteady Semiclassical Mechanics [33], [34], [22], but radiation damping introduces a fundamental change in our notions about wave scatter. Classical experience teaches a clear distinction between scattering, which focuses attention on the far field (and is associated with the continuous part of the spectrum) and resonance, which concerns only the free modes without significant far field (associated with discrete eigenvalues). For waves about an island, such a distinction collapses because the main far field is that of the free modes and the main mechanism of scattering becomes that of radiation excitation and damping of free modes. The analysis of the scattering transient therefore requires an extension of unsteady Semiclassical Mechanics into complex phase space [35] beyond the scope of this account. In any case, the main effect of the scattering transient appears likely to be a runup over the front side of the island, which can be estimated at least roughly by Green's Law [4]. The 'rear' or 'lee' side of the island, by contrast, lies in the scattering shadow.

If I_σ is wide, on the other hand, because the solitary wave S is narrow and high, then there may be one or two values of $|n|$ for which eigenvalues of very small radiation damping exist that have real part within $I(\sigma)$ and have a substantial average of $|A(\sigma)|^2$ over the narrow frequency band of their spectral concentration. For such an $|n|$, the solitary wave does excite and amplify a free-mode pair of long life so as to contribute a notable response very different from scatter. Before the character of that response is described, however, a note of caution is appropriate. Since the solitary wave under consideration must be high, its amplitude must be rather close to the limiting value. By itself, this does not [32] discredit the application of linear theory, but a real ocean is not of uniform depth and the solitary wave must experience considerable shoaling, before its body comes close to the island, during which the wave's shape is modified and its amplitude, boosted - possibly beyond the amplitude limit of exact solitary - wave theory [30] and even the limit of Moderate Nonlinearity, so that wave-breaking results and the energy-conserving, Classical Equations fail to apply. Those Equations can therefore inspire confidence for the response to only a small subset of high solitary waves. The predictions of (7), all the same, will be seen now to have relatively plausible, qualitative meaning for a much larger subset of solitary waves.

To this end, begin by focusing attention on just one, positive integer n for which an eigenvalue $\varepsilon\omega_*^2$ of very small radiation damping exists [27] such that $|Re\ \omega_*| = \sigma_*$ lies in the 'body'-interval I_σ where $|A(\sigma)|$ is not small. In fact, suppose temporarily that there were no other such n or $\varepsilon\omega_*^2$ in order to estimate the particular, persistent free mode corresponding to this n and $\varepsilon\omega_*^2$ which the solitary wave excites. For simplicity, also fix r at some value inside the mode's wave ring around the island shore. Recall from Section V that the temporary disregard of all but the one eigenvalue makes $|\Psi_n|$ exponentially small outside the narrow frequency band of the corresponding spectral concentration; however, $\varepsilon\omega_*^2$ defines a real pair $\sigma_*, -\sigma_*$ of frequencies, so that a spectral concentration is centered on each of those two frequencies. Hence, (7) yields - on our temporary supposition -

$$2\pi\zeta_n \sim A(\sigma_*)e^{-i\sigma_*t}\int_0^\infty \Psi_n(r;\sigma)d\sigma + A(-\sigma_*)e^{i\sigma_*t}\int_{-\infty}^0 \Psi_n(r;\sigma)d\sigma.$$

Now, the path of the first of these integrals passes within an exponentially small distance above the complex point $(\sigma_*, Im\omega_*)$ so that its value is roughly $(-\pi i)$ times the residue of Ψ_n at ω_* and similarly, for the second integral. Since $A(\sigma)$ is even, and if some exponentially small terms are neglected against unity, the estimate becomes

$$\zeta_n \sim \frac{1}{2i} A(\sigma_*)[e^{-i\sigma_*t}w_n(r;\sigma_*)/\Delta_n'(\sigma_*) + e^{i\sigma_*t}w_n(r;-\sigma_*)/\Delta_n'(-\sigma_*)],$$

where the prime denotes differentiation with respect to ω. However, since n and σ_* have both been chosen positive, the two terms in the square bracket do not contribute to the same free mode. On the other hand, since w_n and Δ depend [27] only on n^2 and ω^2, the same estimate results for ζ_{-n} and $\Delta_n'(\sigma)$ is odd in σ, so that the anticlockwise and clockwise, persistent modes of frequency σ_* excited by the solitary wave are approximately

$$\begin{pmatrix} \zeta_a \\ \zeta_c \end{pmatrix} \sim A(\sigma_*)w_n(r;\sigma_*)sin \begin{array}{c} (n\theta - \sigma_*t) \\ (-n\theta - \sigma_*t) \end{array} \Big/ \Delta_n'(\sigma_*). \tag{8}$$

Of course, the temporary supposition may not be tenable, and the persistent response to the solitary wave is generally a sum of mode pairs (8). However, the number of terms in the sum is quite small because almost all eigenvalues of very small radiation damping lie outside the interval I_σ in which $|A(\sigma)|$ is appreciable. It should be noted also that the estimate leading from (7) to (8) is rough in many ways. For instance, it ignores the small distance between ω_* and the real frequency axis and thereby, omits all information on the gradual excitation and subsequent damping of the free-mode pair. This leaves the precise interpretation of $t = 0$ unclear.

Despite these qualifications and uncertainties, it pays to look a little further into a typical, persistent free-mode pair associated with an eigenvalue $\varepsilon\omega_*^2$. On linear theory, the pair (8) combines to a (near -) standing wave

$$\begin{aligned} \zeta_a + \zeta_c &= \zeta_n e^{in\theta} + \zeta_{-n} e^{-in\theta} \\ &\sim -2A(\sigma_*)w_n(r;\sigma_*)\cos(n\theta)\sin(\sigma_*t)/\Delta'(\sigma_*) \end{aligned} \tag{9}$$

with interesting dependence on the angle θ. The incident wave $S(t - x/c)$ travels in the direction of x increasing, so that the 'front' of the island facing the incident wave corresponds to $|\theta| = \pi$, and the 'rear' or 'lee' of the island, to $\theta = 0$; both are, for any $|n|$, among the angles at which the clockwise and anticlockwise members of the mode pair interfere constructively so as to double the local crest height. The dependence on θ is all the more notable when the body of the solitary wave is so narrow that more than one persistent, free-mode pair is excited substantially. As noted above, this can occur only for very few pairs and those can involve only a small number of contiguous values of $|n|$. At any nonzero $|\theta| \le \pi$, some of those pairs must interfere destructively with each other to at least some extent. The absolute maximum of crest height of the persistent waves must therefore occur just at the lee of the island or bank.

This observation is especially relevant when the scope of the discussion is now extended to modifications, common in the contemporary literature, of the incident waves from strict solitary waves [30] to a two-parameter family $A_0 S(t - x/c)$ of incident 'solitary' waves which share the shape and speed of the strict ones, but have an independent amplitude parameter $A_0 a^2$. When $A_0 \neq 1$, they do not share the invariance property over water of uniform depth, but the far field notion so far appealed to is only an abstraction used to fix the ideas upon a consistent start of the analysis. Besides, neither the laboratory nor most directly numerical approaches to the Classical Equations have a far field. Moreover, this modification brings advantages of flexibility because such 'solitary' waves can be made to specify equally well incident waves of depression as of elevation, and the two can be combined to incident 'N- waves' (but this will not be considered here). It is a short further step to a similarly flexible, two-parameter family of incident waves $A_0 G(t - x/c)$, where G denotes a strict Gaussian. The present discussion applies to all such incident waves because the transform of $A_0 S(t - x/c)$ is $A_0 A(\sigma_*)$ and hence, (7)-(9) are changed only by the replacement of A by $A_0 A$. Similarly, the transform of a strict Gaussian is another strict Gaussian, and if one is narrow and high, then the other is broad and low, which is just the relation between S and A appealed to in the present discussion.

The main motivation for the use of such incident waves in place of a strict far field tends to be the liberation from the amplitude limitation of strict solitary waves by means of factors A_0 with $|A_0|$ well in excess of unity. Of course, this enhances the likelihood of wave breaking during the shoaling before the 'solitary' wave reaches the neighborhood of the island. However, it also raises the possibility of modeling incident waves which direct far more energy flux into frequency bands of strongly persistent free modes than is plausible for strict solitary waves, because such modes have large frequency $|\sigma_*| = O(\varepsilon^{-1/2})$ [27]. In particular, it raises the possibility that, while every one of all the free modes excited by a 'solitary' wave has a strength within the scope of linear theory, the constructive interference of all the persistent free modes at the lee of the island documented in (9) generates local crest heights there which are well beyond Moderate Nonlinearity. If this causes local wave breaking, moreover, then the attendant dissipation will compromise the strength of each free mode before it can continue its travel around the island and therefore, the most violent part of the wave response will remain confined to the lee-side, and will be approximated roughly by the appropriate sum of terms represented by (9) with $\theta = 0$ and $A_0 A(\sigma_*)$ in place of $A(\sigma_*)$.

In sum, the response to a very short, incident wave train exhibits none of the extreme amplification and great frequency-sensitivity characterizing the quasiresonant response (Sections IV, V) to a monochromatic wave train. An incident wave with but a single crest or trough may cause no more than a brief scattering transient with run-up at the island's front estimated roughly by Green's Law [4] and with rather quiet water in the scattering 'shadow' on the lee-side of the island. However, if such an incident wave is steep-sided, it will also cause a more persistent response of free modes interfering constructively most of all just at the lee of the island. Moreover, if such an incident wave has also a sufficiently large amplitude, then this persistent response can, much by contrast to the scattering transient, cause havoc at the lee-shore of the island [23].

VIII Time-development of Persistent Response

The big difference between the nature of the response found in Sections IV, V and in Section VII suggests consideration of an intermediate model which can interpolate to some extent between those extremes and can also help a little towards repairing the omission, so far, of any attention to the development in time of wave response around an island or submerged bank. A simple such model is a semi-infinite train of plane waves of surface elevation

$$\zeta_\infty = A_0 H(\tau) \exp(-i\sigma\tau), \quad \tau = t - x/c \tag{10}$$

in the far field, where H denotes Heaviside's unit step and the constants A_0 and c denote the amplitude and phase velocity, respectively. An added attraction is that such a half-train might come closer to approximating an approaching tsunami or storm than either of the monochromatic or solitary-wave models considered so far.

In some ways, however, this ζ_∞ is a poor model of water waves and the distinction must be kept in mind between the questions which can, and those which cannot, be answered credibly by the linear theory of such a crude model. In particular, even trains of wavelength as long as that of tsunamis accumulate appreciable effects of dispersion over long travel distances [12] so that the head of such a train [36] looks, even on the linear theory, rather different from the ζ_∞ just proposed. On the linear theory, of course, this flaw can be removed by adding a correction which is the response to a far-field wave of little more than a couple of wave periods. However, such a response correction is similarly brief and can add little to the lessons of Section VII, namely that a relatively brief disturbance in the far field generates a relatively minor response near the island unless the amplitude $|A_0|$ in (10) is large. This model is therefore helpful only for questions on the development of the response after the initial transient.

Those questions are less important for tsunamis, where the larger amplitude of the head of the wave train is more relevant than the cumulative effect of a long train. For storm waves, on the other hand, the more persistent response is of primary importance and it is useful to explore the circumstances under which such wave trains can endanger offshore structures. An instance of such waves was captured on film, many years ago, when a storm forced abandonment of a radar tower on the lee slope of Browns Bank, East of Boston.

A closer look (in the Appendix) shows the magnitude of the more persistent response to grow linearily in time, at first, which confirms that its largest local amplitudes must always be expected to occur on the lee side of a bank or island, as noted in Section VII. It is also seen in the Appendix that the approach to full quasiresonance requires both a frequency σ in a band of spectral concentration (Sections IV, V) and a long wait for the ultimate violence of the wave response.

In conclusion, linear theory gives illuminating information on the main predictions of the Classical Equations of surface waves of Ideal Water and shows such waves around islands and submerged banks or reefs to differ in important and striking ways from the waves more familiar in Oceanography. From a practical point of view, it indicates two properties of incident waves most likely to generate a response capable of inflicting serious damage: One is the obvious one of large amplitude and the second is substantial energy flux into at least one of the narrow frequency bands favoring the amplified excitation of free wave modes. Separately, these properties generate quite different types of response. The most distinctive effect of large amplitude is scattering with major run-up on the island's front side. That of excited free modes, by contrast, is constructive interference of mode pairs, especially on the lee-side of the island, deep in the scattering shadow. The first is more typical of tsunamis and the second, of storm waves; both properties can combine, however, both for tsunamis and storm waves, and have thus, on rarer occasions, caused disaster.

APPENDIX

The far field (10) may be represented as the inverse of its Fourier transform with respect to $\tau = t - x/c$ by

$$\zeta_\infty = \frac{A_0}{2\pi i} \int_{-\infty+i0}^{\infty+i0} e^{-i\omega(t-x/c)} \frac{d\omega}{\sigma - \omega} = A_0 \sum_{n=-\infty}^{\infty} \zeta_{n\infty} e^{in\theta},$$

$$\zeta_{n\infty} = \frac{i^{n-1}}{2\pi} \int J_n(\omega r/c) e^{-i\omega t} \frac{d\omega}{\sigma - \omega},$$

at least in the distributional sense, with ω interpreted as $\mathrm{Re}\,\omega + iO$, so that $\zeta_\infty = 0$ for $t < x/c$. The corresponding Fourier component ζ_n of the surface elevation is then

$$\zeta_n = \frac{1}{2\pi i} \int_{-\infty}^{\infty} \frac{w_n(r; \omega)}{\Delta_n(\omega)} e^{-i\omega t} \frac{d\omega}{\sigma - \omega},$$

where, as in Section VII, w_n and Δ_n denote the normalized eigenfunction and the characteristic function, respectively; both are here interpreted as functions of n and ω, for simplicity of notation, even though they really depend on n^2 and $\varepsilon\omega^2$, so that the 'eigenvalues' ω_k, at which $\Delta(\omega_k) = 0$, come in pairs symmetrical to the imaginary ω-axis, as in Sections IV, V, VII. Since w_n and Δ_n depend only on n^2, so does ζ_n, and the joint contribution of ζ_n and ζ_{-n} to the total response is

$$2\zeta_n \cos(n\theta).$$

For $t > 0$, the path of the last integral can be deformed to enclose just a branch cut along the negative imaginary ω-axis [18]; by Watson's Lemma, that new integral R_n represents a short-lived transient of wave scatter [18]. The deformation, however, collects also residues at all the poles of the integrand, so that the last formula becomes (for $t > 0$)

$$\zeta_n = \frac{w_n(r; \sigma)}{\Delta_n(\sigma)} e^{-i\sigma t} - \sum_k \frac{w_n(r; \omega_k)}{(\sigma - \omega_k)\Delta_n'(\omega_k)} e^{-i\omega_k t} + R_n, \qquad (11)$$

where the sum is over all the 'eigenvalues' w_k at the roots of $\Delta(\omega)$. The first term on the right of (11) is a forced wave and the sum is a superposition of all the free modes.

Since all the 'eigenvalues' have Im $\omega_k = -\eta_k < 0$, the sum in (11) decays by radiation damping, and

$$\zeta_n \rightarrow w_n(r, \sigma) e^{-i\sigma t} / \Delta(\sigma) \text{ as } t \rightarrow \infty \tag{12}$$

(because σ is real), which gives the ultimate response and represents just the quasiresonant phenomena of extreme amplification and frequency-sensitivity discussed in Sections IV, V. Moreover, almost all the terms in the sum of (11) represent rapidly damped, free modes, which form part of the initial, brief, scattering transient poorly represented by (10), (11). The response persisting beyond that transient is therefore

$$_p\zeta_n = \frac{w_n(r; \sigma)}{\Delta_n(\sigma)} e^{-i\sigma t} - \sum_{q=1}^{2m} \frac{w_n(r; \omega_q)}{(\sigma - \omega_q)\Delta_n'(\omega_q)} e^{-i\omega_q t}$$

collecting only the free modes of very small damping rate $\eta_q = -\text{Im } \omega_q > 0$, and therefore, m will hardly exceed 2 or 3. Moreover, the 'eigenvalues' ω_q occur in pairs because Δ_n really depends on $\varepsilon\omega^2$, and if $d\Delta_n/d(\varepsilon\omega^2) = \Delta_n^\dagger(\varepsilon\omega^2)$, and $|\text{Im } \omega_q/\text{Re } \omega_q|$ is neglected against unity, the last equation reduces to

$$_p\zeta_n = \frac{w_n(r; \sigma)}{\Delta_n(\sigma)} e^{-i\sigma t} - \sum_{q=1}^{m} \frac{w_n(r; \omega_q)}{\varepsilon(\sigma^2 - \omega_q^2)\Delta_n^\dagger(\varepsilon\omega_q^2)} e^{-i\omega_q t} \tag{13}$$

over only those q for which σ and Re ω_q share the same sign. (The factor ε^{-1} in this sum might give a misleading impression because the 'eigenvalues' retained in (13) have $\omega_q^2 = O(\varepsilon^{-1})$; this denominator is small only if $|\sigma - \text{Re } \omega_q|$ is small for some q.) Clearly also, the magnitude of each term in (13) is initially linear in t.

For the interpretation of (13), it is relevant that, for given n, the real frequency axis consists of narrow, "loud" intervals of spectral concentration (Sections IV, V) in which $|\Delta(\sigma)|$ is very small, alternating with wide, "quiet" intervals in which $|\Delta(\sigma)|$ is large. If the value of n is one for which σ lies in a quiet interval, then none of the denominators in (13) is small, so that none of the forced and free modes for this n and $-n$ are amplified substantially by the island topography. Since a reasonably close approach to the limit (12) is seen to require effective decay of all the free modes, it takes a time of order η_*^{-1}, where η_* is the smallest damping rate of free modes for this n.

On the other hand, if n is such that σ lies in a loud interval, say, of ω_1, then the forced mode (12) is strongly amplified by the island or submerged bank, but also, is almost cancelled out initially by the free mode associated with ω_1. The other free modes in (13) are all of similar structure, so that the simplest approach to representative information is to ignore all but one of them, labeled $q = 2$, say, and preferably chosen as that among them with the smallest damping rate, to assure a correct estimate of the time needed for close approach to the quasiresonant limit (12). To make the initial cancellation apparent, the three retained terms may be recombined to $(Q_1 + Q_2) \exp(-i\sigma t)$ where, with $\sigma - \omega_1 = \delta$,

$$Q_1 = \frac{w_n(r;\sigma)}{\Delta_n(\sigma)} - \frac{w_n(r;\omega_1)}{\delta\Delta_n'(\omega_1)}e^{i\delta t},$$

$$\sim (1 - e^{i\delta t}[1 + O(\delta)])w_n(r;\sigma)/\Delta_n(\sigma),$$

$$Q_2 = \frac{w_n(r;\omega_2)}{(\varepsilon\sigma^2 - \varepsilon\omega_2^2)\Delta_n^{\pm}(\varepsilon\omega_2^2)}e^{i(\sigma-\omega_2)t},$$

since $|\delta|$ is exponentially small in ε (Section V). As long as t is not large, it follows that

$$Q_1 \sim [t + O(\varepsilon^{1/2})]w_n(r,\omega_1)/[i\Delta_n'(\omega_1)],$$

while $Q_2 = O(1)$ in ε. It is also seen that the time for a reasonably close approach to the full, quasiresonant response (12) is of order η_1^{-1}, after all, even if $\eta_2 < \eta_1$, because $|Q_2|$, like any further terms in (13), never gets large.

Finally, it must be recalled that the full solution is

$$\zeta = A_0 \sum_{-\infty}^{\infty} \zeta_n e^{in\theta},$$

that the joint contribution of ζ_n and ζ_{-n} to it is $2\zeta_n\cos(n\theta)$, and that the constructive interference resulting from this mode pairing is maximal at the lee of the island or submerged bank.

Acknowledgements. This work was supported in part by Grants DMS-9123227 and 9424508 from the National Science Foundation.

Fig. 1 is from [19] in the Physics of Fluids, American Institute of Physics, New York, 1968.

Figs. 2 and 3 are from [18] in the Journal of Fluid Mechanics, Cambridge University Press, London, 1967.

Fig. 4 is from [6] in the Journal of Fluid Mechanics, Cambridge University Press, London, 1983.

Fig. 5 is from [25], Geophysical and Astrophysical Fluid Dynamics, Gordon and Breach Science Publ., London, 1983.

Author's address: Richard E. Meyer, Center for the Mathematical Sciences, University of Wisconsin-Madison, 1308 W. Dayton St., Madison, WI 53715-1149.

Bibliography

[1] W. G. Van Dorn, Tsunamis, in *Advances in Hydroscience*, vol. 2, Academic Press, N.Y. 1965.

[2] J. B. Keller, J. Fluid Mech. **4**, (1958), 607.

[3] G. F. Carrier and H. P. Greenspan, J. Fluid Mech. **4**, (1958), 97.

[4] G. Green, On the motion of waves in a variable canal of small depth and width, Camb. Trans. vi. (1837).

[5] F. P. Shepard, *Submarine Geology*, Harper & Row, N.Y., 1963.

[6] Y. Renardy, J. Fluid Mech. **132**, (1983), 105.

[7] R. L. Miller, J. Geophys. Res. **73**, (1968), 4497.

[8] R. E. Meyer, J. Geophys. Res. **75**, (1970), 687.

[9] J. J. Mahony and W. G. Pritchard, J. Fluid Mech. **101**, (1980), 809.

[10] J. W. Miles, J. Fluid Mech. **212**, (1990), 365.

[11] R. E. Meyer, J. C. Strikwerda and J.-M. Vanden-Broeck, Wave Motion **17**, (1993), 11.

[12] G. F. Carrier, J. Fluid Mech. **24**, (1966), 641.

[13] E. N. Pelinovsky and R. KH. Mazova, Nat. Hazards **6**, (1992), 227.

[14] C. C. Lautenbacher, J. Fluid Mech. **41**, (1970), 655.

[15] F. Ursell, Proc. Camb. Phil. Soc. **47**, (1951), 347.

[16] F. Ursell, Proc. Roy. Soc. London Ser. A **214**, (1952), 79.

[17] M. C. Shen's Theorem, see pp. 406--408 and 410 in *Waves on Beaches*, (R. E. Meyer, ed.), Academic Press, N.Y. 1972.

[18] M. S. Longuet-Higgins, J. Fluid Mech. **29**, (1967), 781.

[19] M. C. Shen, R. E. Meyer and J. B. Keller, Phys. Fluids **11**, (1968), 2289.

[20] J. B. Keller, SIAM Rev. **27**, (1985), 485.

[21] R. E. Meyer, Lec. in Appl. Math. **13**, Amer. Math. Soc., (1971), 189.

[22] R. E. Meyer, Adv. Appl. Mech. **19**, (1979), 53.

[23] H. Yeh, P. Liu, M. Briggs and C. Synolakis, Nature **372**, (1994), 353.

[24] W. Summerfield, Phil. Trans. Roy. Soc. London **A272**, (1972), 361.

[25] B. J. S. Barnard, W. G. Pritchard and D. G. Provis, Geophys. Astrophys. Fluid Dyn. **24**, (1983), 23.

[26] R. Smith and T. Sprinks, J. Fluid Mech. **72**, (1975), 373.

[27] C. Lozano and R. E. Meyer, Phys. Fluids **19**, (1976), 1075.

[28] R. E. Meyer and J. F. Painter, J. Eng. Math. **13**, (1979), 33.

[29] J. J. Stoker, *Water Waves*, §10.9, Interscience Publ., N.Y., 1957.

[30] K. O. Friedrichs and D. H. Hyers, Commun. Pure Appl. Math. **7**, (1954), 517.

[31] F. Ursell, Proc. Camb. Phil. Soc. **49**, (1953), 685.

[32] R. E. Meyer, pp. 17--21 in Progress in Applied Mechanics (Prager Anniversary Volume) Macmillan Co., New York, 1963.

[33] M. C. Shen and J. B. Keller, SIAM J. Appl. Math. **28**, (1975), 857.

[34] D. Hector, J. Cohen and N. Bleistein, Stud. Appl. Math. **51**, (1972), 121.

[35] R. E. Meyer, SIAM J. Appl. Math. **51**, (1991), 1585.

[36] H. Jeffreys and B. S. Jeffreys, *Methods of Mathematical Physics*, Camb. Univ. Press, Cambridge, 1956, p. 517.

Review of Tsunami Simulation with a Finite Difference Method

F. Imamura

Disaster Control Research Center, Tohoku University

1. Introduction

The aim of this chapter is to review the tsunami simulation using a finite difference methods (FDM) and to summarize their advantages and problems especially regarding a runup which is a challenging problem in hydrodynamics (Liu, Synolakis & Yeh, 1991).

Looking the history of FDM being applied to long waves such as tidal waves and tsunamis, Hansen (1956)'s simulation of a tide in the north sea with the linear long waves theory descretized by Eularian forward scheme would be the first attempt. As for tsunamis, a near-fields tsunami in Tokyo bay (Isozaki and Unoki , 1964) and a far field tsunami of the 1960 Chilean tsunami (Ueno, 1965) were carried out. After developing the earthquake mechanism to determine tsunami initial conditions, many historical tsunamis have been successfully reproduced by Aida (1969) with the FDM tsunami simulations. Numerical simulation has made much progress during 30 years and is now used as one of the most effective means in the practical design of tsunami defence works and structures (Shuto, 1991).

In a runup of tsunamis problem, it is not easy to express moving boundary for a wave front with the fixed coordinate system in the Eulerian description which is used for most of tsunami simulations with FDM. If the equations in the Lagrangian description are used, the moving condition can be expressed with no approximation (Shuto and Goto, 1978). It is clear, however, that Lagrangian methods are well applicable only to one-dimensional problems, but poorly to any two-dimensional practical problem with complicated topography. That is why the numerical model with the Eulerian description has still been taking important role in tsunami research and prevention work. However, some problems in the accuracy of computation by the Eulerian description are pointed out. Therefore, a boundary condition of a runup is still one of unsolved problems.

This chapter describes the governing equation, numerical scheme, expressing bottom friction, stability condition, and boundary condition especially focusing on a runup problem by reviewing the recent numerical codes for tsunamis. Through the comprehensive review, we could provide information what condition is to assure the stable and to obtain accurate results, and how multiple wave interactions can be included to predict the coastal effects accurately.

2. Governing Equations

2.1. Shallow water theory

Tsunamis mainly generated by the deformation of sea bottom due to earthquakes belong to long waves. Ninety percent of all tsunamis during the period of 1790 - 1990 were caused by earthquakes. In a theory for long waves, the vertical acceleration of water particles are negligible compared to the gravitational acceleration except for a oceanic propagation of tsunami or propagation with a long distance on a gentle slope such as in river and on a continental shelf [Kajiura, 1963]. Consequently, the vertical motion of water particles has not effect on the pressure distribution. The following equation with integrated form, which is called the shallow water theory, is widely used as the governing equation for tsunami propagation:

$$
\begin{cases}
\frac{\partial \eta}{\partial t} + \frac{\partial M}{\partial x} + \frac{\partial N}{\partial y} = 0 \\
\frac{\partial M}{\partial t} + \frac{\partial}{\partial x}\frac{M^2}{D} + \frac{\partial}{\partial y}\frac{MN}{D} + gD\frac{\partial \eta}{\partial x} + \frac{\tau_x}{\rho} = A\left(\frac{\partial^2 M}{\partial x^2} + \frac{\partial^2 M}{\partial y^2}\right) \\
\frac{\partial N}{\partial t} + \frac{\partial}{\partial x}\frac{MN}{D} + \frac{\partial}{\partial y}\frac{N^2}{D} + gD\frac{\partial \eta}{\partial y} + \frac{\tau_y}{\rho} = A\left(\frac{\partial^2 N}{\partial x^2} + \frac{\partial^2 N}{\partial y^2}\right)
\end{cases}
\tag{1}
$$

where x and y are the horizontal coordinates, z the vertical coordinate which are shown in Fig.1, η the vertical displacement of water surface above the still water surface, g the gravitational acceleration, D the total water depth given by $h + \eta$, τ_x and τ_y the bottom frictions in the $x-$ and $y-$ directions, A the horizontal eddy viscosity which is assumed to be constant in space and time. In many cases, the shear stress on a surface wave can be neglected. M and N are the discharge fluxes in the $x-$ and $y-$ directions defined by

$$
M = \int_{-h}^{\eta} u\,dz = \bar{u}D, \quad N = \int_{-h}^{\eta} v\,dz = \bar{v}D
\tag{2}
$$

For a stage of tsunami generation with area of the several tens of kilometers wide and long in a deep sea of several kilometers, the depth -to- length ration is on the order of 10^{-2} and the wave steepness is of order 10^{-3}, suggesting that the linear long wave theory neglecting nonlinear terms in Eq.(1) is a good first-order approximation.

2.2. Bottom friction

Although a bottom friction term affects a dynamics of a runup process and a propagation in the shallow region, its modeling is still in the process being developed, and a consensus on a proper form of roughness coefficient is not yet reached. The commonly used bottom friction is Manning formula which is given by

$$
\begin{cases}
\frac{\tau_x}{\rho} = \frac{gn^2}{D^{7/3}} M \sqrt{M^2 + N^2} \\
\frac{\tau_y}{\rho} = \frac{gn^2}{D^{7/3}} N \sqrt{M^2 + N^2}
\end{cases}
\tag{3}
$$

where n is the Manning's roughness which are familiar among civil engineers being given in Table 1 [for example, Linsley and Franzini, 1979]. The instability of friction is discussed in 3.3. The friction coefficient f and Manning's roughness n are related by

$$n = \sqrt{\frac{fD^{1/3}}{2g}} \qquad (4)$$

It is noted that f becomes rather larger as the total depth is shallower, as long as n is almost constant. Another formula used for bottom friction model is the Chésy formula with different types of roughness coefficient (Packwood and Peregrine, 1981). There formula including the Manning roughness gives the approximation under the steady flow, so that these are not applicable to an unsteady flow during a runup especially accompanying a turbulence. Study of designing of a proper roughness coefficient instead of commonly used Manning's coefficient is required.

2.3. A note on the convection term

The other expression of the shallow water equation by using the averaged velocities in $x-$ and $y-$ directions is often introduced :

$$\begin{cases} \frac{\partial \eta}{\partial t} + \frac{\partial [\bar{u}D]}{\partial x} + \frac{\partial [\bar{v}D]}{\partial y} = 0 \\ \frac{\partial \bar{u}}{\partial t} + \bar{u}\frac{\partial \bar{u}}{\partial y} + \bar{v}\frac{\partial \bar{u}}{\partial y} + g\frac{\partial \eta}{\partial x} + \frac{gn^2}{D^{1/3}}\bar{u}\sqrt{\bar{u}^2 + \bar{v}^2} = 0 \\ \frac{\partial \bar{v}}{\partial t} + \bar{u}\frac{\partial \bar{v}}{\partial y} + \bar{v}\frac{\partial \bar{v}}{\partial y} + g\frac{\partial \eta}{\partial y} + \frac{gn^2}{D^{1/3}}\bar{v}\sqrt{\bar{u}^2 + \bar{v}^2} = 0 \end{cases} \qquad (5)$$

We should be careful to estimate the value of convection terms in the above equation that the terms in Eq.(5) are not same as those in Eq.(1). For example, the convection terms in the momentum equation in the x-coordinate in Eq.(1) divided by D can be modified as

$$\frac{1}{D}\left\{ \frac{\partial}{\partial x}\left(\frac{M^2}{D}\right) + \frac{\partial}{\partial y}\left(\frac{MN}{D}\right) \right\} = \bar{u}\frac{\partial \bar{u}}{\partial x} + \bar{v}\frac{\partial \bar{u}}{\partial x} + \frac{\bar{u}}{D}\left\{ \frac{\partial [\bar{u}D]}{\partial x} + \frac{\partial [\bar{v}D]}{\partial y} \right\}$$

$$= \bar{u}\frac{\partial \bar{u}}{\partial x} + \bar{v}\frac{\partial \bar{u}}{\partial x} - \frac{\bar{u}}{D}\frac{\partial \eta}{\partial t} \qquad (6)$$

The above equation shows that the convection terms in Eq.(1) involves additional term in the right side of Eq.(6), which can be canceled by the other in the acceleration of Eq.(5). Therefore total balance of term in the momentum equation in both equations are same, however each term is different, meaning it is not possible to compare each inertia terms in the Eq. of momentum.

3. Numerical Scheme

3.1. FDM for tsunami simulation

From the early stage of developing numerical simulation, the *finite difference method* based upon the *Taylor expansion series* has one of most fundamental and standard numerical methods. In the finite-difference approach, a continuous problem domain is descretized so that the dependent variables are considered to exist only at discrete points. Then derivatives are approximated by differences resulting in an algebraic representation of the partial differential equation (PDE) using the Taylor expansion. Until now numerous numerical schemes in FDM have been proposed. It should be reminded that several consideration regarding a scheme, grid size, stability condition, discretization errors and so on are required to know whether the solution obtained can be a good approximation to the exact solution of the original PDE.

Firstly, we discuss a numerical scheme in marching problems in which dependent values are varied in time. The numerical schemes in marching problems can be categorized into two; an *explicit scheme* for which only one unknown appears in the difference equation to permit evaluation in term of known quantities and an *implicit one* for which several unknown would appear in the equation, requiring the simultaneous solution of several equations involving the unknowns.

Three different numerical schemes; the staggered leap-frog, the Crank-Nicholson and two-step Lax-Wendroff scheme, are widely applied to simulate long waves. Numerical code of TUNAMI (Tohoku University Numerical Analysis Model for Investigation of tsunamis), models by Aida(1969) and Satake(1995) are based on the *staggered leap-frog scheme* which is an explicit scheme with the second order approximation. However an explicit scheme does not require much CPU time in calculation for one time step, most of them have each criteria to obtain stable results, restricting the grid size in spatial and time. *Crank Nicolson* scheme is most basic implicit one. Since large CPU time and memory are necessary in this scheme, the original form of the Crank Nicolson has not used at the present. The *two-step Lax Wendroff* scheme has been used for the shock waves or bore on a sloping beach because of an advantage to reduce the numerical oscillation caused by discontinuity by the numerical dissipation effect, providing stable and smoothed results (Hibbered & Peregrine, 1979, Packwood, 1980, and Kobayashi & Greenwald, 1987).

3.2. Staggered leap-frog scheme

Let us introduce the example of scheme (Imamura, 1995). The linearized long wave equation without bottom frictions in a one dimensional propagation is selected here as the governing equation in the deep sea region, which is given by

$$\frac{\partial \eta}{\partial t} + \frac{\partial M}{\partial x} = 0, \quad \frac{\partial M}{\partial t} + C_o^2 \frac{\partial \eta}{\partial x} = 0 \tag{7}$$

Where C_o is the propagation velocity of the linear long waves.

The discretized equations by the staggered leap-frog scheme is obtained by

$$
\begin{cases}
\dfrac{\left[\eta_{j+\frac{1}{2}}^{n+\frac{1}{2}} - \eta_{j+\frac{1}{2}}^{n-\frac{1}{2}}\right]}{\Delta t} + \dfrac{[M_{j+1}^n - M_j^n]}{\Delta x} + O(\Delta x^2) = 0 \\[3mm]
\dfrac{[M_j^{n+1} - M_j^n]}{\Delta t} + g\dfrac{\left(h_{j+\frac{1}{2}} + h_{j-\frac{1}{2}}\right)}{2}\dfrac{\left[\eta_{j+\frac{1}{2}}^{n+\frac{1}{2}} - \eta_{j-\frac{1}{2}}^{n+\frac{1}{2}}\right]}{\Delta x} + O(\Delta x^2) = 0
\end{cases}
\tag{8}
$$

where Δx and Δt are the grid sizes in the directions of x and t, and $O(\Delta x^2)$ is the truncation error of the second order approximation, which is the difference between the partial derivative and its finite differential representation. For dealing with discrete values in numerical computations, $\eta(x, y)$ and $M(x, t)$ are expressed for the case of the leap frog scheme as

$$
\begin{cases}
\eta(x, t) = \eta\left((j + \frac{1}{2})\Delta x, (n + \frac{1}{2}\Delta t)\right) = \eta_{j+\frac{1}{2}}^{n+\frac{1}{2}} \\[2mm]
M(x, t) = M(j\Delta x, n\Delta t) = M_j^n
\end{cases}
\tag{9}
$$

The point schematics for the numerical scheme are illustrated in Fig.2. Note that Eq.(8) staggered in space and time is different from the other type of the *leap-frog* scheme [Meisnge and Arakawa, 1976] staggered in only space with the second order of approximation in space and only the first order in time.

3.3. Scheme for bottom friction

A friction term also becomes a source of instability if it is discretized with an explicit scheme [Ramming and Kowalik,1980]. To make discussion of instability simply, let us take the following linearized momentum equation without convection terms:

$$
\frac{\partial M}{\partial t} + gD\frac{\partial \eta}{\partial x} + \frac{gn^2}{D^{7/3}}M|M| = 0
\tag{10}
$$

The explicit form of Eq.(10) is :

$$
M^{n+1} = \left[1 - \frac{\Delta t g n^2 |M|}{D^{7/3}}\right]M^n - gD\Delta t\frac{\partial \eta}{\partial x}
\tag{11}
$$

When a velocity become large or a total depth is small in a shallow region, the absolute of coefficient (amplification factor) of the first term on the right-hand side in Eq.(11) become more than unity, which amplifies proceeding velocity and leads numerical instability. In order to overcome this problem, an implicit scheme to set a friction term can be basically introduced. For example, a simple implicit form is

$$
M^{n+1} = M^n / \left[1 + \frac{\Delta t g n^2 |M|}{D^{7/3}}\right] - gD\Delta t\frac{\partial \eta}{\partial x} / \left[1 + \frac{\Delta t g n^2 |M|}{D^{7/3}}\right]
\tag{12}
$$

which ensures numerical stability, because the amplification factor on the right-hand side in Eq.(12) is always less than unity. However an effect of friction in a shallow water become so large that numerical results would be fairly dumped. Another implicit form, a combined implicit one for the friction term is given by

$$M^{n+1} = M^n \frac{\left[1 - \frac{\Delta t g n^2 |M|}{2D^{7/3}}\right]}{\left[1 + \frac{\Delta t g n^2 |M|}{2D^{7/3}}\right]} - gD\Delta t \frac{\partial \eta}{\partial x} \frac{1}{\left[1 + \frac{\Delta t g n^2 |M|}{2D^{7/3}}\right]} \tag{13}$$

This scheme also gives a stable result. The amplification factor is kept to be less than unity and is larger than that in Eq.(12), meaning that a numerical dissipation effect would be larger. The above scheme causes a numerical oscillation at a wave front because the amplification factor could be negative in a shallow region with small total depth, meaning that a water mass is pulled backward by unrealistically large friction term. We should select the best scheme among some implicit schemes to apply the bottom friction term with Manning's roughness.

4. Stability

Since tsunami propagation is categorized into marching problems, the concept of stability is necessary to obtain a reasonable results. In our experiences of numerical simulations, a numerical result is un-expected diverged depending on grid size and time step, which is caused by instability of numerical scheme. In order to avoid such an instability, a condition to assure the stability is specified in each scheme. Generally speaking, a stable scheme is one for which errors from any sources (round-off, truncation and so on) are not permitted to grow in the sequence of numerical procedures as the calculation proceeds from one marching step to the next.

Fourier or von Nuemann analysis is widely applied to obtain the stability condition for a given numerical scheme as long as the numerical scheme is linear (Anderson et a., 1984). Let ε represent the rounding-off error in the numerical solution. The numerical solution, F, actually calculated by using a numerical scheme may be written as follows:

$$F = D + \varepsilon \tag{14}$$

where D the exact solution of a finite difference equation. The computed numerical solution F must satisfy a given finite difference equation. For example, substitute the central difference scheme into the convection equation,

$$\frac{\partial F}{\partial t} + C \frac{\partial F}{\partial x} = 0 \tag{15}$$

yields

$$D_j^{n+1} + \varepsilon_j^{n+1} = D_j^n + \varepsilon_j^n - C \frac{\Delta t}{2\Delta x} \left[D_{j+1}^n + \varepsilon_{j+1}^n - D_{j-1}^n - \varepsilon_{j-1}^n\right] \tag{16}$$

Since the exact solution D must also satisfy the difference equation, the same is true of

$$\varepsilon_j^{n+1} = \varepsilon_j^n - C\frac{\Delta t}{2\Delta x}\left[\varepsilon_{j+1}^n - \varepsilon_{j-1}^n\right] \tag{17}$$

It, therefore, is found that D and ε must both satisfy the same difference equation. This means that the numerical error and the exact numerical solution both possess the same growth property in time, which is very important result for stability analysis and make possible to study their progress analytically. For nonlinear problems, the above relationship can not be obtained and the finite difference equation for numerical errors involves D in the previous time step, indicating not only a numerical scheme but also the initial value or condition strongly affect the behavior of errors. That is the reason why the nonlinear stability analysis is not well-understood. The error can be expressed in a Fourier series as follows :

$$\varepsilon(x,t) = \sum_{m}^{\infty} e^{at} e^{ik_m x} \tag{18}$$

where k is real but a may be complex, and i means the complex index only in this section. If Eq.(18) is substituted into Eq.(17), we obtain

$$e^{a\Delta t} = 1 - \frac{r}{2}\left[e^{ik_m\Delta x} - e^{-ik_m\Delta x}\right] = 1 - r\cos\beta \tag{19}$$

where $r = C\Delta t/\Delta x$ the ration of the wave celerity to the numerical speed which is called the Courant number. $\beta = k_m\Delta x$ is the non-dimensional wave number. The coefficient of $|e^{a\Delta t}|$ implies the amplification factor which is the ratio of the value in one time step to that in the previous one. It is clear that if $|e^{a\Delta t}|$ is less than or equal to one, the error will not grow from one marching step to the next one, which is called the stability condition. However, $|e^{a\Delta t}|$ in Eq.(19) is not always satisfied with the stability condition, as a result this numerical scheme lead instable results. The stability condition in Eq.(19) is that r is less than or equal to one.

5. Truncation error and accuracy of simulation

5.1. Truncation error

It is our experiences that even when a computations is stably carried out, the computed results often behave against what is expected from the partial differential equation we used. These errors are introduced by a Truncation Error (TE) inevitably cause by discretization with a numerical scheme, and the result in the damping of wave height or numerical oscillations behind the main wave are found. For the analysis of the TE to know the accuracy of a given numerical scheme, the von Nuemann analysis can be also be applied. Because the finite difference equations for the exact solution of FDM are the same as those for numerical errors in linear problems. For stability analysis, only absolute value of the error (amplification factor) expressed by Fourier series is concerned, however, for the analysis of the TE, a phase angle (ϕ) as well as amplitude $|G|$ defined by

$$e^{a\Delta t} = |G|e^{i\phi} \tag{20}$$

should be discussed. $|G|$ affects on the damping of wave height (*dissipation*) and ϕ change a wave celerity depending on a wave number (*dispersion*). For example, the staggered leap-frog scheme has

$$e^{a\Delta t} = e^{ik_m C_1 \Delta t}$$

$$\text{where} \quad C_1 = \frac{C_o}{K\beta} \sin^{-1}[K \sin \beta] \quad \text{and} \quad \beta = \frac{K_m \Delta x}{2} \tag{21}$$

The above relationship indicates that the scheme involves only numerical dispersion effect (wave celerity is varied by the wave number) as shown in Fig.3 while it does not have any dissipation on a amplitude. Moreover, it is shown that the magnitude of numerical dispersion strongly depends on the non-dimensional wave number, β. Imamura and Goto (1989) obtains the amplification coefficient in the other schemes and show that the TE including dissipation and dispersion become larger as β is increased. Therefore, in order to reduce the TE, the high resolution keeping β small is required. In the other word, good accuracy of numerical solutions are obtained as long as there are enough grid points per wave length. Shuto et al.(1986) suggested the critical resolution of at least 20 spacial grid points through numerical experiments in FDM and FEM. Note that required resolution depends on the numerical scheme and method.

5.2. High order approximation method

There are two ways to obtain high accurate solutions; reducing the TE in a given scheme or keeping high resolution in a whole computational region. The study of the first approach is devoted by Kowalik(1993), and Sayama et al.(1987). Kowalik(1993) introduced the fourth-order scheme in time as well as space for the linear long wave equation by eliminating a TE of third derivative terms in the three-time-level leap-frog scheme. The study of Sayama et al. (1987) is more challenging, in which the TE of third derivative are eliminated and the physical dispersion terms in the linear Boussinesq equation are added. Since the both possess the same third derivative with different coefficients, the TE is to be converted into the dispersion term. As a result, the procedure of the new method is the same as the linear Boussinesq equation, however the numerical results are expected to be more accurate.

Another type of study is proposed by Imamura and Shuto (1990) that a physical dispersion term is replaced by a numerical dispersion inevitably introduced in the linear long wave theory by the discretization. Setting equal the coefficients of the thirds terms in equation of the linear Boussinesq and the modified differential equation from the FDM of the linear long wave theory, it makes possible that the linear long wave theory gives the same results as the linear Boussinesq equation, thereby saving much the CPU time and computer memory.

As for the latter approach keeping high resolution, Titov (1989) proposed the numerical modeling by using variable grid in a computational region to keep a resolution even in the shallow region where a wave length is shortened approaching to a shore. On the other hand Shuto et al. (1990) and Liu et al. (1993) developed a nested multiple grid system that a smaller grid system in a shallower region is nested in a large grid system in a deeper region. At a certain time level, discharge and water level in each grid systems are calculated and the values of those along the boundaries between two grid systems are exchanged. Repeating the procedure, the calculation in the all systems is proceeded.

6. Numerical errors

Two numerical errors ; rounding-off and truncation errors, are very important to study the accuracy of a numerical simulation. Any computed solution may be affected by rounding to a finite number of digits in the arithmetic operations, which caused the rounding-off error. The magnitude of rounding-off error is essentially proportional to the number of grid points in the problem domain. On the other hand, the truncation error is caused by discretization operation to replace the particle differentiation by the finite difference representation. As shown in Eq.(3), the magnitude of truncation error is expressed by n-th power of Δx. Therefore, refining the grid may decrease the truncation error but increase the rounding-off error as shown in Fig.4.

The example of tsunami simulation in the one-dimensional problems done by Imamura and Goto (1986) shows that the magnitude of the rounding-off error ranging from 10^{-4} to 10^{-2} % in the case of simple precision with 7 digits number is much smaller than the truncation error of 10^{-1} to 10^{1} %. As long as the numerical scheme is satisfied with the stability condition, the rounding-off error can be negligible in the tsunami simulation.

For additional information, there is another numerical error; a discretization error which is defined as the truncation error plus the error introduced by the treatment of boundary condition. For example, a numerical oscillation would be generated by the notched geometry of coastal line which is poorly approximated by the rectangular grids. Moreover, poor approximation of topography causes large error in estimation of tsunami heights, even if a higher-order numerical scheme is used. So far any proper criteria to judge topographical approximation is not introduced because it involves error and limitation in surveying and natural geometry itself is complicated like a fractal geometry.

7. Boundary conditions for a runup

7.1. Condition of wave front and its estimation

Runup is taken into consideration only in nonlinear computations but not in linear computations because the linear theory is not allowed to be applied in a shallow region including a land area. Whether a cell is dry or submerged is judged as follows :

$$D = \begin{cases} h + \eta > 0, & \text{then the cell is submerged} \\ h + \eta \leq 0, & \text{then the cell is dry} \end{cases} \quad (22)$$

Figure 5 shows the example in the case of staggered scheme in space and time that the cell at $j + 1/2$ is dry while one at $j - 1/2$ is submerged or wet. The wave front should be located between the "dry " and " submerged " cells. Generally, discharge across the boundary between the two cells is calculated if the ground height in the dry cell is lower than the water level in the submerged cell (case A in Fig.5). In other cases, discharge is considered to be zero (case B in Fig.5).

This is a unique approach for a runup condition to introduce a "slot" at every grid points into the equation of continuity (Bundgaard and Warren, 1991). A "slot" is a very narrow channel shown in Fig.6 setting every gird points be " wet" point at any time during a simulation, therefore the concept of wave front between dry and wet cell is not necessary. When a water surface reaches at a certain level, it is considered to be submerged or wet. However, there is problem to determine the proper slot size.

The previous studies regarding determination of a wave front or shoreline boundary point, are summarized as follows (ref.Fig.7). As a simplest method, Titov and Synolakis (1993) determine a wave front as the intersection of the beach with the horizontal projection of the last "wet" point. Another method is to use the equation of continuity or motion. Sielecki and Wurtele (1970) proposed the method by using the extrapolation of the sea level at the first dry point based on the continuity equation. On the other hand, Kim & Shimazu (1982) applied an equation of motion neglecting friction and inertia. This is not reasonable because Matsumoti (1983) theoretically shows the importance of both at a wave front. He determines the wave front by Whitham's assumptions which is a first good approximations in the tip region of flow on a dry bed. Kirkgöz (1983) uses the equation of motion with the including the shear stress on a bottom and inertia, and determine the height of front wave in a uprush zone.

7.2. Calculation of discharge

In the case of the staggered scheme in space and time, we don't need to determine the exact location of a wave front which is considered to be just located between a "dry" and "wet" two cell points. And the calculation the equation of motion and continuity are alternatively proceeded. However, there is one problem how we can calculate the discharge between the two cells, which strongly affect the water level at the next time step. The following is summary of methods to estimate the inflow discharge near a wave front shown in Fig.8. Iwasaki and Mano (1979) assume that the line connecting the water level and the bottom height gives the surface slope to the first-order approximation. Then Eq. of momentum without convection terms is directly used to calculate a discharge. Hibberd and Peregrine (1979) give a provisional water level in the dry cell on a linearly extrapolated water surface. Then, the discharge calculated with this provisional water level gives the total amount of water into the dry cell and the water depth in the cell. If necessary, the computation will be repeated with the water level thus modified. Aida (1977) and Houston & Butler (1979) evaluate the discharge into the dry cell with broad-crested weir formulas in which the water depth above the bottom of the dry cell is substituted. The coefficient of the discharge to $\sqrt{gh}$ should be determined by a flow condition such as the Froud number ($= u/\sqrt{gh}$). Imamura (1995) evaluates the discharge by applying directly the equation of momentum keeping the zero total depth on a first "dry" cell. In this model, the total depth at a point of the discharge is given by the difference between the ground level on a dry cell and the water level on a wet one, which could overcome the problem in Iwasaki and Mano (1979) that the discharge is evaluated even when level in submerged cell is lower than a ground level in a dry cell as shown the case-B in Fig.5.

These approximations are convenient to handle but introduce numerical errors in estimation of wave front, which is studied by Goto and Shuto (1983). The runup height computed with the Iwasaki-Mano method agrees with the theoretical solution with a 5

$$\Delta x/\alpha gT^2 < 4 \times 10^{-4} \tag{23}$$

With the Aida method, the condition is given by

$$\Delta x/\alpha gT^2 < 10^{-3} \tag{24}$$

in which $\alpha$the angle of slope, g is the gravitational acceleration, x is the spatial grid length, and T is the wave period.

8. Initial Conditions and others

The present program is only for tsunamis. No wind waves and tides are included here. The still water level is given by tides and is assumed constant during tsunamis are computed for a short time such as one or two hours. Accordingly, no motion is assumed up to the time n. It means in sea,

$$\eta_{i+\frac{1}{2},j+\frac{1}{2}}^{n-\frac{1}{2}} = 0, \quad \text{and} \quad M_{i,j}^n, N_{i,j}^n = 0 \tag{25}$$

For a runup computation on land, the initial water level η is equal to the ground height h.

$$\eta_{i+\frac{1}{2},j+\frac{1}{2}}^{n-\frac{1}{2}} = h_{i+\frac{1}{2},j+\frac{1}{2}} \tag{26}$$

It should be kept in mind that values of h take negative sign on land.

For tsunami simulation in a deep sea including a tsunami source, there are two kinds of initial condition with and without dynamic effects of fault motion (rupture velocity and rising time). If such dynamic effects can be negligible for tsunami initial propagation, the final deformation of the sea bottom caused by the fault is given as the initial water surface. On the other hand, in order to include such effect, the modified mass conservation equation should be used instead of that in Eq.(1).

$$\frac{\partial \eta}{\partial t} + \frac{\partial M}{\partial x} + \frac{\partial N}{\partial y} = \frac{\partial \xi}{\partial t} \tag{27}$$

where is the deformation of sea bottom.

As for boundary conditions of water overflows on structures, the Hom-ma formula is used when water overflows breakwaters and sea walls in the computation region. Discharge overflowing the structures is given by

$$Q = \begin{cases} \mu h_1 \sqrt{gh_1}, & \text{if } h_2 \leq \frac{2}{3}h_1 \\ \mu' h_1 \sqrt{g(h_2 - h_1)}, & \text{if } h_2 > \frac{2}{3}h_1 \end{cases} \tag{28}$$

where h_1 and h_2 are the water depths in front of and behind structures measured above the top of structures shown in Fig.9, μ=0.35 and μ'=2.6m.

Reference

Aida,I. (1969),Numerical experiments for the tsunami propagation -the 1964 Niigata tsunami and the 1968 Tokachi-oki tsunami, Bull.Earthq.Res.Inst., Vol.47, pp.673-700.

Aida,I.(1977), Numerical experiments for inundation of tsunamis, Susaki and Usa, in the Kochi prefecture, Bull.Earthq.Res.Inst., Vol.52, pp.441-460 (in Japanese).

Anderson,Da., J.C.Tannehill and R.H.Pletcher.(1984), Computational fluid mechanics and heat transfer, McGraw-Hill Book Company, 599p..

Bundgaard, H.I. and I.R. Warren (1991), Modelling of tsunami generation and run-up, Sci. Tsunami Hazard, pp.23-29.

Goto,C. and N.Shuto (1983), Numerical simulation of tsunami propagation and run-up, in Iida and T.Iwasaki (eds.), Tsunamis: Their Science and Engineering, Terra Science Pub.Co., pp.439-451.

Hansen,W. (1956),Theorie zur errechnung des wasserstands und der stromungen in randme eren nebst anwendungen, Tellus, 8, pp.287-300.

Hibberd,S. and D.H.Peregrine (1979), Surf and run-up on a beach: uniform bore, J.Fluid Mech., Vol.95, pp.322-345.

Houston, J.R. and H.L.Boutler (1979), A numerical model for tsunami inundation, WES Rech. Rep. HL-79-2.

Imamura,F and C.Goto (1986), Truncated error of tsunami numerical simulation by the finite difference method, Proc. of JSCE, vol.375, no.II-6, pp.241-250.

Imamura,F. and N.Shuto (1990), Tsunami propagation by use of numerical dispersion, Proc.Int.Smy.comp. Fluid Dynamics, Nogoya, pp.390-395.

Imamura,F (1995), Tsunami numerical simulation with the staggered leap-frog scheme, manuscript for TUNAMI code, School Civil Eng., Asian Insti. Tech.,45.p.

Isozaki,I. and S.Unoki (1964), The numerical computation of the tsunami in Tokyo bay caused by the Chilean earthquake in May, 1960, Studies on Oceanogr. Dedicated to Prof. Hidaka in commemoration of his Sixtieth Birthday, 389p..

Iwasaki,T. and A.Mano (1979), Two-dimensional numerical simulation of tsunami run-ups in the Eulerian description, Proc. 26th Conf.Coastal Eng., JSCE, pp.70-72 (in Japanese).

Kajiura,K. (1963), The leading wave of a tsunami, Bull.Earthq.Res.Inst., 41, pp.535-571.

Kim,S.K. and Y.Shimazu(1982), Simulation od tsunami runup and estimate of tsunami disaster by the expected great earthquake in the Tokai District, Central Japan, J.Earth Sci., Nagoya Univ., Vol.30, pp.1-30.

Kirkgz,M.S., (1983), Breaking and run-up of long waves, In K.Iida and T.Iwasaki, eds, Tsunamis: their science and engineering, Terra Sci., Tokyo, pp.467-478..

Kobayashi,N. and J.H. Greenwald (1987), Wave reflection and run-up on rough slopes, J.Waterway, Port , Coastal and Ocean Engng Div., ASCE, Vol.113, pp.282-298.

Kowalik,Z.(1993), Solution of the linear shallow water equations by the fourth-order leapfrog scheme, J.Geophys.Res.Let., Vol.98, pp.10205-102-9.

Linsley,R.K. and J.B.Franzini (1979),Water-Resources Engineering, 3rd edition, McGraw-Hill Kogakukusha.Ltd., 716pp.

Liu,P.L-F, C.E.Synolakis and H.Yeh (1991), Report on the international workshop on long-wave run-up, J.Fluid Mech., Vol.229, pp.675-688.

Liu,P.L-F, S.B.Yoon, S.N.Seo and Y.S.Cho (1993), Numerical simulations of tsunami inundation at Hilo, Hawaii, Proc. IUGG/IOC Int. Tsunami Symp., Wakayama, pp.

Matsutomi, H. (1983), Numerical analysis of the run-up of tsunamis on dry bed, In K.Iida and T.Iwasaki, eds, Tsunamis: their science and engineering, Terra Sci., Tokyo, pp.479-493.

Mesinger,F. and A.Arakawa (1976),Numerical methods used in a atmospheric models,GARP Publ. Ser.,17, 64pp.

Packwood,A.R. (1980), Surf and runup on beaches, Ph.D. thesis, University of Bristol, Bristol, UK.

Packwood,A.R. and D.H.Peregrine (1981), Surf and runup on beaches: models of viscous effect, Rep.AM-1-07, School Math., University of Bristol, Bristol, 34PP.

Ramming,H.G. and Z.Kowalik, 1980, Numerical modeling of marine hydrodynamics - Applications to dynamic physical prosses, Elsevier Oceanography Series 26, Elsevier Scientific Pub. Co.Ltd., 360p..

Satake,K. (1995), Linear and non-linear computations of the 1992 Nicaragua earthquake tsunami, topical issue of "Tsunamis : 1992-94", Pure and Applied Geophysics, Vol. 144, No.3/4, pp.455-470.

Sayama,J., F.Imamura, C.Goto and N.Shuto (1987), A study on a highly accurate numerical method for tsunami in Deep ocean, Proc. of Coastal Eng. in Japan, Vol.34, pp.177-181 (in Japanese).

Shuto et al., (1986), A study of numerical techniques on the tsunami propagation and run-up, Sci.Tsunami Hazard, Vol.4, pp.111-124.

Shuto,N, C,Goto and F.Imamura (1990), Numerical simulation as a means of warning for near-field tsunami, Coastal Eng.in Japan, Vol.33, No.2, pp.173-193.

Shuto,N.(1991), Numerical simulation of tsunamis -Its present and near future, Natural Hazards, Vol.4, pp.171-191.

Sielecki,A. and M.G.Wurtele (1970), The numerical integration of the nonlinear shallow-water equations with sloping boundaries, J.Comput.Phys., Vol.6, pp.219-236.

Titov,V.V., (1989), Numerical modeling of tsunami propagation by using variable grid, Proc. of IUGG/IOC Int. Tsunami Symp., Novosibirsk, pp.46-51.

Titov,V.V. and C.E.Synolakis (1993), A numerical study of wave runup of the September 1, 1992 Nicaraguan tsunami., Proc. of IUGG/IOC Int. Tsunami Symp.,Wakayama, pp.627-635.

Ueno,T.(1965), Numerical computations for the chilean earthquake tsunami, Oceanogr. Mag., Vol.17, pp.87-94.

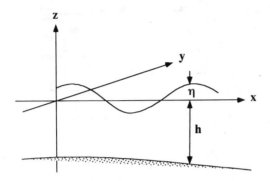

Figure 1 Coordinate sysytem

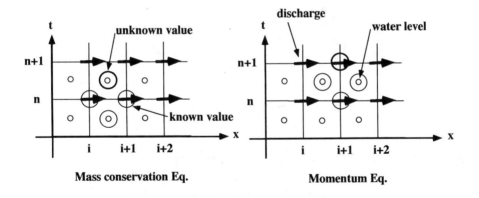

Figure 2 Arrangement of points for the computaions with the staggered leap-frog scheme

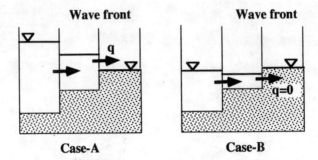

Figure 5 Boundary condition at a wave front

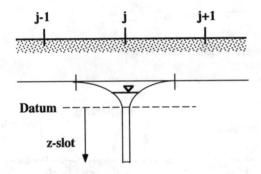

Figure 6 Description of a slot

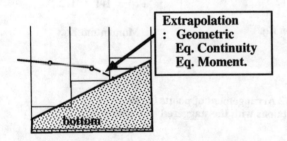

Figure 7 Estimation of wave front

Figure 7 Various moving boundary condition

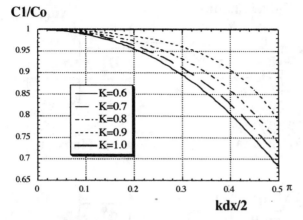

Figure 3 Numerical dispersion effect for the staggered leap-frog scheme with various Courant number (K) as a parameter

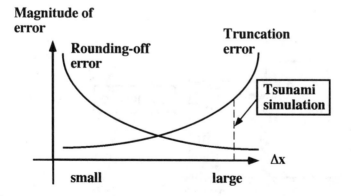

Figure 4 Comparison of magnitude between rounding-off and truncation errors

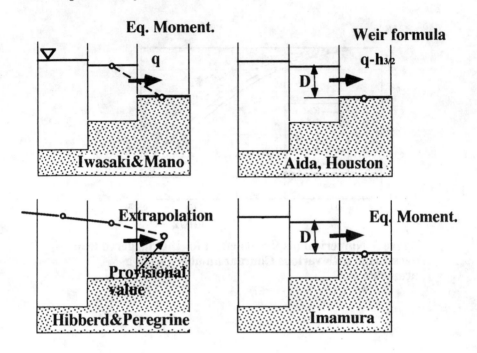

Figure 8 Various moving boudary condition

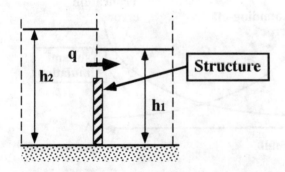

Figure 9 Overflows on structure

Review of Finite Element Methods for Tsunami Simulation

Roy A. Walters[a] and Toshimitsu Takagi[b]

1. Introduction

The finite element method has developed as a numerical technique that can sucessfully solve a variety of partial differential equations that arise in different fields of science. After extensive application in structural mechanics, this method was applied to fluid dynamics problems, which are described by differential equations such as the Navier-Stokes and Reynolds equations, as well as more restrictive equations such as the shallow-water equations, Boussinesq equations, and others. The focus of this review is the finite element methods which are applicable to long-water-wave dynamics, especially tsunami propagation with wave runup.

In the last two decades, there has been an increasing growth of applications of finite element methods to fluid dynamics problems[96,93]. This is a result of several factors, some of which are: the methods have reached a degree of maturity and have become more acceptable, there is a desire to study more complicated domains which is a strong point of finite element methods, and computer power has increased to the level where large-scale, high-resolution problems are feasible.

However, in a review such as this, it is not possible to cover all aspects of this field. In particular, the focus will be on useful methods to examine the tsunami propagation and wave runup problem. At this time, three-dimensional, free-surface simulations using Navier-Stokes equations are not practical for technical as well as resource reasons. In the future, a coupled wave propagation model and a three-dimensional nearshore model to describe wave runup may be desirable. For the present, however, the subject matter is restricted to long-wave phenomena that can be described by the shallow-water equations or some modified form of them.

An important enabling technology for successful model simulations is an automated grid generation procedure. The finite element grid is usually constructed of triangular or rectangular elements that conform to the boundary geometry which can change in time, and have element sizes that depend on both the equations being solved and the topography. It is imperative to have the grid generation carried out by an automated method that ensures grid quality and speeds the process. Recent improvements in grid generation techniques and computer technology have made this

[a]U.S. Geological Survey, 1201 Pacific Avenue, Tacoma, WA
[b]INA Corporation, Tokyo, Japan

possible and a number of public and commercial packages are available.

As yet, there are few tsunami simulation models that use the finite element method. Kawahara et al[43] and Kawahara and Nakazawa[41] applied a two-step explicit scheme to an analysis of the tsunamis resulting from the Tokachi-oki and Kanto earthquakes in Japan. Kawahara et al[40] and Kodama et al[47] present a treatment of an open boundary condition for the shallow-water equations and showed results for the tsunami caused by the Kanto Earthquake. Myers et al[63] present a simulation of the 1993 Hokkaido tsunami. In addition, there is some work with solitary waves[35,60]. All these analyses are based on the shallow-water equations. On the other hand, Houston[29] presented an analysis of the tsunami in the Hawaiian islands, that is based on the Helmholz equation.

In tsunami analysis, it is necessary to compute not only wave propagation but also wave runup because most damage to property and human life is caused by the wave runup. Modeling wave propagation involves choosing the correct mathematical description and using accurate numerical methods. Modeling wave runup with moving boundaries is more a numerical problem than formulation problem. There are two basic approaches: the edge of the grid moves with the waterline and the elements deform either locally or globally[51,53], or elements are active or inactive depending whether they have dry nodes[39,28,83,81,82,44,36]. The former is the more accurate method to solve this problem but can involve an overwhelming amount of computational overhead. The latter is simpler and directly analogous to the strategies used in finite difference methods, but presents an oversimplification of the phenomenon and can lead to inaccuracies. Some techniques for dealing with this problem are discussed in the sections on moving boundaries (Sections 2.5.3 and 3.9.2).

The following sections are organized so as to flow from the more general aspects of the problem to the details. In the next section, the mathematical description of the problem is presented. Following this is a presentation of the numerical methods along with a description of the numerical implementation of the boundary conditions and and some matrix solution strategies. The section after this contains information on errors, particularly related to analysis of the discrete dispersion relation, convergence rates, and error estimates. After this, an overview of grid generation is presented. Finally, this review lists some applications and ends with concluding remarks.

2. Mathematical Formulation

2.1. General Remarks

In general, there is a much greater range of strategies for solving the shallow-water equations when using finite element methods than when using finite difference methods. This is related to the early development of solution methods in both fields. In the finite difference method, a major problem with computational noise was solved with the C-grid, a fully staggered finite difference computational cell[59] (see Section 4 for more details). Thus, most models tend to use the primitive shallow-water equations combined with finite difference stencils that use the C-grid. However, in the finite element method, no equivalant of the C-grid has been found (except in one-dimension) so that other strategies have developed to solve this problem. The basic choice is to use primitive equations with modified elements or procedures, or to modify the governing equations and use standard elements. This was some of the impetus behind the formulation of a wave equation and other forms of the governing equations. In addition, recasting the dependent variables, such as in the TELEMAC model (see Section 3.7), results in an easy description of moving boundaries. Hence, the reader will encounter several forms of the governing equations, each of which has particular advantages. All of the methods described here are applicable to the tsunami problem. The particular method adopted will depend on the details of the actual problem being solved.

2.2. Primitive Equations

2.2.1. Shallow-Water Equations

The shallow-water equations are derived by vertically integrating the Navier-Stokes or Reynolds equations[70] that are derived in a rotating frame of reference. In vector form, they consist of the continuity equation

$$\frac{\partial \eta}{\partial t} + \nabla \cdot (h + \eta)\overline{\mathbf{u}} = 0, \tag{1}$$

and the horizontal components of the momentum equation

$$\frac{\partial (H\overline{\mathbf{u}})}{\partial t} + \nabla \cdot (H\overline{\mathbf{u}\mathbf{u}}) + \mathbf{f} \times (H\overline{\mathbf{u}}) - \nabla \cdot (HA_h \nabla \overline{\mathbf{u}}) + gH\nabla \eta - \frac{\tau_s}{\rho} + \frac{\tau_b}{\rho} = H\overline{\mathbf{F}} \tag{2}$$

where bottom boundary conditions are given as

$$A_v \frac{\partial \mathbf{u}}{\partial z} = \frac{\tau_b}{\rho} = C_f |\mathbf{u}| \mathbf{u} \qquad (z = -h), \tag{3}$$

essential boundary conditions on η are set at open boundaries, and $(\overline{\mathbf{u}} \cdot \hat{\mathbf{n}}) = 0$ (no normal flow, where $\hat{\mathbf{n}}$ is the unit vector normal to the boundary) is applied on land boundaries. For the tsunami problem, surface friction effects caused by wind and baroclinic effects may be ignored.

The various terms appearing in these equations are defined as follows. The surface elevation relative to mean sea level is given as $\eta(x, y, t)$, $\mathbf{u}(x, y, z, t)$ is the horizontal velocity, and $\overline{\mathbf{u}}(x, y, t)$ is the vertical or depth average of $\mathbf{u}$. The water depth from mean sea level is given as $h(x, y)$, while $H(x, y)$ is the total water depth, $H = h + \eta$; $\mathbf{f}$ is the Coriolis vector, and g is the gravitational acceleration. The surface stress is denoted as τ_s, and τ_b denotes the bottom stress; C_f is the bottom stress coefficient, and ρ is a reference density. $\mathbf{F}$ represents body forces such as density gradient forces, and

∇ is the horizontal gradient operator $(\partial/\partial x, \partial/\partial y)$. $A_h(x, y, z, t)$ is the coefficient for the horizontal component of viscous stresses. In many models, A_h is usually assumed to vanish. However, this term can be used as a shock-capturing term to dissipate small-scale features near fronts or other steep gradients.

Generally, the coefficient C_f is determined by Manning's or Chezy's formula as follows:

$$Manning's \quad formula \quad : \quad C_f = \frac{gn^2}{(h+\eta)^{1/3}} \tag{4}$$

$$Chezy's \quad formula \quad : \quad C_f = \frac{g}{C^2} \tag{5}$$

where n is the Manning coefficient and C is the Chezy coefficient.

In order to close (2), the advection term and bottom stress term must be expressed in terms of the depth-averaged variables. Eq. (2) is typically used in a two-dimensional (horizontal) model where the variations in the vertical are known or are insignificant. Thus a coefficient α can be defined that takes into account the autocorrelation of the vertical variation in $\mathbf{u}$ and can be used to express the advection term as a function of $\overline{\mathbf{u}}$ [70]. A common procedure is to set $H\overline{\mathbf{uu}} = H\alpha\overline{\mathbf{u}}\,\overline{\mathbf{u}}$ and $\alpha = 1$. In the bottom stress term, $\mathbf{u}$ is replaced by $\overline{\mathbf{u}}$ and the coefficient C_f changed accordingly.

There are several options for the formulation of the momentum equation. The equation can be stated in conservative or non-conservative form by applying the continuity equation to the first two terms in (2).

With the assumptions stated above for tsunami simulation, the momentum equation is expressed as

$$\frac{\partial \overline{\mathbf{u}}}{\partial t} + \overline{\mathbf{u}} \cdot \nabla \overline{\mathbf{u}} + \mathbf{f} \times \overline{\mathbf{u}} - \nabla \cdot (A_h \nabla \overline{\mathbf{u}}) + g\nabla\eta + \frac{C_f}{(h+\eta)}C_f|\overline{\mathbf{u}}|\overline{\mathbf{u}} = 0 \tag{6}$$

and the continuity equation remains as in Eq. (1).

2.2.2. Boussinesq Equation

The Boussinesq equations resemble the shallow-water equations but contain high-order terms to account for dispersion[68,35]. In this way, dispersion of gravity waves can be accounted for without resorting to three-dimensional wave potential theory. Following the development by Peregrine[68,17], the Boussinesq equations can be written as:

$$\frac{\partial \overline{\mathbf{u}}}{\partial t} + \overline{\mathbf{u}} \cdot \nabla \overline{\mathbf{u}} + \mathbf{f} \times \overline{\mathbf{u}} + g\nabla\eta = \frac{\partial}{\partial t}\left[\frac{h}{2}\nabla(\overline{\mathbf{u}} \cdot \nabla h) + \frac{h^2}{3}\nabla(\nabla \cdot \overline{\mathbf{u}})\right] \tag{7}$$

and the continuity equation is the same as in Eq. (1). These equations are correct to $O(\epsilon^2)$ [68,17], where $\epsilon = \eta/h$.

From a modeling perspective, these equations can be solved in the same manner as the shallow-water equations, but with an additional loading term. The element interpolation functions must be of sufficiently high order to approximate the derivatives in this new term, but otherwise there is no new complications in the discretization. The complications arise in the time discretization which results in a set of equations that require a matrix solution. Some strategies for treating this are mentioned in the section on solution methods.

2.3. Wave Equation

A number of methods that use modified forms of the governing equations have emerged to solve the shallow-water problem. Of the several computational forms of these equations that have been explored in the literature, one of the most effective is the wave continuity equation formulation[66,76,52,49,55,93,86,87]. This form of the equations has been studied in the context of both time stepping and harmonic expansion schemes. Among its many advantages are computational efficiency, freedom from grid scale noise and a demonstrated robustness on field scale problems.

The continuity and momentum equations can be combined to give a wave equation that replaces the continuity equation. In this way, the solution for surface elevation and velocity are decoupled[52,55]. Differentiating Eq. (1) with respect to time and replacing the second term with terms from the momentum equation, the wave equation can be conveniently written as

$$\frac{\partial^2 \eta}{\partial t^2} + \gamma_0 \frac{\partial \eta}{\partial t} - \nabla \cdot (gH\nabla\eta) = \nabla \cdot \mathbf{R}, \tag{8}$$

where

$$\mathbf{R} = \nabla \cdot (H\overline{\mathbf{u}}\overline{\mathbf{u}}) + \mathbf{f} \times H\overline{\mathbf{u}} - \nabla \cdot (HA_h\nabla\overline{\mathbf{u}}) - \gamma_0 H\overline{\mathbf{u}} - \frac{\tau_s}{\rho} + \frac{\tau_b}{\rho} - H\overline{\mathbf{F}} \tag{9}$$

There are several variations of the equations above. Perhaps the most common is the addition of a penalty term that is a coefficient times the continuity equation[55,48]. This is included in Eq. (8) where the coefficient is γ_0.

2.4. TELEMAC

The TELEMAC model is directed towards several kinds of equations occurring in free surface hydraulics: shallow-water equations and mild slope equations[20,25,26]. With total water depth, H, rather than η as the dependent variable, the shallow-water equations can be expressed as follows:

$$\dot{H} \quad + \quad \mathbf{u} \cdot \nabla H + H\nabla \cdot \mathbf{u} = 0 \ , \tag{10}$$
$$\dot{\mathbf{u}} \quad + \quad \mathbf{u} \cdot \nabla \mathbf{u} + g\nabla H - \nabla \cdot (\nu\nabla\mathbf{u}) = \mathbf{S} - g\nabla h \tag{11}$$

where $\mathbf{S}$ is the source term that includes bottom friction, Coriolis force, wind stress, and other processes. A unique feature of this model is that the celerity of shallow-water waves $c = \sqrt{gh}$ is chosen as a new variable instead of the water depth H. With this transformation, the shallow water Eqs. (10) and (11) become

$$\dot{c} \ + \ \mathbf{u} \cdot \nabla c + \frac{c}{2} \nabla \cdot \mathbf{u} = 0 \ , \tag{12}$$

$$\dot{\mathbf{u}} \ + \ \mathbf{u} \cdot \nabla \mathbf{u} + 2c\nabla c - \nabla \cdot (\nu \nabla \mathbf{u}) = \mathbf{S} - g\nabla h \tag{13}$$

These equations are solved with a fractional step method with two steps: an advection step, and a step that includes propagation, diffusion, and sources. Further discussion is left to the Numerical Approximation section (Section 3.7). Using characteristics to handle the advection in both the continuity and momentum equations, this model gives excellent results for shocks such as hydraulic jumps. There is a rather heavy conputational overhead associated with characteristics, however.

A real asset of this model is the ability with which it simulates wetting and drying. Because c is used as one dependent variable, c and thus H go gracefully to zero when drying occurs, so no special treatment of moving boundaries or dry areas needs to be made.

2.5. Boundary and Initial Conditions

2.5.1. General Remarks

The boundary condition can be classified as shown in Figure 1. On the boundary S_1, where the velocity is given, the boundary condition is:

$$\mathbf{u} = \hat{\mathbf{u}} \qquad on \qquad S_1 \tag{14}$$

where the overhat denotes the prescribed value on the boundary. If only the normal component of velocity on the boundary is assumed to be zero, the so called "slip condition", the velocity is specified as follows:

$$\mathbf{u}_n = \hat{\mathbf{u}}_n = 0 \qquad on \qquad S_1' \tag{15}$$

where the subscript n denotes normal direction to the boundary.

On the boundary S_2, where the water elevation is given, the boundary condition is

$$\eta = \hat{\eta} \qquad on \qquad S_2. \tag{16}$$

In tidal models, the water elevation on the open boundary S_2 is given as a sum of periodic functions of time.

On the boundary $S_1' + S_2$, surface traction (stresses) are specified by

$$\tau = A_h \nabla \overline{\mathbf{u}} \cdot \mathbf{n} = \hat{\tau} \qquad on \qquad S_1' + S_2. \tag{17}$$

where A_h is a horizontal viscosity coefficent.

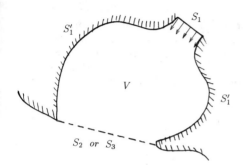

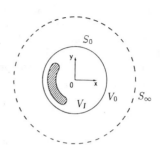

Fig. 1. Definitions for boundary conditions   Fig. 2. Definitions for open boundary

2.5.2. Open Boundary

In many problems, there are open boundaries which divide a domain into inner and outer regions as shown in Figure 2. For some problems, an incident wave amplitude may be specified at the open boundary. Nonetheless, any outgoing wave at the open boundary must be treated in such a way that it is not reflected back into the inner region. There are two general classes of methods that are used to accomplish this: A coupling of the inner region to a simple model for the outer region[67,12], and a direct application of a radiation boundary condition at the open boundary[74,32].

As an example of the first method, Pearson and Winter[67] use the method of characteristics to provide radiation boundary conditions to a part of a two-layer inlet governed by one-dimensional shallow-water theory. They solve for the inner problem using standard techniques, then solve for the boundary nodes by applying characteristics, a method which is largely independent of the spatial and temporal approximations.

Kawahara[40] derives an analytic solution for waves that propagate outward from the boundary. The analytic solutions are expressed as the sum of sinusoidal functions where, in his analysis, only outgoing waves through the open boundary are considered. These solutions are of the form

$$\left\{ \begin{array}{c} \overline{\mathbf{u}} \\ \overline{\eta} \end{array} \right\} = \overline{\eta}_0 \left\{ \begin{array}{c} \sqrt{g/h}\mathbf{k}/k \\ 1 \end{array} \right\} exp(-j\omega t)exp(j\mathbf{k}\cdot\mathbf{x}) \qquad (18)$$

where ω is angular frequency, k_i are the components of wave number $\mathbf{k}$, $j = \sqrt{-1}$, and $\overline{\eta}_0$ is an unknown constant. The wave number k at the open boundary can be determined by a wave period T and wave velocity C of the inner wave incident on the boundary. Hence, some information about the wave must be obtained before this analysis can be applied. The unknown constant in Eq. (18) is determined by requiring continuity of the wave in the inner and the outer regions[40]. The procedure to apply this technique to the finite element method as well as other methods to construct special finite elements, will be dicussed in the Numerical Approximation Section (Section 3).

The second technique is to apply radiation conditions directly. Commonly, the Sommerfeld condition

$$\frac{\partial \eta}{\partial t} \pm c \frac{\partial \eta}{\partial n} = 0 \tag{19}$$

is used at the open boundary[32], where $\frac{\partial \eta}{\partial n}$ is the derivative of η normal to the boundary. This is an approximation of the general condition derived by Sommerfeld[77] for the linear wave equation given by Eq. (8) with $\gamma_0 = 0$ and $H = constant$.

Johnson et al[32] derive a generalized radiation boundary condition for the wave equation in the form:

$$\frac{\partial^2 \eta}{\partial t^2} - c^2 \frac{\partial^2 \eta}{\partial x^2} + c_1 \eta + c_2 \frac{\partial \eta}{\partial x} + c_3 \frac{\partial \eta}{\partial t} + c_4 \frac{\partial^2 \eta}{\partial t \partial x} = 0 \tag{20}$$

The resulting radiation condition is

$$\frac{\partial^2 \eta}{\partial t^2} + (\frac{c^4}{2} \pm c) \frac{\partial^2 \eta}{\partial x \partial t} + \frac{c_1}{2} \eta + \frac{c_2}{2} \frac{\partial \eta}{\partial x} + \frac{c_3}{2} \frac{\partial \eta}{\partial t} = 0 \tag{21}$$

For the telegraph equation given by Eq. (8), this becomes

$$\frac{\partial^2 \eta}{\partial t^2} + c \frac{\partial^2 \eta}{\partial x \partial t} + \frac{\gamma_0}{2} \frac{\partial \eta}{\partial t} = 0 \tag{22}$$

Through a series of test cases, they demonstrated the utility of this expression. The Numerical Approximation section (Section 3) contains information on using this condition within the context of the finite element spatial approximation.

2.5.3. Moving Boundary

In many cases, the land boundary is considered to be a vertical wall. Thus the boundary conditions are a constraint on the normal component of velocity at a fixed boundary such as described by Eq. (15). However, for the runup problem, two additional conditions are required: a boundary that is a vertical wall but moves with time, and a boundary that floods and drys.

The moving wall type of boundary condition is applicable to the moving paddle in the runup benchmark test cases, and to a landslide or other earth movement into water. In general terms, this condition is expressed as

$$\hat{n} \cdot u_b = \hat{n} \cdot u \qquad on \qquad S_1(t) \tag{23}$$

where $\hat{n}$ is the normal to the boundary and u_b is the boundary velocity. This is the more general case for Eq. (15).

For the wetting/drying type of moving boundary, the line where $H = 0$ is tracked and the grid follows this line. This may be stated as

$$\frac{DH}{Dt} = \frac{\partial H}{\partial t} + \mathbf{u} \cdot \frac{\partial H}{\partial \hat{n}} = 0 \tag{24}$$

A sufficient condition is given by Eq. (23). In addition, the location of the boundary x_b is tracked using

$$H = 0 \qquad x_b = x_0 + \int_0^t u_b dt \tag{25}$$

This condition is similar to that for the moving wall; however, here the boundary velocity must be calculated rather than specified directly. Hence this condition is nonlinear because it depends on the solution for sea level and velocity.

In practice, there are many methods in which to enforce the moving boundary conditions. These methods are strongly linked to the form of the governing equations and to the numerical approximation used. Therefore, the details are left to the Numerical Approximation section (Section 3.9.2).

2.5.4. Initial Condition

As initial conditions, the water velocity and the water elevation are given as:

$$\mathbf{u} = \hat{\mathbf{u}}^0, \quad \eta = \hat{\eta}^0 \qquad at \quad t = t_0 \tag{26}$$

where, $\hat{\mathbf{u}}^0$ and $\hat{\eta}^0$ are the prescribed values of the water velocity and the water elevation at $t = t_0$. In the tsunami wave analysis, the initial water elevation is given by the bottom deformation estimated by the formulation, for instance presented by Mansinha and Smylie[58]. The initial water velocity in the computed area is usually assumed to be negligible.

3. Numerical Approximation

3.1. General Remarks

As mentioned in the previous section, a number of different procedures have evolved to solve the shallow-water equations. However, when using the finite element method, the spatial part of all these equations are solved in the same manner. First, a weighted residual statement is derived, then these equations are solved by discretizing space into a set of finite elements of triangular and/or quadrilateral shape. Most models use a time-marching procedure that is discrete in time. However, finite element methods can also be used in time by increasing the dimensions by one and using space-time elements. In this case, linear elements in time are frequently used.

3.2. Finite Element Method

The major difference between finite difference and finite element methods is the approximation method used for the governing equation. In the finite difference approach, each term is differenced separately with a certain stencil that is accurate to some chosen order. The basis of the finite element approach is to form approximating integral equations where finite elements form the local interpolation of the variables.

There are a variety of approaches to formulating the integral equations which are the foundation of the finite element method. For symmetric operators, the Rayleigh-Ritz procedure based on the calculus of variations has been used frequently[6]. A more general procedure is the method of weighted residuals of which the Bubnov-Galerkin method is a subset[6,70]. The more general procedure is applicable to the runup problem considered herein.

For a brief introduction to these concepts, consider the time-independent equation

$$Lu = f \qquad in \quad \Omega \qquad (27)$$

where the operator L acts on the unknown function u to generate the known function f in the bounded domain Ω. Next replace $u(x)$ with an approximation $\hat{u}(x)$ given by

$$\hat{u}(x) = \sum_{j=1}^{N} u(x_j)\phi_j(x) \qquad (28)$$

where x_j are the locations of the N computational nodes and $\phi_j(x)$ are the functions that are used for interpolation over Ω. These functions are variously known as basis, trial, or interpolation functions. They are chosen in a special way such that $\phi_j(x_i) = 1$ if $j = i$ and 0 otherwise.

Using these approximations, the residual is defined as

$$R = L\hat{u} - f = L\left[\sum_{j=1}^{N} u(x_j)\phi_j(x)\right] - f \qquad (29)$$

If the approximation were exact, the residual would vanish. In the method of weighted residuals, this residual is forced to zero, in an average sense, through selection of the $u(x_j)$. They are calculated by satisfying the constraints that arise when the weighted integrals of the residuals are set to zero.

$$\int_{\Omega} R(x)w_i(x)d\Omega = 0 \qquad (i = 1, 2, ..., N) \qquad (30)$$

In one procedure, the Galerkin or Bubnov-Galerkin method, the weighting functions are taken to be the basis functions ϕ.

For time-dependent problems, an additional term appears in Eq. 27 which can be approximated following the procedure above.

$$\frac{\partial u}{\partial t} = \sum_{j=1}^{N} \phi_j(x)\frac{\partial u_j}{\partial t} \qquad (31)$$

In the method of weighted residuals, this term becomes

$$\sum_{j=1}^{N} \int_{\Omega} w_i(x)\phi_j(x)\frac{\partial u_j}{\partial t}d\Omega = 0 \qquad (i = 1, 2, ..., N) \qquad (32)$$

A term that arises in all transient calculations is

$$M = \int_{\Omega} \phi_i(x)\phi_j(x)d\Omega \qquad (33)$$

where M is known as the mass matrix, and is symmetric and positive definite. This term appears frequently in the various procedures in the next section. Thus the transient problem has been decomposed into spatial approximation functions, and nodal variables that are time-dependent. A variety of procedures can be used to approximate the time derivative. The most common are the use of finite difference approximation in time, finite element approximation in space following the same procedure as above[70], and harmonic methods. These procedures are described in more detail below.

To apply the finite element method, the spatial domain is divided into a number of elements of triangular or quadrilateral shape. These elements cannot be highly distorted because of error constraints. In fact, one of the main characteristics of the finite element method is the use of non-structured grids which gives it great flexibility in the approximation of very complex spatial domains.

Next, the interpolation functions are chosen for the elements. Linear and quadratic functions are commonly used, although higher order methods can have very high convergence rates[88]. For a triangle with linear basis functions, the spatial interpolation is a plane surface in the element defined by the values of the variable at the three vertices. Interpolation functions employed in a number of finite element analyses for the shallow-water equations are listed in the references[37]. In unsteady flow problems, a low-order interpolation function, such as a linear polynomial based on three-node triangular finite elements is commonly used.

After the elements and interpolation functions are defined, the weighted residual on each element is calculated for the particular form of the governing equation used. This leads to highly modular and flexible code. For instance, the element subroutine can be changed from a wave problem to a diffusion problem with little or no change to the remaining code.

Next, the element contributions are summed (assembled) into a global matrix and this weighted residual statement is set to zero. This leads to N equations for the N unknowns, before the boundary conditions are enforced. This enforcement is dependent on the form of the equations used so the details are left to the sections that follow.

Finally, the global set of equations is solved for the dependent variables. There are two classes of solvers in common use: direct solvers of which frontal solvers are an example, and iterative solvers. Some of these methods with references are contained in sub-section 3.10.

3.3. Primitive Equations- Spatial Approximation

3.3.1. Galerkin Approach

The weighted-residual equations corresponding to the shallow-water equations (1) and (2) are derived by substituting for the operator L in Eq. (29) and forming Eq. (30). Using ψ as the weighting function for the continuity equation and ϕ for the momentum equations, the integral equations become

$$\int_\Omega \frac{\partial \eta}{\partial t}\psi d\Omega - \int_\Omega (h + \eta)\overline{\mathbf{u}} \cdot \nabla\psi d\Omega = -\oint \mathbf{Q} \cdot \mathbf{n}\psi d\Gamma, \qquad (34)$$

where the divergence term in (1) has been integrated by parts, $\mathbf{Q} = H\overline{\mathbf{u}}$, and

$$\int_\Omega \frac{\partial \overline{\mathbf{u}}}{\partial t}\phi d\Omega + \int_\Omega \overline{\mathbf{u}} \cdot \nabla\overline{\mathbf{u}} \; \phi d\Omega + \int_\Omega \mathbf{f} \times \overline{\mathbf{u}} \; \phi d\Omega + g\int_\Omega \nabla\eta \; \phi d\Omega$$

$$- \int_\Omega \frac{1}{H}\nabla \cdot (HA_n\nabla\overline{\mathbf{u}}) \, d\Omega - \int_\Omega \frac{\tau_s}{\rho H}\phi d\Omega + \int_\Omega \frac{\tau_b}{\rho H}\phi d\Omega = \int_\Omega \overline{\mathbf{F}}\phi d\Omega \qquad (35)$$

where the equation is written in non-conservative form with $\alpha = 1$. Here Ω is the computational domain with boundary Γ. For the continuity equation, essential conditions on η or the boundary integral for discharge are specified on all external boundaries. The natural boundary condition is that $\mathbf{Q} \cdot \mathbf{n} = 0$.

In practice, taking $\psi = \phi$ leads to equal-order interpolation, and the equations contain spurious computational modes that must be dealt with through damping or other means. The case where the interpolation functions are different leads to mixed-order interpolation which is discussed in the next sub-section.

Commonly, integration-by-parts is applied to various terms in the integral equations in order to reduce the order of the derivatives and allow lower-order basis functions, and to enable specification of the boundary conditions through boundary integrals. For instance, consider the divergence term in the continuity equation:

$$\int_\Omega \nabla \cdot (H\overline{\mathbf{u}})\psi d\Omega = -\int_\Omega (h + \eta)\overline{\mathbf{u}} \cdot \nabla\psi d\Omega + \oint \mathbf{Q} \cdot \mathbf{n}\psi d\Gamma \qquad (36)$$

In this case, $\mathbf{Q}$ is specified along the boundary where there are no essential conditions on η. A natural condition is that $\mathbf{Q} = 0$ or no flow normal to the boundary. In addition, the gravity term can be integrated by parts so that the boundary forcing for η appears as a line integral. Also, the conservative form of the advection term and lateral stress terms can be integrated by parts. Thus there appears a boundary term for the momentum transfer at the boundaries. A primary criterion for choosing these options is to provide a more convenient method to apply the boundary conditions.

The weighted residual equations for the Boussinesq equations can be obtained in the same way as above. In this case, there is an additional term on the right-hand side of Eq. (35). This term may be rearranged and integrated by parts in order to reduce the order of the required interpolation functions.

3.3.2. Mixed-Order Interpolation

Mixed-order interpolation was developed in an attempt to surpress the computational noise generated in some approximations to the primitive shallow-water equations that use the same order interpolation for both η and $\mathbf{u}$. In practice, η is approximated at one order less than $\mathbf{u}$ which removes the grid-scale noise in η but still requires some viscous damping for $\mathbf{u}$[85,89]. When this form of interpolation is used, η is usually approximated with linear bases and $\mathbf{u}$ with quadratic bases. This requires a six-node triangular element with three vertex nodes and three midside nodes to support the quadratic functions. The most useful quadrilateral element is the nine-node element with four corner nodes, four midside nodes, and one interior node. The eight-node element without the interior node is not recommended because of spurious results along shoreline boundary layers. The problem arises because the interpolation is linear through the center of the element, and quadratic along the edges.

In the weighted residual statement, different order weighting is used. The continuity equation is weighted with the bases for η, *i.e.*, linear bases, and the momentum equation is weighted with the bases for velocity, *i.e.*, quadratic bases. Thus there are two forms for the mass matrix, one using linear bases in the continuity equation, and one using quadratic bases in the momentum equation. This has a tendency to provide less constraints in a relative sense to the continuity equation than the momentum equation.

Most implementations of mixed interpolation for the shallow-water equations use quadratic elements coupled with implicit time integration[45,56,46,91]. This requires a full assembly and solution of a nonstationary matrix at each time-step, leading to rather large computational requirements for practical problems. A basin-scale simulation of a tsunami using this approach is beyond the capabilities of all but the most high-perfomance computers, although local studies of runup would be practical.

3.4. Primitive Equations- Time Approximation

3.4.1. General Remarks

There are many methods for treating the transient nature of problems. It is not practical to cover all the time-approximation methods in use; rather, only the more common methods are mentioned. A variety of methods may be found in the reference list, and in particular, in some of the standard references[6,50,97]. For the purpose of this review, there are four general techniques mentioned: explicit, semi-implicit, implicit, and harmonic methods. Explicit methods generally involve the least computational effort per time-step because the mass matrix is stationary and has other nice properties. However, there are stability constraints on the time-step size that can lead to overly small steps in deep water. Implicit methods have no time-step constraint (other than proper resolution of the signal) but involve the solution of a nonstationary matrix at each time-step. However, the time-step in this case can be sufficiently large that the computational time may be less than that for explicit methods. Cochet et al[13] present a comparison of these methods. Semi-implicit methods were developed to bridge the gap, to remove stability constraints while maintaining a stationary matrix. Finally, harmonic methods were developed for periodic problems and have computational efficiency several orders of magnitude greater than time-stepping methods. Although they are applicable to incident wave fields with line spectra, they are not applicable to the solitary wave found in many tsunami and runup problems. A variety of numerical-integration schemes in time that have been reported in the literature are summarized by Kawahara[37]. A high-order scheme is presented by Holz[27].

3.4.2. Explicit Time-Marching Methods

Consider a simple time-marching scheme, the forward difference in time (Euler scheme). Writing Eqs. (34) and (35) in a more general fashon,

$$M_{ij}\frac{\partial \eta_j}{\partial t} + F_i(\eta, \mathbf{u}) = 0 \tag{37}$$

$$M_{ij}\frac{\partial \mathbf{u}_j}{\partial t} + \mathbf{G}_i(\eta, \mathbf{u}) = 0 \tag{38}$$

Taking a forward difference in time:

$$M_{ij}\eta_j^{n+1} = M_{ij}\eta_j^n + \Delta t F_i^n \tag{39}$$

$$M_{ij}\mathbf{u}_j^{n+1} = M_{ij}\mathbf{u}_j^n + \Delta t \mathbf{G}_i^n \tag{40}$$

These equations are subject to stability constraints which relate Δt and Δx to the Courant number. This particular scheme is unconditionally unstable without friction[50]. However, there are a variety of other explicit schemes listed in the references[50,21].

The leapfrog scheme has three-levels in time and normally approximates friction at the k-1 level for stability. Among its good features, it decouples the equations for η and $\mathbf{u}$, requires no iteration for the full equations, and has only a stationary mass matrix to solve. The phase accuracy is fair in comparison with other schemes. However, this scheme has been observed to have severe difficulties with spurious computational modes[22]. An alternative semi-implicit scheme was proposed which retains the computational advantages of the leapfrog scheme, but reduces the computational noise[22].

The split step scheme decouples the equations by solving for η and $\mathbf{u}$ at different time levels[50,4,14,39,42,83]. In addition, the phase accuracy is better than the leapfrog scheme[50]. However, the full equations are not centered in time unless iteration is used, leading to only $O(\Delta t)$ accuracy.

Although these methods are explicit, there is still a matrix equation to solve because of the presence of the mass matrix. However, the matrix is stationary, symmetric, and positive definite so there are very efficient solvers available. Some of the methods that follow attempt to avoid the matrix solution by lumping the mass matrix on the diagonal (a procedure that makes the resulting equations look like the equivalent finite-difference equations). However, this leads to less accurate phase response in the model so caution must exercised in this regard. Finally, it is sometimes more efficient to use a semi-implicit or implicit time-marching procedure to gain much larger time-step size. However, accuracy usually decreases as Courant number is increased so again caution is required.

3.4.3. Two-Step Selective Lumping Method

In order to avoid a matrix solution and to acquire numerical stability, an explicit time-marching method with a lumping technique has been used by Kawahara et al[39]. It consists of two steps:
The first step is

$$\overline{M}_{ij}\eta_j^{n+\frac{1}{2}} = \widetilde{M}_{ij}\eta_j^n - \frac{\Delta t}{2}F_i^n \tag{41}$$

$$\overline{M}_{ij}\mathbf{u}_j^{n+\frac{1}{2}} = \widetilde{M}_{ij}\mathbf{u}_j^n - \frac{\Delta t}{2}\mathbf{G}_i^n \tag{42}$$

and the second step is

$$\overline{M}_{ij}\eta_j^{n+1} = \widetilde{M}_{ij}\eta_j^n - \Delta t F_i^{n+\frac{1}{2}} \tag{43}$$

$$\overline{M}_{ij}\mathbf{u}_j^{n+1} = \widetilde{M}_{ij}\mathbf{u}_j^n - \Delta t \mathbf{G}_i^{n+\frac{1}{2}} \tag{44}$$

where

$$\widetilde{M}_{ij} = eM_{ij} + (1-e)\overline{M}_{ij}. \tag{45}$$

$\overline{M}_{ij}$ denotes the lumped matrix generated by concentrating non-diagonal elements of the matrix M_{ij} to diagonal ones as follows:

$$M_{ij} = \frac{\Delta}{12}\begin{bmatrix} 2 & 1 & 1 \\ 1 & 2 & 1 \\ 1 & 1 & 2 \end{bmatrix} \Rightarrow \overline{M}_{ij} = \frac{\Delta}{3}\begin{bmatrix} 1 & & \\ & 1 & \\ & & 1 \end{bmatrix} \tag{46}$$

where Δ is area of a element. A parameter e is referred to as a selective lumping parameter, which controls the numerical dumping and numerical stability. This parameter is usually set to be about $0.9 \sim 0.95$.

3.4.4. Semi-Implicit Time-Marching Methods

The semi-implicit procedures were developed in an attempt to create unconditionally stable schemes but yet have a stationary matrix[21,78]. Because the major stability constraint arises from gravity waves, only the gravity wave terms are made implicit[21,57]. This results in a time-invariant matrix that can be re-solved in 1/3 to 1/4 of the time for a full matrix assembly, and there is no iteration required.

3.4.5. Implicit Time-Marching Methods

Approximating the time derivatives with finite differences and weighting by a distance θ through the time interval Δt, Eqs. (35) and (34), that are the general form of Eqs. (34) and (35), can be written as

$$M_{ij}\eta_j^{n+1} - \theta\Delta t F_i^{n+1} = M_{ij}\eta_j^n + (1-\theta)\Delta t F_i^n \qquad (47)$$
$$M_{ij}\mathbf{u}_j^{n+1} - \theta\Delta t \mathbf{G}_i^{n+1} = M_{ij}\mathbf{u}_j^n + (1-\theta)\Delta t \mathbf{G}_i^n \qquad (48)$$

For $\theta \geq 1/2$ the method is implicit and linear stability analysis shows that there are no stability constraints[21]. Fourier analysis also shows that as θ is increased from 1/2 to 1, the numerical damping also increases[21]. This fact is used by models that are fully implicit to increase stability and reduce computational noise through numerical damping. In addition, the method that uses finite elements in time with linear bases in time, is identical to the finite difference method with $\theta = 2/3$.

The challenge with implicit methods is to solve Eq. (48) in an efficient manner. Because the left-hand-side is nonlinear, some form of iteration is required. Typically, a Newton-Raphson iteration is used[45,91]. It has been found that fewer iterations are required if a high-order predictor is used for the dependent variables at the start of the iteration. In particular, the Adams-Bashforth method works well[91].

3.5. Wave Equation

These governing equations are approximated using standard Galerkin techniques. The spatial domain is discretized by defining a set of 2-dimensional elements. A standard Lagrange basis of polynomial degree p is defined on the master element[6] and this basis is used to interpolate variable quantities within each element. Expanding $\eta, \mathbf{u}$ in terms of the finite element basis and numerically integrating produces an algebraic problem for the nodal unknowns. Because the wave equation and the momentum equation are solved separately, there is no requirement that the bases for η and $\mathbf{u}$ be of the same order. In fact, there is a consistency condition that suggests that the bases for $\mathbf{u}$ should be one order less than the bases for η[88].

Using ϕ as the weighting function in the wave equation for η, the weighted residual form of the equations may be expressed as

$$\int_\Omega \frac{\partial^2 \eta}{\partial t^2} \phi d\Omega + \int_\Omega \gamma_0 \frac{\partial \eta}{\partial t} \phi d\Omega + \int_\Omega gH\nabla\eta \cdot \nabla\phi d\Omega$$

$$= -\int_\Omega \mathbf{R} \cdot \nabla\phi d\Omega - \oint(\frac{\partial H\bar{\mathbf{u}}}{\partial t} + \gamma_0 H\bar{\mathbf{u}}) \cdot \mathbf{n}\phi \, d\Gamma \qquad (49)$$

where the divergence terms in (8) have been integrated by parts[55].

With Φ as the weighting function, the momentum equation becomes

$$\int_\Omega \frac{\partial \mathbf{u}}{\partial t} \Phi d\Omega + \int_\Omega [\nabla \cdot (\mathbf{uu})]\Phi d\Omega + \int_\Omega \mathbf{f} \times \mathbf{u} \, \Phi dx + \int_\Omega A_h \nabla\mathbf{u} \cdot \nabla\Phi d\Omega$$

$$= -g\int_\Omega (\nabla\eta - \mathbf{F})\Phi d\Omega + \oint \mathbf{B} \cdot \mathbf{n}\Phi \, d\Gamma \qquad (50)$$

where the stress terms have been integrated by parts and the last term is the associated boundary integral for stress. There are several options in manipulating these equations. Any of the divergence terms can be integrated by parts if this eases the specification of the boundary conditions. Essential conditions on η or the boundary integral in the last term in Eq. (49) are specified on all external boundaries. The natural boundary condition is that the boundary integral is zero.

The wave equation is discretized in time using a 3-level finite difference approximation with time levels $k + 1$, k, and $k - 1$[52,55,48]. All terms are centered at time level k so that the scheme is explicit. The momentum equations are centered between levels $k+1$ and k, with the exception that the dispersion term and the advective term are evaluated at level k. Note that the wave equation and the momentum equation do not need to be solved simultaneously. In practice, the wave equation is solved first, followed by a solution to the momentum equation.

Normally, this would produce a fully 2D matrix problem. However, it is possible to obtain a much simpler problem by applying node-point integration in the horizontal and using linear bases. This effectively diagonalizes the matrix in the horizontal and leaves a (2×2) matrix for each velocity node[55]. This procedure results in very good computational efficiency.

An implicit form of these equations is created by weighting the gravity terms between the time levels[52]. This removes the stability constraint, but requires a matrix solution. Iterative solvers work well in this regard.

3.6. SUPG- Streamline Upwind Petrov-Galerkin

Advection-dominated flows are notorious for their computational difficulties. In compressible, advection dominated flows this has led to a class of stabilized methods[30,31] that are characterized by a transformation to a symmetric form of the equations, and by an operator duality between the weighting functions and the governing equations. Recently, these concepts have been applied to the shallow-water equations[9]. This procedure has great promise in the simulation of the local runup, where bores, hydraulic jumps, and other shock-like features can occur.

First, the shallow-water equations are transformed to a symmetric form. This form has the property that weighted residual formulations based on it automatically possess certain stability properties associated with exact solutions of the governing equations[30]. The shallow-water equations can be expressed in vector form as

$$\mathbf{U}_{,t} + \mathbf{A}_1(\mathbf{U})\mathbf{U}_{,x_1} + \mathbf{A}_2(\mathbf{U})\mathbf{U}_{,x_2} = \mathbf{S}(\mathbf{U}) \tag{51}$$

where $\mathbf{U} = (H, Hu_1, Hu_2)^t$, and the superscript t denotes the transpose. The flux Jacobian matrices are given by

$$\mathbf{A}_1(\mathbf{U}) = \begin{pmatrix} 0 & 1 & 0 \\ gH - u_1^2 & 2u_1 & 0 \\ -u_1 u_2 & u_1 & u_2 \end{pmatrix} \tag{52}$$

$$\mathbf{A}_2(\mathbf{U}) = \begin{pmatrix} 0 & 0 & 1 \\ -u_1 u_2 & u_2 & u_1 \\ gH - u_2^2 & 0 & 2u_2 \end{pmatrix} \tag{53}$$

and $\mathbf{S}(\mathbf{U})$ is given by

$$\mathbf{S}(\mathbf{U}) = (0, gHh_{,x_1} - \tau_{b_1}, gHh_{,x_2} - \tau_{b_2})^t \tag{54}$$

where τ_{b_i} is the bottom friction in the i direction.

This system is hyperbolic and the flux Jacobian matrices are not symmetric, nor can they normally be made so. Then define a change of dependent variables, $\mathbf{U} = \mathbf{U}(\mathbf{V})$, which makes the resulting system symmetric. Bova and Carey[9] make arguments that relate entropy in gas dynamics with total energy of the water column for the shallow-water equations. They then derive the symmetric hyperbolic system

$$\hat{\mathbf{A}}_0\mathbf{V}_{,t} + \hat{\mathbf{A}}_1(\mathbf{V})\mathbf{V}_{,x_1} + \hat{\mathbf{A}}_2(\mathbf{V})\mathbf{V}_{,x_2} = \hat{\mathbf{S}}(\mathbf{V}) \tag{55}$$

where $\mathbf{V} = (H - V^2/2, u_1, u_2)^t$, $V^2 = u_1^2 + u_2^2$, $\hat{\mathbf{A}}_0$ is given by

$$\hat{\mathbf{A}}_0(\mathbf{V}) = \begin{pmatrix} 1 & u_1 & u_2 \\ & H + u_1^2 & u_1 u_2 \\ symm. & & H + u_2^2 \end{pmatrix} \tag{56}$$

and the flux Jacobian matrices are given by

$$\hat{\mathbf{A}}_1(\mathbf{V}) = \begin{pmatrix} u_1 & H + u_1^2 & u_1 u_2 \\ & u_1(3H + u_1^2) & u_2(H + u_1^2) \\ symm. & & u_1(H + u_2^2) \end{pmatrix} \tag{57}$$

$$\hat{\mathbf{A}}_2(\mathbf{V}) = \begin{pmatrix} u_2 & u_1 u_2 & H + u_2^2 \\ & u_2(H + u_1^2) & u_1(H + u_2^2) \\ symm. & & u_2(3H + u_2^2) \end{pmatrix} \tag{58}$$

and $\hat{\mathbf{S}}(\mathbf{V})$ is given by

$$\hat{\mathbf{S}}(\mathbf{V}) = (0, gHh_{,x_1} - \tau_{b_1}, gHh_{,x_2} - \tau_{b_2})^t \tag{59}$$

Then following the second half of the development, they develop a streamline upwind Petrov-Galerkin formulation for the weighted residual statement, based on the approach of Hughes and others[30] for the Euler equation in gas dynamics. The weighted residual formulation is

$$\int_\Omega \hat{\mathbf{W}}^t(\hat{\mathbf{A}}_0 \mathbf{V}_{,t} + \hat{\mathbf{A}}^t \nabla \mathbf{V} - \hat{\mathbf{S}}) d\Omega = 0 \tag{60}$$

where $\hat{\mathbf{A}}^t = (\hat{\mathbf{A}}_1, \hat{\mathbf{A}}_2)^t$ and

$$\nabla \mathbf{V} = \begin{pmatrix} \mathbf{V}_{x_1} \\ \mathbf{V}_{x_2} \end{pmatrix} \tag{61}$$

For the Petrov-Galerkin formulation, the weighting functions are the standard Galerkin trial functions plus a perturbation:

$$\hat{\mathbf{W}} = \mathbf{W} + \tau \hat{\mathbf{A}} \nabla \mathbf{W} \tag{62}$$

This is the same form as the operators in the weighted residual statement. When the intrinsic time scale $\tau = 0$, the form reduces to the standard Galerkin formulation. For large velocities, τ is large and the equation is weighted in an upwind direction. This method has generated impressive results for flows over spillways with hydraulic jumps, constricted flows with hydraulic jumps, and shoaling of a wave. Thus far, it has not been applied to the moving boundary problem. In addition, there are some research questions that need to be resolved in order to create a more efficient solution to transient problems.

3.7. *TELEMAC-2D Model*

The equations for c and $\mathbf{u}$ given by Eqs. (12) and (13) are solved in two fractional steps. In the advective step,

$$\dot{c} + \mathbf{u} \cdot \nabla c = 0 , \tag{63}$$
$$\dot{\mathbf{u}} + \mathbf{u} \cdot \nabla \mathbf{u} = 0 \tag{64}$$

are solved using a characteristic method[16,26]. The characteristic path is calculated at each node and the dependent variables are interpolated at the upwind location using the the finite element grid. In the propagation step (all the other terms in the equation), the time integration is implicit with a weighting factor $\theta \geq 1/2$. Quadrilateral elements with linear bases are used in the spatial discretization of the above equations to arrive at $A\,X = Y$, with:

$$A = \begin{pmatrix} M_1 & B_x & B_y \\ -B_x^T & M_2 & 0 \\ -B_y^T & 0 & M_3 \end{pmatrix} \quad and \quad X = \begin{pmatrix} C \\ U \\ V \end{pmatrix}^{n+1} \tag{65}$$

where C, U and V are vectors containing respectively the unknown wave celerity and components of velocity at time t^{n+1}. M_1, M_2, M_3, B_x and B_y are the following square matrices:

$$M_1 = \frac{2}{\theta_u \Delta t} M \tag{66}$$

$$M_2 = M_3 = \frac{1}{2\theta_u \Delta t} M + \frac{1}{2\theta_c} D \tag{67}$$

where M is the mass matrix, θ_u and θ_c are the time-weighting coefficients for velocity and the celerity, D is the diffusion matrix and B_x and B_y are matrices of the form:

$$B_{xij} = -\int_V \Psi_j \frac{\partial}{\partial x}(c^n \Psi_i) dV \tag{68}$$

$$B_{yij} = -\int_V \Psi_j \frac{\partial}{\partial y}(c^n \Psi_i) dV \tag{69}$$

where Ψ_i and Ψ_j are the basis functions and c^n is the celerity at time t^n.

For initial conditions, $c(\mathbf{x}, 0)$ and $\mathbf{u}(\mathbf{x}, 0)$ are known, and an essential condition is applied for either $c(\mathbf{x}, t)$ and $\mathbf{u}(\mathbf{x}, t)$ on the boundary.

3.8. Harmonic Methods in Time

For a wide range of problems that use steady or periodic forcing, such as astronomical and radiational tides, and steady flows, it is more convenient to solve the governing equations in the frequency domain than in the time domain. This approach can lead to much smaller computational effort and is especially useful for exploratory studies with complicated geometries. This is a technique commonly used in oceanography because many of the variables of interest are characterized by line spectra. The idea was developed in the context of numerical models independently by several groups[38,49,52,66,76]. In addition, a comparison between basic harmonic and time-stepping models for the two-dimensional equations is presented in Walters and Werner[92].

In following this approach, the dependent variables are expressed as periodic functions

$$\eta(x, y, t) = \frac{1}{2} \sum_{n=-N}^{N} \eta_n(x, y) e^{-i\omega_n t} \tag{70}$$

$$\mathbf{u}(x, y, z, t) = \frac{1}{2} \sum_{n=-N}^{N} \mathbf{u}_n(x, y, z) e^{-i\omega_n t} \tag{71}$$

where ω is the angular frequency, and n is the index for the N tidal and residual constituents. Applying (70) and (71) to (1) and (2) and using harmonic decomposition[76], the governing equations become

$$- i\omega_n \eta_n + \nabla \cdot (h\mathbf{u}_n) = \nabla \cdot \mathbf{W}_n \tag{72}$$

and

$$(-i\omega_n + \gamma)\mathbf{u}_n + \mathbf{f} \times \mathbf{u}_n = -g(\nabla \eta_n - \mathbf{T}_n) \tag{73}$$

where $\mathbf{W}$ is the continuity nonlinearity, $\gamma(x, y)$ is the time-independent part of the bottom friction term $C_D|\mathbf{u}|$, and $\mathbf{T}$ contains all the nonlinear terms in the momentum equation, including advection and time-dependent part of bottom stress[87].

The treatment of the nonlinear continuity and advection terms is straightforward as they contain only simple products of the various frequencies. Thus they lead to sums and differences between the frequencies of the N constituents and contribute as source terms in the generation of harmonics. The treatment of the bottom friction is more difficult due to the factor $|\mathbf{u}|$. However, these terms can be treated by a method developed[76] for the two-dimensional shallow water equations and extended to three dimensions with the finite element method[87].

The wave equation form of these equations becomes

$$- i\omega_n \eta_n - \nabla \cdot \left\{ \left(\frac{gh}{q_n^2 + f^2} \right) [q_n (\nabla \eta_n - \mathbf{T}_n) - \mathbf{f} \times (\nabla \eta_n - \mathbf{T}_n)] \right\} = \nabla \cdot \mathbf{W}_n \tag{74}$$

where $q_n = -i\omega_n + \gamma$.

This equation is derived by solving for $\mathbf{u}$ in the depth-averaged momentum equation and substituting that expression into the divergence term in the continuity equation[54,87]. This form of the equation has the same advantages as were mentioned earlier.

Using ϕ as the weighting function in the Helmholz equation for η, the weighted residual form may be expressed as

$$-i\omega_n \int_\Omega \eta_n \phi d\Omega + \int_\Omega \left\{ \left(\frac{gh}{q_n^2 + f^2} \right) [q_n (\nabla \eta_n - \mathbf{T}_n) - \mathbf{f} \times (\nabla \eta_n - \mathbf{T}_n)] \right\} \cdot \nabla \phi d\Omega$$
$$- \int_\Omega \mathbf{W}_n \cdot \nabla \phi d\Omega = - \oint \mathbf{Q}_n \cdot \mathbf{n} \phi d\Gamma \tag{75}$$

where the divergence term in (74) has been integrated by parts, and $\mathbf{W}_n$ and $\mathbf{T}_n$ are treated as known functions from the previous iteration. The wave equation is approximated by replacing ϕ with Lagrange bases of order p and numerically integrating over the spatial domain with sufficiently high order that the integrals are exact.

Similarly, with Φ as the weighting function, the momentum equation becomes

$$-i\omega_n \int_\Omega \mathbf{u}_n \Phi d\Omega + \mathbf{f} \times \int_\Omega \mathbf{u}_n \Phi d\Omega = -g \int_\Omega (\nabla \eta_n - \mathbf{T}_n) \Phi d\Omega \qquad (76)$$

These equations are solved in sequence: first the wave equation is solved using a direct or iterative solver, then velocity is back-calculated from these results. This procedure is iterated until convergence is attained, typically 5 to 10 iterations. Because the convergence is oscillatory, an under-relaxation of velocity speeds convergence.

3.9. Boundary Conditions

3.9.1. Finite Element Formulation of the Open Boundary

The basic concepts behind the treatment of the open boundary have been described in an earlier section (Section 2.5.2). Here, these concepts are put into a finite element context using the shallow-water equations. First, consider a coupling of the inner region to a simple model for the outer region.

One approach is to couple a standard finite element discretization in the inner region, to infinite elements in the outer region. In the approach presented by Chen[12], a circular arc is used to define the open boundary where the two regions connect. The infinite elements are defined by radials drawn from the nodes on the arc to infinity. Then bases are defined in the infinite elements so they reduce to the standard linear bases on the boundary, satisfy the Sommerfeld radiation condition at infinity, and are of a form that can be integrated analytically in the weighted residual equations. Some examples are presented in the cited reference[12].

Using a different approach, specific terms in the governing equations are integrated by parts to obtain boundary integrals, then analytic solutions are substituted into these integrals[40,79]. Consider the primitive form of the shallow-water equations given in Eqs. (34) and (35). Following Kawahara[40], the equations are linearized and the divergence term in the continuity equation and the gravity term in the momentum equation are integrated by parts.

$$\int_\Omega \phi \dot{\eta} d\Omega - h \int_\Omega \nabla \phi \cdot \mathbf{u} d\Omega + h \int_\Gamma \phi \bar{\mathbf{u}} \cdot \hat{\mathbf{n}} d\Gamma = 0 \qquad (77)$$

and

$$\int_\Omega \phi \dot{\mathbf{u}}_i d\Omega - g \int_\Omega \phi \eta d\Omega + g \int_\Gamma \phi \bar{\eta} n_i d\Gamma = 0 \qquad (78)$$

Substituting the analytic solution described in Eq. (18) into the boundary integration terms in Eqs. (78) and (77) yields a closed solution. In Kawahara[40], the two-step explicit finite element method is applied to the equations above. A figure in the cited reference demonstrates the application of this procedure with two open boundary configurations.

In the second case, the direct application of radiation boundary conditions, the basic idea is to integrate specific terms in the governing equation by parts to obtain boundary integrals, then substitute the expression for the boundary condition into these integrals. A similar procedure as that used above is applied to the wave equation (Eq. (8)), except that the boundary condition is used directly rather than an analytic solution. Expressing Eq. (49) as a telegraph equation[32], and integrating the Laplacian by parts,

$$\int_\Omega \frac{\partial^2 \eta}{\partial t^2} \phi d\Omega + \int_\Omega \gamma_0 \frac{\partial \eta}{\partial t} \phi d\Omega + \int_\Omega c^2 \nabla \eta \cdot \nabla \phi d\Omega = - \oint c^2 \nabla \eta \cdot \hat{n} \phi d\Gamma \qquad (79)$$

Using the generalized radiation boundary condition given by Eq. (22) and integrating with respect to time, the boundary integral above may be written

$$- \oint c^2 \nabla \eta \cdot \mathbf{n} \phi d\Gamma = \oint c \left(\frac{\partial \eta}{\partial t} + \frac{\gamma_0}{2} \eta \right) \phi d\Gamma = c \left(\frac{\eta_b^{k+1} - \eta_b^{k-1}}{2 \Delta t} + \frac{\gamma_0}{2} \eta_b^k \right) \oint \phi_b d\Gamma \qquad (80)$$

where subscript b refers to values on the boundary. Results using this approach are contained in Johnson et al[32]

3.9.2. Moving Boundary

As indicated earlier, there are two types of techniques for the application of moving boundaries (Figure 3). One is a deforming grid approach in which the grid is deformed according to the moving boundary[51,53], and the other is a fixed-grid approach where elements along the boundary are dry or wet (inactive or active) depending on whether they contain dry nodes[39,28,81,82]. The moving-grid technique takes advantage of the flexibility of the spatial discretization offered by the finite element approach. However, the details of the grid deformation must be incorporated efficiently, otherwise there can be far too much computational overhead associated with grid regeneration. The fixed-grid technique is a simpler procedure, but conserves neither mass or momentum unless something more sophisticated than switching elements on and off is used. An example is the technique used by Holz and Nitsche[28] where the boundary is tracked through a fixed grid, and mass in the partially dry elements is accounted for.

Deforming grids are used in a wide class of problems; for instance, the study of deforming solids. They have been used less in flow problems[51,53,61]. The moving-grid problem proceeds forward in two steps: the grid deforms either locally or globally with fixed element connections as the boundaries move, then the grid is reconnected when the elements have deformed to the point where they do not provide an accurate description of the spatial domain. Normally, many deformation steps take place, then the grid is regenerated and that grid is used as an initial condition for further steps. Grid regeneration can be a time-consuming process; therefore, it is usually preferable to limit grid deformation to a local area along the boundary and only regenerate this strip of elements.

The implementation of moving boundaries is a straightforward procedure. The location of the boundary is given by Eqs. (23) and (25). The basis or interpolation functions for the deforming elements are written as functions of time as well as space;

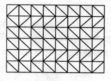

(a) Initial grid

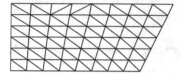
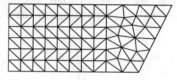

(b) Deforming grid: global(left), local(right)

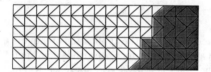
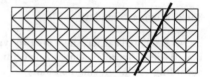

(c) Fixed-grid with switching(left) and tracking(right). The shaded elements are inactive.

Figure 3: Methods to approximate a moving boundary

that is, $\Phi = \Phi(\mathbf{x}, t)$. Then a typical weighted residual term with a time derivative becomes

$$\int_\Omega \frac{\partial \eta}{\partial t} \Phi_i d\Omega = \int_\Omega \frac{\partial \eta_j}{\partial t} \Phi_j \Phi_i d\Omega + \int_\Omega \frac{\partial \Phi_j}{\partial t} \eta_j \Phi_i d\Omega \qquad (81)$$

Then expressions for $\frac{\partial \Phi_j}{\partial t}$ are developed[51] and the problem is closed. Lynch and Gray[53] present an application of this method to wave runup on a sloping beach and show the difference in results for no-flux and moving boundary conditions. In addition, an application of this approach to the paddle movement in Benchmark Problem 3 is presented in this volume.

In the fixed-grid approach, nodes are checked for water depth to determine if they are wet or dry and the element calculations handled accordingly, that is, dry nodes are identified by $H \le \varepsilon_h$, where ε_h is a minimum water depth determined from a computational stability criterion or other means. Examples of this technique are contained in the Benchmark Problems in this volume.

For example, an element with a dry node such as shown in Figure 4 can be omitted from the calcualtions and a slip condition applied at the wet nodes. In this case, there

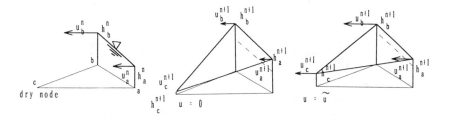

(a) Water boundary element   (b) Non-slip condition   (c) Slip condition

Fig. 4.   Illustration for moving boundary condition[82]

is neither conservation of mass nor of momentum.

On the other hand, some methods include this element in the computation, because the nodes **a** and **b** keep the water depth greater than ε_h at the n-th time step. However this is not true for node **c**, so the problem then is how to treat the boundary condition at the dry node **c**. In one approach[82], a slip condition on the dry node is imposed, and the following procedure is applied in each time step as shown in Figure 4. Here $u^n_{i(dry)} = \tilde{u}_i$ as well as for all the fractional steps, where

$$\tilde{u}_i = \sum_{i=1}^{k} u_i \, / \, k \qquad (\text{if} \quad u_i > 0 \text{ , k is the number of wet nodes)} \qquad (82)$$

such that $\tilde{u}_i$ is imposed on the dry node as a transient value. This technique is flexible, because the wet and dry regions can be changed freely if the region estimated to be located on the moving boundary is discretized with reasonable fine elements. Figures 5 and 6 show the application of this moving boundary condition to a broken dam problem and wave run-up on a sloping beach.

Holz and Nitsche[28] carry this procedure one step further to account for mass conservation in the partially-dry elements. They follow the shoreline by interpolation of the water depth from the vertices. They then adjust the grid and integrate only over the flooded part of the element.

3.10. Matrix Solution Methods

All of the weighted residual statements discussed above lead to a system of equations for the nodal values of the dependent variables. The method of solution impacts the computational resources necessary to solve a particular problem such that there can be 2 or 3 orders of magnitude difference in the run times for different model equa-

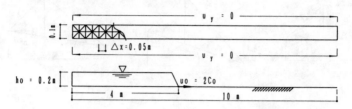

(a) Computational model for 1-D broken dam flow

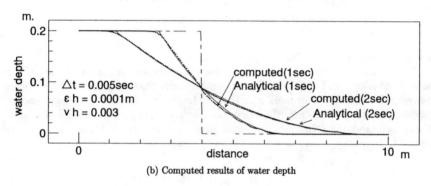

(b) Computed results of water depth

Fig. 5. Verification in a broken dam flow problem

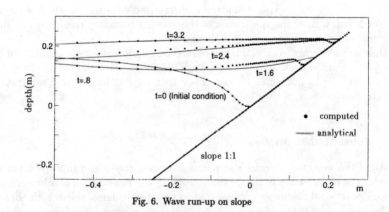

Fig. 6. Wave run-up on slope

tions and different matrix solution methods. Thus the solution method is of primary importance for model development.

The primitive equation formulation of the shallow-water equations is much more computationally intensive than the wave equation formulation. One reason is that the former involves a simultaneous solution for all dependent variables - η and the u components - whereas with the latter η and u are decoupled which leads to smaller matrices. A common method of solving the non-linear system of equations arising from the primitive equations is to use an implicit finite difference time-stepping method coupled with a Newton-Raphson iteration on the resulting matrix. The resulting efficiency then depends on the matrix solution algorithm used. Because the continuity equation is not necessarily diagonally dominant, some form of pivoting must be used to avoid an ill-conditioned matrix. This method of solving the shallow-water equations is by far the most computationally intensive option. For periodic problems, there can be an order of magnitude or better improvement in run times when using the harmonic formulation of the problem because of the reduced number of iterations.

Using the wave equation formulation, η is solved for first followed by a back-calculation of u. For the explicit time-stepping method[52], a matrix system is solved and no iteration is required. Moreover, for the case where node-point integration is used, the matrix becomes diagonal so that no matrix solution is required, and the velocity is back-calculated in a simple manner. The implicit version of the wave equation requires a matrix solution[50], although a simple iterative solver appears adequate. For periodic problems, a useful approach is to use the harmonic form of the wave equation. First η is calculated from the wave equation and u is back-calculated from the momentum equation which involves a simple 2×2 block diagonal matrix. The procedure requires iteration. The wave equation is diagonally dominant so that pivoting is not necessary.

Three useful matrix solution algorithms for finite element analysis are frontal solvers, banded matrix solvers, and iterative solvers. All of these methods take advantage of the sparse nature of the matrix. In particular, frontal solvers take advantage of the geometric nature of finite element grids and can solve the global matrix on a smaller submatrix by sweeping a front through the grid. A modified form of the frontal solver that reduces storage and run time has been used extensively for this problem[84]. Standard iterative solvers have been used with success with the time-marching form of the wave equation[94,48]. In addition, iterative methods modified for complex matrices have been used successfully with the harmonic form, although the break even point is rather high when compared with the frontal solver[5].

4. Convergence and Dispersion Analysis

One of the basic types of analysis that has been applied to the discretized shallow-water equations is Fourier or dispersion analysis. Its purpose is to find stability constraints, and assess the accuracy of the approximate dispersion relation, $\omega = \omega(k, \Delta t, \Delta x)$. Another type of analysis is aimed toward a quantitative assessment of

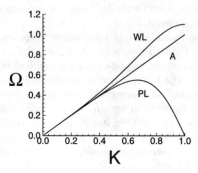

Fig. 7. Dispersion relation for continuum (A), primitive equations with linear bases (PL), and wave equation with linear bases (WL). $\Omega = \omega \Delta x / (\pi c)$, $K = k \Delta x / \pi$.

convergence rates and error estimates. The existing literature on these subjects is sufficiently large that it is not practical to cover all the concepts nor go into very much detail. The intent here is to state the basic concepts as applied to finite element approximations, and provide references to literature where more details can be obtained.

Early attempts to use finite element methods with the shallow-water equations led to numerous problems with spurious oscillation modes caused by a folded dispersion relation for the discrete problem[21,69,90]. A similar problem was encountered in the early days of finite difference analysis with the A- and B-grids. The A-grid places **u** and η at each of the corner nodes of a computational cell, equivalent to the quadrilateral finite element with linear bases. It has a computational mode with internode ($2\Delta x$) oscillations. The B-grid places **u** at each of the corner nodes and η at the center of a computational cell, equivalent to the quadrilateral finite element with linear bases for velocity and piecewise-constant bases for η. It has a computational mode with "checkerboard" oscillations, that is, values for η that alternate up and down between adjacent elements. A Fourier analysis of the linearized, primitive shallow-water equations shows why problems were encountered[50,21,89,18]. In the continuum, the dispersion relation is given by $\omega = ck$, where $c = \sqrt{gH}$. This shown as A in figure 7 and can be compared to the relation for the primitive shallow-water equations with linear bases which is shown as PL. As may be seen, $\omega = 0$ at the grid Nyquist frequency. An eigenmode analysis[89] show that this mode is indeed the oscillatory mode. Further, Platzman[69] showed that because the dispersion relation was folded, long-wave forcing would excite two modes: the long-wave mode that is the desired solution, and a short-wave spurious mode. The amplitude of the short-wave mode depends on the details of the grid, and is not really predictable for complicated grids.

Three general methods have been used to mitigate the effects of numerical oscillations: (1) damping the short-wave modes, (2) creation of elements that do not support these modes, and (3) modification of the governing equations. Note that in the finite difference method, this problem was solved with the C-grid. The C-grid places η at the center of a computational cell and one velocity component at the center of each side, thereby staggering all the dependent variables. In the finite element context, no element has been found that has these same properties and is practical to apply.

Damping can include the use of spatial filters[57,15], use of excessive artificial or real viscosity[65], or the use of dissipative time-stepping methods. Lynch[50] has used Fourier analysis to examine a large number of time-stepping procedures and has extended this concept to examine amplitude errors. It is instructive to compare the many figures he presents. In general, it is difficult to localize the effects of damping to small wavelengths only. Hence damping becomes a method of last resort.

The attempt to find elements that do not support spurious modes has not been very successful[89], although some exist[19]. Most candidate elements in two- or three-dimensions suffer from convergence problems, or are too difficult to apply. In one-dimension, however, there is an element that is similar to a C-grid and performs well[89]. A useful approach is the use of mixed-interpolation elements where velocity has quadratic bases and sea level has linear bases[45,56,46,91]. Although the modes are removed in sea level, they still occur in the velocity solution[89,85].

The modification of the governing equations has led in several directions. One of these is the wave equation formulation, another is SUPG, and another is TELEMAC. The dispersion relation for the wave equation with linear bases is shown in figure 7 as WL[50,85,62]. Note that this approach has eliminated the spurious mode entirely and provides a more accurate dispersion relation. This has been verified by a large number of simulations by many groups [50,21,66,49,85,86,48,55,87] At this time, the dispersion relation for SUPG and TELEMAC have not been derived; however, numerical results indicate that they do not contain spurious computational modes[9,16,20,25,26]

Convergence theory for finite element methods is well established[6,11]. A standard way to describe errors is through the global L^2 error norm. Theoretical considerations lead to a relation between the L^2 error norm and the node spacing

$$L^2 = C \Delta x^{(p+1-r)} \tag{83}$$

where C is a problem-dependent constant, Δx is a node spacing, p is the order of the polynomial bases, and r is a parameter that accounts for reduction in order from derivatives and grid irregularity. Typical linear bases lead to convergence rates of the order Δx^2, a well known fact. In addition, it is found that for regular grids, high orders of p lead to very high convergence rates[3,64,88]. Raising the order of the element bases is an effective alternative to reducing the size of Δx and can lead to much higher convergence rates. Note that the mixed-interpolation element does not have this advantage because velocity is dependent on gradients so that the order remains at Δx^2, rather than Δx^3.

However, the global error norm does not tell the whole story when using highly irregular grids encountered in real-world problems. Local errors from irregular geometry or abrupt depth variations can seriously degrade the global convergence rates[88,7].

For this reason and others, there has been considerable recent research into error estimators whose purpose is to guide adaptive refinement of the grid[3,64,95]. A particularly useful procedure is to use a combination of adaptive grid refinement and higher order p convergence to provide very accurate solutions in complex domains[3,64,88]. Moreover, error estimates are obtained as an outcome of the procedure.

5. Grid Generation Methods

One of the primary advantages of using finite elements is the flexibility in the discretization of regions with highly irregular geometry. In studying the behaviour of tides, storm surges, or tsunamis in coastal waters by means of numerical models, there are several advantages in being able to vary the size and orientation of elements in the grid on which the model is based. For instance, it is possible to obtain a much better fit to the coastlines when the edges of grid elements can be made of arbitrary length and direction. Moreover, the use of irregular grids also facilitates better model design in other practical and theoretical respects. Considerable computing time is wasted when a grid with uniform element size is used in situations where water depth varies substantially over the area modelled. If an explicit difference scheme is used, the Courant stability criterion requires that $c\Delta t/\Delta x \leq 1$, where c = wave speed, Δt = time step, and Δx is some linear measure of element size, such as length or width. Since c increases with water depth and it is impractical to use different sizes of time step at different parts of the grid, having Δx uniform over the whole grid leads to unnecessarily frequent computation everywhere except at the point of maximum depth. Most implicit schemes are stable even if the Courant criterion is violated, but truncation error generally increases with the local value of $c\Delta t/\Delta x$; so again, if Δx is uniform over the grid, some unnecessary computing is carried out over the shallower parts of the model domain.

In addition, there are other grid design criteria that are related to errors in the flow calculations. These errors can arise from small scale features in the flows such as shears, or from truncation errors associated with abrupt changes in depth. The result, then, is that it is desirable to construct grids with variable element size, and this size variation should depend on some general weighting function that controls node density.

It is no exaggeration to say that the amount of effort required to generate conputational grids has been a serious impediment to modeling areas with realistic geometry. A basic difficulty with grid generation lies in achieving an acceptable compromise between competing design requirements: (a) boundary line segments of the model grid should fit the boundaries of the problem domain with sufficient accuracy, (b) element area should be proportional to some scalar field defined over the grid, and (c) elements should not depart far from a regular shape for the sake of accuracy in subsequent calculations.

Purely manual construction of an irregular grid obeying any of the constraints described in the preceeding section is difficult and time-consuming, even leaving aside the complicated book-keeping required to specify the geometry of the finished grid in a form suitable for subsequent use in numerical models. Simpson[75] and Thacker[80] reviewed many widely varied methods for handling all or part of the grid-preparation process by computer. Simpson classified the available methods for producing triangular grids into four types:

1. curvilinear coordinate mapping of simply shaped regions with regular grids into regions with complicated boundaries.

2. generating grids by subdividing initially coarse grids.

3. advancing front methods in which grids are constructed inwards from the model boundaries.

4. vertex triangulation methods in which vertices (nodes) are distributed through the model domain and then connected appropriately by a triangulation algorithm.

The last two methods have proven to be very flexible and are in common use. The difference between them is that advancing front methods generally provide a better grid along the boundaries, whereas node generation provides a better grid in the interior. Better is in the sense of meeting the design criteria above.

A program for irregular grid design needs at least the following five main functions:

- input, verification, and adjustment of coastline and bathymetric data;

- preparation of an interpolation grid for depth that covers the domain to be modelled;

- production of an irregular model grid with nodes suitably positioned for accurate and efficient numerical modelling;

- interactive checking and editing, including trimming and joining, of model grid;

- display and plotting of interpolation and model grid, and optionally model output.

So far, triangular elements have generally been preferred over quadrilateral elements for shallow-water models because of the ability to completely tile an area. The choice depends mainly on the numerical scheme to be used later with the grid. If higher-order elements are used, they are usually generated in a post-processing step from the simple elements with linear bases, *i.e.*, six-node triangles from three-node triangles and nine-node quadrilaterals from four-node quadrilaterals.

It is common practice to make the individual elements in triangular grids as nearly equilateral as possible, on the grounds that this should reduce truncation errors in the subsequent numerical solutions. On the other hand, some people feel that this condition is not important[75]. The arguments presented are necessarily dependent on the numerical scheme used. Foreman[18] found when comparing particular schemes

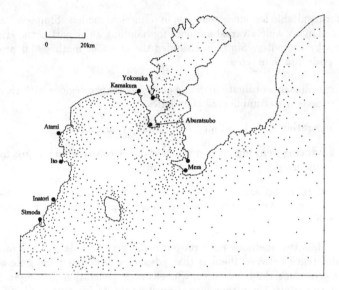

(a) Input data for boundary configuration and depth

Total number of element : 75,301
Total number of node   : 38,842

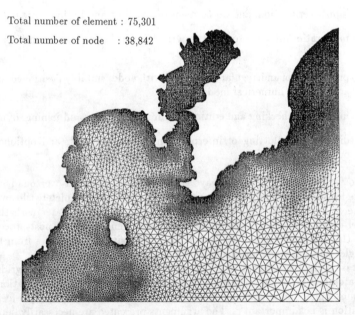

(b) Finite element idealization

Fig. 8.   Automatic grid generation [33]

on regular grids composed of equilateral or isosceles triangles that the equilateral grids allowed isotropic wave propagation, whereas in grids of isosceles triangles, wave speed varied with directon. A good choice is to aim at making all triangles as nearly equilateral as possible, on the assumption that this is likely to prove desirable for some numerical schemes and seems unlikely to impair the qualities of others.

An additional criteria is that the size of the elements should vary smoothly over the spatial region. That is, it is undesirable to have small elements adjacent to large elements as this configuration increases the truncation errors and hence reduces the accuracy of most numerical approximation schemes. This a particular problem with grids that are generated manually. In practical terms, area ratios between adjacent elements should not exceed approximately 2. It is desirable that the grid generation programs include a variety of tests to verify the quality of the grid.

One such interactive-graphics program for generating a set of suitably-spaced nodes for a model grid is outlined in Henry[23] and Henry and Walters[24]. The triangulation algorithm used was devised by Renka[71] and later modified to account for convex domains[8]. This particular implementation uses a generalized weighting function to control node density. The default for the function is the Courant number scaled by a constant. In addition, there are extensive editing features.

Another automatic grid generation method for the shallow-water analysis presented by Kashiyama et al[34]. The key feature of this approach is that the finite element grid is generated so that the Courant number on each element is nearly constant in the whole domain. Furthermore, the bathymetric data at each node is obtained by interpolation using data similar to that given by charts. This program uses a method to place nodes on a element-by-element basis, and avoids the storage of a fine mesh as in the previous scheme. The computed results applied to the offshore region of Kanto area including Tokyo bay in Japan are shown in Figure 8.

In the last section, adaptive grid refinement was mentioned. After the initial grid generation (above) and model runs have been made, there are additional steps that need to be taken to reduce the computational errors. It is not usually feasible to regenerate the entire grid so some method of local refinement is required. There are several ways to approach this by subdividing elements[8], dividing nodes[24], or bisecting the longest element side[72,73]. In the end, an iterative procedure between adaptive refinement and estimation of model errors provides a procedure for achieving acceptable error levels.

6. Applications

6.1. Puget Sound Tsunami

Using several types of evidence, researchers have concluded that a large earthquake occurred on the Seattle fault approximately 1000 years ago[10,2]. They have inferred a vertical displacement of about 7 meters on this fault which passes under Puget Sound just south of Seattle, Washington.

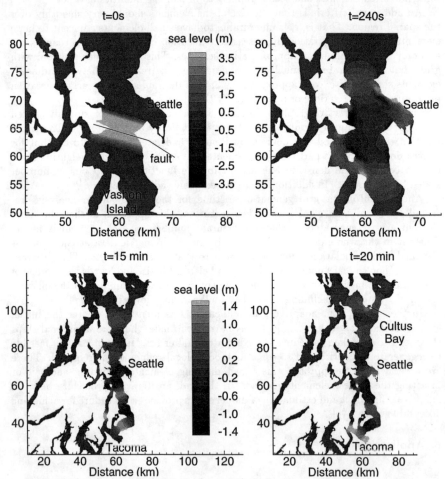

Fig. 9. Initial sea level pattern for tsunami (upper left). Sea level pattern after 240 seconds (upper right), after 15 minutes (lower left), and after 20 minutes (lower right).

The propagation of a tsunami that may have been generated by a this earthquake was simulated using a model based on the wave equation formulation discussed earlier. The model is based on Eqs. (49) and (50) with $\mathbf{f} = A_h = 0$. Integral-lumping was used in both the wave equation and the momentum equation.

A grid of triangular elements was constructed[24] for the entire Puget Sound region south of Juan de Fuca Strait, an area of approximately 75 by 155 km. Grid resolution varies from 200 to 700 m along the boundaries to several kilometers in the interior and was created using a criterion of constant Courant number. This resolution provides a reasonable resolution of the numerous islands and channels, yet can accomodate high resolution subareas for other specific studies, such as local runup.

Figure 9 shows the initial displacement for the free surface for one faulting scenario, and the wave pattern after 240 seconds, 15 minutes, and 20 minutes. At 240 seconds, the tsunami is beginning to run up along the Seattle waterfront and the north end of Vashon Island. At 15 minutes, two leading waves (one on each side of Vashon Island) are propagating southward. There is also a secondary wave behind these. Before 20 minutes elapsed time, these 2 waves meet near Tacoma. At 20 minutes, these waves set up an oscillation at Tacoma and start a wave propagating into south Puget Sound. At this time, the northward propagating wave has reached Cultus Bay where tsunami deposits have been identified[2].

6.2. *Tokachi-oki Tsunami*

The propagation of Tokachi-oki Tsunami presented by Kawahara et al[40] is shown in Figures 10 to 12. An earthquake with magnitude 7.9 and an epicenter located approximately 120 km south of Erimo Cape in Hokkaido occured on May 16,1968. The earthquake was accompanied by a rather strong tsunami which inflicted considerable damage along the Pacific coast of the Hokkaido and Tohoku districts. It was known from the tide gage records that the tsunami began with a distinct down wave in the region from the west of Eriomo cape through Muroran and Hakodate to Hachinohe[33].

The model used to simulate this tsunami is based on the shallow water equations given by Eqs. (6) with $f = 0$ and (1). The two step explicit scheme was used for the time-integration. Radiating open boundary conditions are realized by using the analytical solutions of the linear shallow water equations linked with the computed velocity and water elevation as described earlier[40]. Figure 10 shows the depth variation in this area. The finite element grid is composed by 32,386 elements and 16,547 nodes. Figure 11 shows the initial displacement for the free surface, which was obtained from the numerical experiments by Aida[1]. Finally, figure 12 shows the process of tsunami propagation. At 6 min., the front of the tsunami wave has arrived at the Erimo cape in Hokkaido, and at the 18 min. the wave has approached the Sanriku coast with an increase in water elevation.

7. Concluding Remarks

This review has presented an overview of finite element methods that are appli-

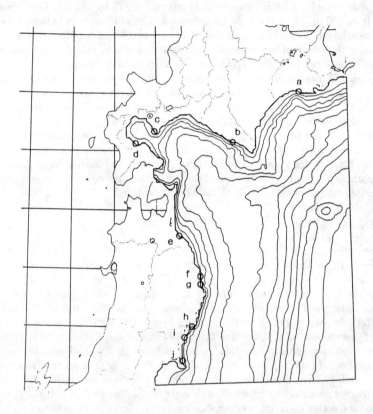

Figure 10: Water depth in the area of the Tokachi-oki tsunami.

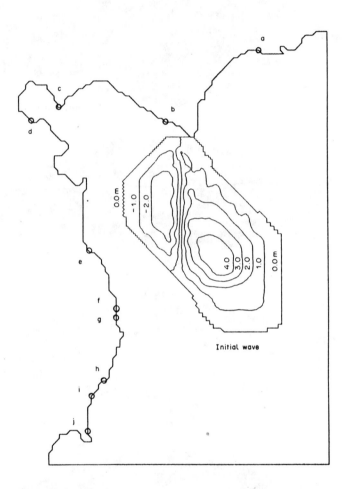

Figure 11: Initial displacement for the free surface

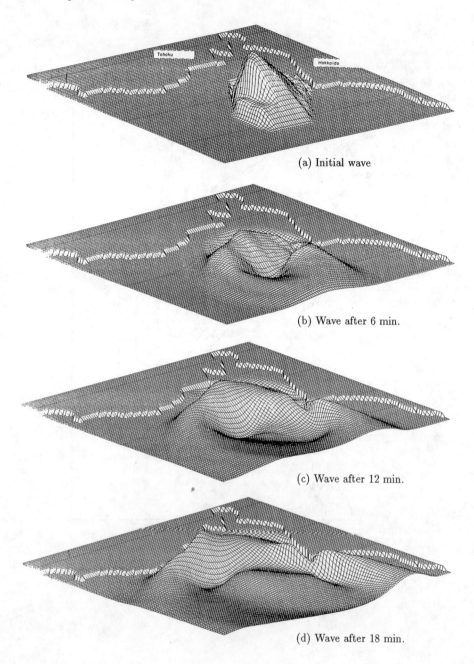

(a) Initial wave

(b) Wave after 6 min.

(c) Wave after 12 min.

(d) Wave after 18 min.

Fig. 12. Process of the tsunami propagation

cable to the tsunami and wave runup problems, and has focused on 2-dimensional (horizontal) simulation using the shallow-water equations. Important attributes of the finite element method are its ability to model irregular geometry with irregular subregions with different properties, and its modular structure which lends it to flexible programming. For instance in estuarine and coastal problems with complicated geometry; irregular boundaries, exterme variations in bottom topography, and the presence of numerous islands can be accomodated easily. For these and other reasons, these methods are being used with increasing frequency.

There are several areas of active research related to issues of efficiency and complicated flows. A mass matrix must be solved at each time step (except for lumped mass matrix), so efficient methods of solution are necessary. In addition, there are methods such as SUPG that are under investigation for high speed flows with shocks and other features. Finally, there is further work needed in accurate and efficient application of radiation and moving boundary conditions.

8. References

1. Aida, I.,Numerical Experiments for the Tsunami Propagation- the 1964 Niigata Tsunami and the 1968 Tokachi-oki Tsunami, *Bull. Earthq. Res. Inst.* **47** (1969) 673.

2. Atwater, B. F., and A. L. Moore, A tsunami about 1000 years ago in Puget Sound, Washington, *Science* **258** (1992) 1614-1617.

3. Babuska, I. and M. Suri, The $h - p$ Version of the Finite Element Method: An Overview, *CMAME* **21** (1990) 5.

4. Baker, A. J., M. O. Soliman and D. W. Pepper, A Time-Split Finite Element Algorithm for Enviromental Release Prediction, in *Proc. 2d Conf. Finite elements Water Resources*, eds. C. A. Brebbia et al (1978) 4.53-4.65.

5. Barragy, E. J., R. A. Walters, and G. F. Carey, Tidal Simulation Using Conjugate Gradient Methods, in *Finite Element Modeling of Environmental Problems*, ed. G. F. Carey (Wiley, New York, 1995) 115-136.

6. Becker, E., G. F. Carey, and J. T. Oden, *Finite Elements: An Introduction*, (Prentice-Hall, Englewood Cliffs, New Jersey, 1981).

7. Blain, C. A., J. J. Westerink, and R. A. Luettich, Domain and Grid Sensitivity Studies for Hurricane Storm Surge Prediction, in *Comutatinal Methods in Water Resources X*, eds. A. Peters et al (1994) 1275-1282.

8. Bova, S.W. and G.F. Carey, Mesh Generation Using Fractal Concepts and Iterated Function Systems, *Int. J. for Numerical Methods in Engineering* **33** (1992) 287-305.

9. Bova, S. W., and G. F. Carey, An Entropy Variable Formulation and Petrov-Galerkin Methods for the Shallow Water Equations, in *Finite Element Modeling of Environmental Problems*, ed. G. F. Carey (Wiley, New York, 1995) 115-136.

10. Bucknam, R. C., E. Hemphill-Haley, and E. B. Leopold, Abrupt uplift within the past 1700 years at southern Puget Sound, Washington, *Science* **258** (1992) 1611-1614.

11. Carey, G. F. and J. T. Oden, *Finite Elements: Mathematical Aspects* (Prentice-Hall, Englewood Cliffs, New Jersey, 1986).

12. Chen, H. S., Infinite elements for combined diffraction and refraction, in *Finite Element Analysis in Fluids*, eds. T. J. Chung and G. R. Karr (University of Alabama Press, Huntsville, Alabama, 1989) 5.138-5.147.

13. Cochet, J. F., D. Dhatt and G. Touzot, Comparison of Explicit and Implicit Methods Applied to Finite Element Models of Tidal Problems, in *Proc. 3d Int. Conf. Finite Element Meth. Flow Prob.* **II**, eds. D. H. Norrie et al (1980) 13-122.

14. Connor, J. J. and J. Wang, Finite Element Modeling of Hydrodynamic Circulation in *Numerical Methods in Fluid Dynamics*, eds. C. A. Brebbia and J. J. Connor (Pentech Press, Plymouth, 1977) 355-387.

15. Cullen, M. J. P., On the Use of Artificial Smoothing in Galerkin and Finite Difference Solutions of the Primitive Equations, *Q. J. R. Meteor. Soc.* **102** (1976) 77-93.

16. Daubert, O., J. M. Hervouet and A. Jami, Description of Some Numerical Tools for Solving Incompressible Turbulent and Free Surface Flows, *Int. J. for Numerical Methods in Engineering* **27** (1989) 3-20.

17. Dunbar, D. S., P. H. LeBlond, and D. O. Hodgins, Evaluation of tsunami levels along the British Columbia coast (Seaconsult Marine Research, Ltd, Vancouver, B.C., 1988) 132pp.

18. Foreman, M. G. G., A Two-Dimensional Dispersion Analysis of Selected Methods for Solving the Linearized Shallow Water Equations. *J. Comp. Phys.* **56** (1984) 287-323.

19. Fortin, M., Old and new finite elements for incompresible flows, *Int. J. for Numerical Methods in Fluids* **1** (1981) 347-364.

20. Galland, J.-C., N. Goutal, and J.-M. Hervouet, TELEMAC: A new numerical model for solving shallow water equations, *Advances in Water Resources* **14** (1991) 83-95.

21. Gray, W. G. and D. R. Lynch, Time-stepping schemes for finite element tidal model computations, *Advances in Water Resources* **1** (1977) 83-95.

22. Gray, W. G. and D. R. Lynch, On the control of noise in finite element tidal computations: a semi-implicit approach, *Computers and Fluids* **7** (1979) 47-67.

23. Henry, R. F., Interactive Design of Irregular Triangular Grids. in *Numerical Methods for Transport and Hydrologic Processes* (Elsevier, 1988) 445-450.

24. Henry, R. F. and R. A. Walters, A geometrically-based, automatic generator for irregular triangular networks. *Comm. in Numerical Methods in Engineering.* **9** (1993) 555-566.

25. Hervouet, J.-M., TELEMAC:a fully vectorised finite element software for shallow water equations, *Computer Methods and Water Resources* (1991) 7-11.

26. Hervouet, J.-M., J.-M. Janin and C. Moulin, Preconditioning and Solving Linear Systems for the Computation of Free Surface Flows, in *Computational Methods in Water Resources X*, eds. A. Peters et al (1994) 1409-1416.
27. Holz, K. P., A Higher-Order Time Integration Scheme for Open Channel Flow, *Comput. Meth. Appl. Mech. Eng.* **8** (1976) 117-124.
28. Holz, K. P., and G. Nitsche, Tidal Wave Analysis for Estuaries with Inter-Tidal Flats, in *Finite Elements in Water Resources*, ed. S.Y.Wang (University of Mississippi Press, University, 1980) 5.113-5.126.
29. Houston, J. R., Interaction of Tsunamis with the Hawaiian Islands Calculated by a Finite Element Model, *J.Phys. Oceanography* **8** (1978) 93-102.
30. Hughes, T. J. R., L. P. Franca, and M. Mallet, A New Finite Element Formulation For Computational Fluid Dynamics: I. Symmetric Forms of the Compressible Euler and Navier-Stokes Equations and the Second Law of Thermodynamics, *Computer Methods in Applied Mechanics and Engineering* **54** (1986) 223-234.
31. Hughes, T. J. R., and M. Mallet, A New Finite Element Formulation For Computational Fluid Dynamics: III. Generalized Streamline Operator for Multidimensional Advective-Diffusive Systems, *Computer Methods in Applied Mechanics and Engineering* **58** (1986) 305-328.
32. Johnson, M., K. D. Paulson, and F. E. Werner, Radiation boundary conditions for finite element solutions of generalized wave equations, *Int. J. for Numerical Methods in Fluids* (1990).
33. Kajiura, K., T. Hatori, I. Aida and M. Koyama, A survey of tsunami accompanying the Tokachi-oki Eartquake of May, 1968, *Bull. Earthq. Res. Inst.* **46** (1968) 1369-1396 (in Japanese).
34. Kashiyama, K. and T. Okada, Automatic Mesh Generation Method for Shallow Water Flow Analysis, *Int. J. for Numerical Methods in Fluids* **15** (1992) 1037-1057.
35. Kato, S., A. Anju, and M. Kawahara, A Finite Element Study of Solitary wave by Boussinesq Equation, in *Computatinal Methods in Water Resources X*, eds. A. Peters et al (1994) 1067-1072.
36. Kato, S., Anju, A., and M. Kawahara, A treatment for open boundary condition in Boussinesq equation, *Advances in Hydro-Science and Engineering* **II** (1995) 1481-1486.
37. Kawahara, M., Finite-element applications in fluid mechanics, Lake and Harbor motion, in *Finite Element Handbook*, eds. H. Kardestuncer et al (1987) 3.81-3.94.
38. Kawahara, M. and K.Hasegawa, Periodic Galerkin Finite Element Method of Tidal Flow, *Int. J. for Numerical Methods in Engineering* **12** (1978) 115-127.
39. Kawahara, M., H. Hirano, K. Tsubota and K. Inagaki, Selective Lumping Finite Element Method for Shallow Water Flow, *Int. J. for Numerical Methods in Fluids* **2** (1982) 89-112.
40. Kawahara, M., T. Kodama, M. Kinoshita, Finite element method for tsunami

wave propagation analysis considering the open boundary condition, *Computers Mathematics with Applications* **16** (1988) 139-152.

41. Kawahara, M. and S. Nakazawa, Finite Element Method for Unsteady Shallow Water Wave Equation, in *U.S.-Japan Seminar on Interdisciplinary Finite Element Methods*, eds. J.F.Abel et al (Cornell University Press ,Ithaca, N.Y., 1979) 267-283.

42. Kawahara, M., S. Nakazawa, S. Ohmori, and T. Takagi, Two Step Explicit Finite Element Method for Storm Surge Propagation Analysis, *Int. J. for Numerical Methods in Engineering* **15** (1980) 1129-1148.

43. Kawahara, M., N. Takeuchi and T. Yoshida, Two Step Explicit Finite Element Method for Tsunami Wave Propagation Analysis, *Int. J. for Numerical Methods in Engineering* **12** (1978) 331-351.

44. Kawahara, M.,T. Umetsu, Finite element method for moving boundary problems in river flow, *Int. J. for Numerical Methods in Fluids* **6** (1986) 365-386.

45. King, I. P., and W. R. Norton, Recent Application of RMA's Finite Element Models for Two Dimensional Hydrodynamics and Water Quality, in *Proc. 2d Conf. Finite Elements Water Resources*, eds. C. A. Brebbia et al (1978) 2.81-2.99.

46. King, I. P., A Three Dimensional Finite Element Model for Stratified Flow, in *Finite element Flow Analysis*, ed. T.Kawai (University of Tokyo Press, Tokyo, 1982) 513-520.

47. Kodama, T., T. Imai, M. Kawahara, A computational technique dealing with the open boundary condition and its application to tsunami wave propagation analysis, *Computer Modelling in Ocean Engineering* **91** (1991) 385-394.

48. Kolar, R. L., J. J. Westerink, M. E. Catekin, and C. A. Blain, Aspects of nonlinear simulation using shallow-water models based on the wave continuity equation, *Computers and Fluids* **23** (1994) 523-538.

49. Le Provost, C., G. Rougier and A. Poncet, Numerical modeling of the harmonic constituents of the tides, with application to the English Channel, *J. Phys. Oceanography* **11** (1981) 1123-1138.

50. Lynch, D. R., *Finite element solution of the shallow water equations* (Princeton Univ., PhD Dissertation, Univ. Microfilms International, 1978).

51. Lynch, D. R., and W. G. Gray, Finite Element Simulation of Shallow Water Problems with Moving Boundaries, in *Proc. 2d Conf. Finite elements Water Resources*, eds. C. A. Brebbia et al (1978) 2.23-2.42.

52. Lynch, D. R. and W. G. Gray, A wave equation model for finite element tidal computations, *Computers and Fluids* **7** (1979) 207-228.

53. Lynch, D. R. and W. G. Gray, Finite Element Simulation of Flow in deforming Regions, *J. of Comp. Phys* **36** (1980) 135-153.

54. Lynch, D. R. and F. E. Werner, Three dimensional hydrodynamics on finite elements. Part 1: linearized harmonic model, *Int. J. for Numerical Methods in Fluids* **7** (1987) 871-909.

55. Lynch, D. R. and F. E. Werner, Three dimensional hydrodynamics on finite el-

ements. Part 2: nonlinear time stepping model, *Int. J. for Numerical Methods in Fluids* **12** (1991) 507-533.

56. MacArthur, R. C., I. P. King, and W. R. Norton, Application of the Finite Element Method to Vertically Stratified Hydrodynamic Flow and Water Quality, in *Finite Elements in Water Resources*, eds. S.Y. Wang et al (University of Mississippi Press, University, 1980) 5.92-5.102.

57. Malone, F. D. and J. T. Kuo, Semi-Implicit Finite Element Methods Applied to the Solution of the Shallow Water Equations, *Journal of Geophysical Research* **86** (1981) 4029-4040.

58. Mansinha, L. and D. E. Smylie, The displacement fields of inclined faults, *Bull. Seism. Soc. Amer.* **61** (1971) 1433-1440.

59. Messinger, F., and A. Arakawa, Numerical methods in atmospheric models, *GARP Publication Series* **17** (WMO/ICSU, 1976) 64.

60. Mitchell, A. R., S. W. Schoombie, Finite element studies of solitons, in *Numerical Methods in Coupled System* (Wiley, Chichester, 1984) 465-488.

61. Mueller, A. C., and G. F. Carey, Continuously deforming finite elements, *Int. J. for Numerical Methods in Engineering* **21** (1985) 2099-2126.

62. Mullen, R., and T. Belytschko, Dispersion analysis of finite element semidiscretizations of the two-dimensional wave equation, *Int. J. for Numerical Methods in Engineering* **18** (1982) 11-29.

63. Myers, E. P. and A. M. Baptista, Finite Element Modeling of the July 12, 1993 Hokkaido Nansei-Oki Tsunami, *Pure and Applied Geophysics* **144** (1995) 1070-1103.

64. Oden, J. T., W. Wu, V. Legat, An *hp* adaptive strategy for finite element approximations of the navier stokes equations, *Int. J. for Numerical Methods in Fluids* **20** (1995) 831-851.

65. Partridge, P. W., and C. A. Brebbia, Quadratic Finite Elements in Shallow Water Problems, *Proc. ASCE* **102**(HY9) (1976) 1299-1313.

66. Pearson, C. E. and D. F. Winter, On the calculation of tidal currents in homogeneous estuaries. *J. Phys. Oceanography* **7** (1977) 520.

67. Pearson, C. E. and D. F. Winter, On tidal motion in a stratified inlet with particular reference to boundary conditions, *J. Phys. Oceanography* **14** (1984) 1307-1314.

68. Peregrine, D. H., Long waves on a beach, *J. Fluid Mech.* **27** (1967) 815-927.

69. Platzman, G. W., Some response characteristics of finite element tidal models, *J. Comp. Phys.* **40** (1981) 36-63.

70. Pinder, G. F. and W. G. Gray, *Finite Elements in Surface and Subsurface Hydrology* (Academic Press, 1977).

71. Renka, R. J., A Storage-efficient Method for Construction of a Thiessen Triangulation, ORNL/CSD-101 (Oak Ridge National Laboratory, 1982).

72. Rivara, M. C., Algorithms for refining triangular grids suitable for adaptive and multigrid techniques, *Int. J. for Numerical Methods in Engineering* **20** (1984) 745-765.

73. Rivara, M. C., and C. Levin, A 3-D refinement algorithm suitable for adaptive and multigrid techniques, *Comm. in Applied Numerical Methods* **8** (1992) 281-290.

74. Røed, L. P., and C. K. Cooper, Open boundary conditions in numerical ocean models, in *in Advanced Physical Oceanographic Numerical Modeling*, ed. J.J. O'Brien (D. Reidel Publ. Comp., Boston, 1986) 411-436.

75. Simpson, R. B., A Survey of Two Dimensional Finite Element Mesh Generation. *Proc. 9th Manitoba Conf. Num. Math. and Computing* (1979) 49-124.

76. Snyder, R. L., M. Sidjabat, and J. H. Filloux, A study of tides, setup, and bottom friction in a shallow semi-enclosed basin. Part II: Tidal model and comparison with data *J. Phys. Oceanography* **9**: (1979) 170-188.

77. Sommerfeld, A., *Partial Differential Equations in Physics* (Academic Press, New York, 1949).

78. Staniforth, A. N., and H. L. Mitchell, A Semi-Implicit Finite-Element Barotropic Model, *Month. Weat. Rev.* **105** (1977) 54-169.

79. Tanaka, T., Y. Ono, and T. Ishise, The Open Boundary Value Problems in Ocean Dynamocs by Finite Elements, in *Finite Elements in Water Resources*, eds. S.Y. Wang et al (University of Mississippi Press, University, 1980) 5.47-5.63.

80. Thacker, W. C., A Brief Review of Techniques for Generating Irregular Computational Grids, *Int. J. for Numerical Methods in Engineering* **15** (1980) 1335-1341.

81. Umetsu, T., Applications of moving boundary simulation for river flow due to configulation bank environment, *Finite Elements in Fluids* (1993) 806-815.

82. Umetsu, T., A boundary condition technique of moving boundary simulation for broken dam problem by Three-step explicit finite element method, *Advances in Hydro-Science and Engineering*, **II** (1995) 394-399.

83. Umetsu, T., and M. Kawahara, Two-step explicit finite element method for the spill way flow using moving boundary technique, in *Proc. Int. Conf. Computer Modelling in Ocean. Engineering* (Venice, Italy, 1988) 597-608.

84. Walters, R. A. "The frontal method in hydrodynamics simulations", *Computers and Fluids*, **8**, (1980) 265-272.

85. Walters, R. A., Numerically induced oscillations in finite element approximations to the shallow water equations, *Int. J. for Numerical Methods in Engineering* **3** (1983) 591 604.

86. Walters, R. A., A model for tides and currents in the English Channel and southern North Sea, *Advances in Water Resources* **10** (1987) 138-148.

87. Walters, R. A., A three-dimensional, finite element model for coastal and estuarine circulation. *Continental Shelf Research* **12** (1992) 83-102.

88. Walters, R. A. and E. Barragy, Application of $h - p$ methods to the shallow water equations. *Int. J. for Numerical Methods in Fluids* (In press).

89. Walters, R. A. and G. F. Carey, Analysis of spurious oscillation modes for the shallow water and Navier-Stokes equations, *Computers and Fluids* **11** (1983)

51-68.

90. Walters, R. A. and G. F. Carey, Numerical noise in ocean and estuarine models, *Advances in Water Resources* **7** (1984) 15-20.

91. Walters, R. A. and R. T. Cheng, A Two Dimensional Hydrodynamic Model of a Tidal Estury, in *Proc. 2d Conf. Finite elements Water Resources*, eds. C. A. Brebbia et al (1978) 2.3-2.21.

92. Walters, R. A. and Werner, F. E., "A comparison of two finite element models using the North Sea data set", *Advances in Water Resources* **12** (1989) 184-193.

93. Westerink, J. J. and W. G. Gray, Progress in surface water modelling. *Reviews of Geophysics* (Supplement April, 1991) 201-217.

94. Westerink, J. J., R. A. Luettich, A. M. Baptista, N. W. Scheffner, and P.Farrar, Tide and Storm Surge Predictions Using a Finite Element Model, *Jounal of Hydraulic Engineering* **118** (1992) 1373-1390.

95. Westerink, J. J. and P. J. Roache, Issues in convergence studies in geophysical flow computations, in *Quantification of uncertainty in computational fluid dynamics, FED* **213** (ASME, NY, 1995).

96. Zienkiewicz, O. C., R. Lohner, K. Morgan and S. Nakazawa, Finite elements in fluid mechanics - a decade of progress, in *in Finite Elements in Fluids* **5**, eds. R.H.Gallagher et al (Wiley, Chichester, 1984) 1-26.

97. Zienkiewicz, O. C. and R. L. Taylor, *The Finite Element Method, Forth Edition*, (McGraw-Hill Book Company, 1991) 807.

NUMERICAL PREDICTION OF SOLITARY WAVE RUNUP ON VERTICAL WALLS BY FINITE-AMPLITUDE SHALLOW-WATER WAVE MODEL

Nobuhisa Kobayashi[1] and Yukiko Tega[2]

Abstract

The existing one-dimensional time-dependent numerical model based on the finite-amplitude shallow-water equations is modified to predict solitary wave runup on a vertical wall. The computed results for Benchmark Problem No. 3 are compared with the measured time series of the free surface elevation. The comparison indicates that the finite-amplitude shallow-water equations for nondispersive waves can not predict the sharp crest of shoaling solitary waves because of the premature steepening and breaking of the wave front. The numerical model is also compared with the six runs of the experiment by Ramsden and Raichlen (1990) in which the spatial and temporal variations of the free surface elevation were measured for a solitary bore impacting on a vertical wall. The transformation from a breaking solitary wave to an incident bore on the gentle slope in front of the vertical wall is predicted fairly well. The numerical model underpredicts the free surface elevation during the solitary bore impact on the vertical wall probably because of the vertical fluid acceleration and air entrainment neglected in the one-dimensional model.

1. INTRODUCTION

One-dimensional time-dependent models based on the finite-amplitude shallow-water equations were developed and applied to predict bore runup on a beach (Hibberd and Peregrine, 1979; Yeh et al., 1989) and flow over a vertical containment dyke (Greenspan and Young, 1978). The finite-amplitude shallow-water equations based on the conservation equations of mass and horizontal momentum are relatively simple and can be applied to breaking waves although energy dissipation due to wave breaking is not modeled explicitly and occurs numerically (Kobayashi and Wurjanto, 1992). Various numerical methods were developed for flows with shocks (e.g., Richtmyer and Morton, 1967; Anderson et al., 1984), but no attempt has been made to compare the different numerical methods for their capabilities of computing breaking waves.

Kobayashi (1995) summarized practical applications of the one dimensional time-dependent numerical models for breaking waves on coastal structures and beaches. Generally, these models predict the free surface elevation fairly accurately (within about 20% errors). The one-dimensional models predict only the depth-averaged ve-

---

[1]Prof. and Assoc. Dir. Ctr. for Appl. Coast. Res., Ocean Engrg. Lab., Univ. of Delaware, Newark, DE 19716.

[2]Grad. Student, Dept. of Civ. Engrg., Univ. of Delaware, Newark, DE 19716.

locity, which may represent the horizontal velocities measured below the wave trough level reasonably well (Cox et al., 1995). However, a vertically two-dimensional model will be required to predict the detailed vertical variations of the fluid velocities and shear stress under breaking waves (Kobayashi and Johnson, 1995). On the other hand, Kobayashi and Karjadi (1995, 1996) extended their one-dimensional model to predict the free surface elevation and depth-averaged cross-shore and alongshore velocities in the swash and surf zones under obliquely incident regular and irregular waves.

For solitary waves considered in this paper, Kobayashi and Karjadi (1993, 1994) developed a computer program called SBREAK to predict the runup of breaking solitary waves on beaches. SBREAK was shown to be in good agreement with the data of Synolakis (1987) on breaking solitary wave runup on a smooth uniform slope with a limited calibration of the bottom friction factor which is the only empirical parameter involved in SBREAK. The representative solitary wave period and associated surf similarity parameter were introduced in SBREAK to examine the similarity and difference between solitary and regular waves on smooth uniform slopes. The characteristics of the computed solitary wave breaking, decay and reflection as a function of the surf similarity parameter were shown by Karjadi and Kobayashi (1994) to be qualitatively similar to those of regular waves (Battjes, 1974).

The computer program SBREAK is modified herein to predict runup of solitary waves on a vertical wall. The computed results for Cases A, B and C of Benchmark Problem No. 3 are compared with the time series of the free surface elevation measured by Wave Gages 7 and 9 as well as the visually measured runup on the vertical wall. The finite-amplitude shallow-water equations for nondispersive waves will be unlikely to accurately simulate solitary wave propagation on the horizontal bottom from the wavemaker to the toe of the compound slope in this experiment, unlike a dispersive Boussinesq wave model (e.g., Kobayashi et al., 1989; Zelt, 1991). As a result, the wavemaker is assumed to have been controlled using the theory of Goring (1978) and produced perfect solitary waves at the toe of the compound slope which are specified as input to the modified SBREAK. In order to further assess the capabilities and limitations of the numerical model, comparisons are also made with the six runs of Ramsden and Raichlen (1990) who measured the spatial and temporal variations of the free surface elevation of a solitary bore impacting on a vertical wall in a tilting flume.

2. NUMERICAL MODEL

2.1. Governing Equations

The wave motion on a gentle impermeable slope is computed for the normally incident wave train specified at the seaward boundary of the computation domain as shown in Fig. 1 for the case of a gentle slope in front of a vertical wall. The prime

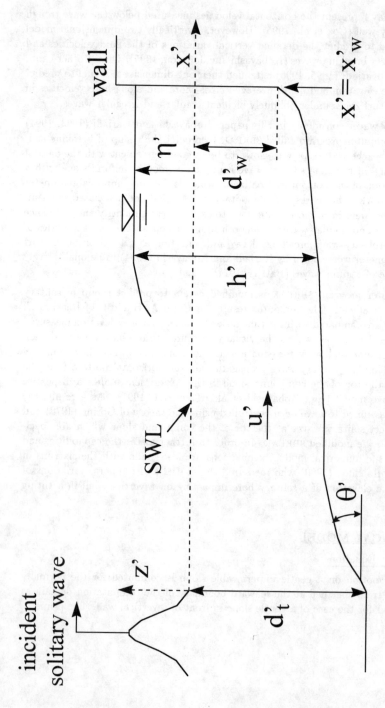

Figure 1: Definition Sketch for Numerical Model.

indicates the dimensional variables in the following. The symbols shown in Fig. 1 are as follows: x' = horizontal coordinate taken to be positive landward with $x' = 0$ at the seaward boundary of the computation domain; x'_w = horizontal distance between the seaward boundary and the vertical wall; z' = vertical coordinate taken to be positive upward with $z' = 0$ at the still water level (SWL); d'_t = water depth below SWL at the seaward boundary; d'_w = water depth below SWL at the vertical wall; θ' = local angle of the slope in front of the vertical wall which may vary along the slope; η' = free surface elevation above SWL; h' = water depth above the impermeable slope; and u' = depth- averaged horizontal velocity.

For finite-amplitude shallow-water waves over the gentle impermeable slope, the vertically-integrated equations for the mass and x'-momentum may be expressed as (Kobayashi et al., 1987; Kobayashi and Wurjanto, 1992)

$$\frac{\partial h'}{\partial t'} + \frac{\partial}{\partial x'}\left(h'u'\right) = 0 \tag{1}$$

$$\frac{\partial}{\partial t'}\left(h'u'\right) + \frac{\partial}{\partial x'}\left(h'u'^2\right) = -gh'\frac{\partial \eta'}{\partial x'} - \frac{1}{2}f'\mid u'\mid u' \tag{2}$$

where t' = time; g = gravitational acceleration; and f' = friction factor related to the shear stress acting on the slope. The friction factor f' is assumed constant, although it can be varied spatially (Kobayashi and Raichle, 1994). The range of f' used in the previous applications was $f' = 0.005$–0.05 for smooth slopes. Use is made of $f' = 0.005$ in the subsequent computations on the basis of the calibration made for breaking solitary wave runup on smooth uniform slopes (Kobayashi and Karjadi, 1994). Wave runup on a smooth vertical wall is expected to be insensitive to the friction factor f' for the smooth slope in front of the wall as long as the toe depth d'_w is positive.

The following dimensionless variables and parameters based on the assumption of finite-amplitude shallow-water waves are introduced to normalize Eqs. (1) and (2):

$$t = t'/T'_r \;\;;\;\; x = x'/\left[T'_r\left(gH'_r\right)^{1/2}\right] \;\;;\;\; x_w = x'_w/\left[T'_r\left(gH'_r\right)^{1/2}\right] \tag{3}$$

$$u = u'/\left(gH'_r\right)^{1/2} \;\;;\;\; z = z'/H'_r \;\;;\;\; h = h'/H'_r \;\;;\;\; \eta = \eta'/H'_r \;\;;\;\; d_t = d'_t/H'_r \tag{4}$$

$$d_w = d'_w/H'_r \;\;;\;\; \sigma = T'_r\left(g/H'_r\right)^{1/2} \;\;;\;\; \theta = \sigma\tan\theta' \;\;;\;\; f = \sigma f'/2 \tag{5}$$

where T'_r = representative wave period; H'_r = representative wave height; σ = dimensionless parameter expressing the ratio between the characteristic horizontal and vertical length scales; θ = dimensionless gradient of the slope in front of the vertical wall; and f = normalized friction factor. The present numerical model assumes that $\sigma^2 \gg 1$ and $(\cot\theta')^2 \gg 1$ in the computation domain (Kobayashi and Wurjanto, 1992). The representative wave period and height used for the normalization can be taken as the period and height used to characterize the incident wave for a particular problem.

Substitution of Eqs. (3)–(5) into Eqs. (1) and (2) yields

$$\frac{\partial h}{\partial t} + \frac{\partial}{\partial x}(hu) = 0 \tag{6}$$

$$\frac{\partial}{\partial t}(hu) + \frac{\partial}{\partial x}\left(hu^2 + \frac{1}{2}h^2\right) = -\theta h - f\,|\,u\,|\,u \tag{7}$$

where θ and f express the effects of the slope and friction, respectively. For a uniform slope, θ in Eq. (7) can be replaced by the surf similarity parameter, $\xi = \theta/(2\pi)^{1/2}$ (Battjes, 1974). In terms of the normalized coordinate system, the slope is located at

$$z = \int_0^x \theta dx - d_t \quad ; \quad 0 \le x \le x_w \tag{8}$$

which reduces to $z = (\theta x - d_t)$ for a uniform slope.

The initial time $t = 0$ for the computation marching forward in time is taken to be the time when the specified incident wave train arrives at the seaward boundary located at $x = 0$ as shown in Fig. 1. The initial conditions for the computation are thus given by $\eta = 0$ and $u = 0$ at $t = 0$ in the region $0 \le x \le x_w$. It is noted that $(h - \eta)$ is the normalized depth below SWL above the bottom elevation expressed in Eq. (8).

In order to derive an appropriate seaward boundary condition, Eqs. (6) and (7) are expressed in the following characteristic forms

$$\frac{\partial \alpha}{\partial t} + (u + c)\frac{\partial \alpha}{\partial x} = -\theta - \frac{f\,|\,u\,|\,u}{h} \quad ; \quad \frac{dx}{dt} = u + c \tag{9}$$

$$\frac{\partial \beta}{\partial t} + (u - c)\frac{\partial \beta}{\partial x} = \theta + \frac{f\,|\,u\,|\,u}{h} \quad ; \quad \frac{dx}{dt} = u - c \tag{10}$$

with

$$c = h^{1/2} \quad ; \quad \alpha = u + 2c \quad ; \quad \beta = -u + 2c \tag{11}$$

where α and β are the characteristic variables.

Assuming that $u < c$ in the vicinity of the seaward boundary where the normalized water depth below SWL is d_t, α and β represent the characteristics advancing landward and seaward, respectively, in the vicinity of the seaward boundary. The total water depth at the seaward boundary is expressed in the form (Kobayashi et al., 1987)

$$h = d_t + \eta_i(t) + \eta_r(t) \quad \text{at } x = 0 \tag{12}$$

where η_i and η_r are the free surface variations normalized by H_r' at $x = 0$ due to the incident and reflected waves, respectively. The incident wave train is specified by prescribing the variation of η_i with respect to $t \ge 0$. The normalized reflected wave train η_r is approximately expressed in terms of the seaward advancing characteristic β at $x = 0$

$$\eta_r(t) = \frac{1}{2} d_t^{1/2} \, \beta(t) - d_t - C_t \quad \text{at } x = 0 \tag{13}$$

where β is given by Eq. (10). The nonlinear correction term C_t in Eq. (13) was introduced by Kobayashi et al. (1989) to predict regular wave set-down and set-up on a beach. For a solitary wave, this term may be neglected because of its transient nature and $C_t = 0$ is used in the subsequent computations.

The landward boundary condition at the vertical wall for the case of no overtopping is the no flux condition

$$u = 0 \qquad\qquad \text{at } x = x_w \tag{14}$$

where it is assumed that the water depth $h > 0$ and no shoreline appears in the computation domain $0 \le x \le x_w$.

2.2 Numerical Method

Eqs. (6) and (7) are combined and expressed in the following vector form:

$$\frac{\partial \mathbf{U}}{\partial t} + \frac{\partial \mathbf{F}}{\partial x} + \mathbf{G} = 0 \tag{15}$$

with

$$\mathbf{U} = \begin{bmatrix} m \\ h \end{bmatrix} \; ; \quad \mathbf{F} = \begin{bmatrix} mu + 0.5h^2 \\ m \end{bmatrix} \; ; \quad \mathbf{G} = \begin{bmatrix} \theta h + f \mid u \mid u \\ 0 \end{bmatrix} \tag{16}$$

where $m = uh$ is the normalized volume flux per unit width. The vectors $\mathbf{F}$ and $\mathbf{G}$ depend on the vector $\mathbf{U}$ for given θ and f.

Eq. (15) is discretized using a grid of constant space size Δx and constant time step Δt based on an explicit dissipative Lax-Wendroff method which is a second-order finite difference method (e.g., Richtmyer and Morton, 1967). In the following, the known quantities at the node located at $x = (j-1)\Delta x$ $(j = 1, 2, \ldots, s)$ and at the time $t = (n-1)\Delta t$ with $n = 1, 2, \ldots$ are indicated by the subscript j without a superscript. The integer s indicates the wet node next to the moving waterline at $t = (n-1)\Delta t$ for the case of wave runup on an inclined slope and the node at the specified landward boundary for the case of wave runup on a vertical wall. The unknown quantities at the node j and at the time $t^* = n\Delta t$ are denoted by the subscript j with the superscript $*$ where the asterisk indicates the quantities at the next time level. The values of $\mathbf{U}_1^*$ and $\mathbf{U}_s^*$ are computed using the seaward and landward boundary conditions, respectively. The values of $\mathbf{U}_j^*$ for $j = 2, 3, \ldots, (s-1)$ are computed using the known values of $\mathbf{U}_{j-1}$, $\mathbf{U}_j$ and $\mathbf{U}_{j+1}$ at the time $t = (n-1)\Delta t$ (Kobayashi et al., 1987)

$$\mathbf{U}_j^* = \mathbf{U}_j - \lambda \left[\frac{1}{2} \left(\mathbf{F}_{j+1} - \mathbf{F}_{j-1} \right) + \Delta x \mathbf{G}_j \right] + \frac{\lambda^2}{2} \left(\mathbf{g}_j - \mathbf{g}_{j-1} - \Delta x \mathbf{S}_j \right) + \mathbf{D}_j \tag{17}$$

where $\lambda = \Delta t/\Delta x$. The vector $\mathbf{g}_j$ in Eq. (17) is given by

$$\mathbf{g}_j = \frac{1}{2}(\mathbf{A}_{j+1} + \mathbf{A}_j)\left[\mathbf{F}_{j+1} - \mathbf{F}_j + \frac{\Delta x}{2}(\mathbf{G}_{j+1} + \mathbf{G}_j)\right] \tag{18}$$

with

$$\mathbf{A} = \left[\begin{array}{ccc} 2u & ; & (h - u^2) \\ 1 & ; & 0 \end{array}\right] \tag{19}$$

The vector $\mathbf{S}_j$ in Eq. (17) is defined as

$$\mathbf{S}_j = \left[\begin{array}{c} \Delta x e_j - 0.5\theta_j\,(m_{j+1} - m_{j-1}) \\ 0 \end{array}\right] \tag{20}$$

with

$$\begin{aligned} e_j &= 2f\,|\,u_j\,|\,h_j^{-1}\Big[\left(u_j^2 - h_j\right)(h_{j+1} - h_{j-1})\,(2\Delta x)^{-1} \\ &\quad -u_j\,(m_{j+1} - m_{j-1})\,(2\Delta x)^{-1} - \theta_j h_j - f\,|\,u_j\,|\,u_j\Big] \end{aligned} \tag{21}$$

The vector $\mathbf{D}_j$ in Eq. (17) represents the additional term for damping high frequency parasitic waves, which tend to appear at the rear of a breaking wave, and is given by

$$\mathbf{D}_j = \frac{\lambda}{2}\left[\mathbf{Q}_j\,(\mathbf{U}_{j+1} - \mathbf{U}_j) - \mathbf{Q}_{j-1}\,(\mathbf{U}_j - \mathbf{U}_{j-1})\right] \tag{22}$$

with

$$\mathbf{Q}_j = p_j\mathbf{I} + \frac{1}{2}\,q_j\,(\mathbf{A}_j + \mathbf{A}_{j+1}) \tag{23}$$

where $\mathbf{I}$ = unit matrix; the coefficients p_j and q_j are given by

$$\begin{aligned} p_j &= \frac{1}{2}\,(c_j + c_{j+1})^{-1}\,[\epsilon_2\,|\,w_{j+1} - w_j\,|\,(v_j + v_{j+1}) \\ &\quad -\epsilon_1\,|\,v_{j+1} - v_j\,|\,(w_j + w_{j+1})] \end{aligned} \tag{24}$$

$$q_j = (c_j + c_{j+1})^{-1}\,(\epsilon_1\,|\,v_{j+1} - v_j\,| -\epsilon_2\,|\,w_{j+1} - w_j\,|) \tag{25}$$

with

$$c = h^{1/2} \quad ; \quad v = u + c \quad ; \quad w = u - c \tag{26}$$

where ϵ_1 and ϵ_2 are the positive damping coefficients determining the amount of numerical damping of high frequency parasitic waves at the rear of a breaking wave. The values of $\epsilon_1 = \epsilon_2 = 1$ or $\epsilon_1 = \epsilon_2 = 2$ have been used in previous computations. The increase of ϵ_1 and ϵ_2 tends to improve numerical stability with negligible effects on computed results (Kobayashi and Wurjanto, 1992). Use is made herein of $\epsilon_1 = \epsilon_2 = 1$ employed in the previous computations for solitary wave runup on uniform smooth slopes (Kobayashi and Karjadi, 1993).

The numerical stability criterion for this explicit finite difference method is given by (Packwood, 1980)

$$\frac{\Delta t}{\Delta x} < (\mid u_m \mid + c_m)^{-1} \left[\left(1 + \frac{\epsilon^2}{4} \right)^{1/2} - \frac{\epsilon}{2} \right] \tag{27}$$

where u_m = maximum value of u expected to be encountered in the flow field; c_m = maximum expected value of $h^{1/2}$; and ϵ = greatest coefficient of ϵ_1 and ϵ_2.

The values of Δt, Δx, ϵ_1 and ϵ_2 need to be specified, considering the numerical stability criterion and desirable spatial and temporal accuracy as will be discussed in Sections 3 and 4.

2.3 Incident Solitary Wave Profile

The normalized incident wave profile, $\eta_i(t) = \eta_i'(t')/H_r'$, with $t = t'/T_r'$ at the seaward boundary of the computation domain needs to be specified as input where H_r' and T_r' are the representative wave height and period used for the normalization in Eqs. (3)–(5). The temporal variation of $\eta_i(t)$ can be the measured incident wave profile at the seaward boundary in the absence of a vertical wall. If no data on the incident wave profile is available, an appropriate wave theory may be used to specify $\eta_i(t)$ for $t \geq 0$ such that $\eta_i = 0$ at $t = 0$ to be consistent with the assumed initial conditions of no wave action in the region of $0 \leq x \leq x_w$ at $t = 0$.

In this paper, solitary wave theory (e.g., Dean and Dalrymple, 1984) is used to specify the normalized incident wave profile $\eta_i(t)$ at $x = 0$. The solitary wave profile normalized by H_r' and T_r' may be expressed as (Kobayashi and Karjadi, 1994)

$$\eta_i(t) = sech^2 \left[K(t - t_c) \right] \quad \text{for } t \geq 0 \tag{28}$$

with

$$K = \frac{\sqrt{3}}{2} \frac{\sigma}{d_t} \left(1 + \frac{1}{d_t} \right)^{1/2} \tag{29}$$

where t_c = normalized arrival time of the solitary wave crest such that $\eta_i = 1$ at $t = t_c$. The normalized height of the incident solitary wave is unity. The normalized time $t_c = t_c'/T_r'$ depends on the representative wave period T_r'.

In order to compute $\eta_i(t)$ as a function of $t \geq 0$ using Eqs. (28) and (29), the parameter t_c and the representative wave period T_r' included in $\sigma = T_r'(g/H_r')^{1/2}$ for the solitary wave need to be specified. Since the time t' is normalized as $t = t'/T_r'$, the unit duration $(t_c - 0.5) \leq t \leq (t_c + 0.5)$ about the crest arrival time t_c may be selected such that $\eta_i \geq \delta_i$ in this unit duration where δ_i needs to be very small and is given by

$$\delta_i = sech^2 \left(\frac{K}{2} \right) \tag{30}$$

which can be rewritten as

$$K = 2 \, \ell n \left(\sqrt{\delta_i^{-1}} + \sqrt{\delta_i^{-1} - 1} \right) \tag{31}$$

Kobayashi and Karjadi (1993, 1994) adopted the small value of $\delta_i = 0.05$ on the basis of sensitivity analyses. For $\delta_i = 0.05$, Eq. (31) yields $K = 4.36$. Eq. (29) is rearranged to compute the value of σ

$$\sigma = \frac{2K}{\sqrt{3}} d_t \left(1 + \frac{1}{d_t} \right)^{-1/2} \tag{32}$$

The representative wave period $T_r' = \sigma(H_r'/g)^{1/2}$ can thus be estimated for given d_t. On the other hand, the initial value of η_i at $t = 0$ is given by

$$\eta_i(t = 0) = sech^2(K \, t_c) \tag{33}$$

Since the initial condition for the computation is taken to be $\eta = 0$ at $t = 0$ in the region $0 \leq x \leq x_w$, the value of t_c needs to be selected such that $\eta_i(t = 0)$ is essentially zero. Kobayashi and Karjadi (1994) adopted $t_c = 1$ so that $\eta_i(t = 0) = 0.00066$. The selection of $t_c = 1$ makes it easier to interpret the computed temporal variations relative to the crest arrival time $t_c = 1$.

2.4 Reflected Wave Profile

The normalized reflected wave train $\eta_r(t)$ in Eq. (12) at the seaward boundary is computed using Eq. (13) in the following. It is also required to find the unknown value of the vector $\mathbf{U}_1^*$ at $x = 0$ at the time $t^* = n\Delta t$ which can not be computed using Eq. (17).

A simple first-order finite difference equation corresponding to Eq. (10) with $f = 0$ is used to find the value of β_1^* at $x = 0$ and the time $t^* = n\Delta t$

$$\beta_1^* = \beta_1 - \frac{\Delta t}{\Delta x} (u_1 - c_1)(\beta_2 - \beta_1) + \Delta t \theta_1 \tag{34}$$

where $\beta_1 = (-u_1 + 2c_1)$ and $\beta_2 = (-u_2 + 2c_2)$. The right hand side of Eq. (34) can be computed for the know values of $\mathbf{U}_j$ with $j = 1$ and 2 at the time $t = (n - 1)\Delta t$ where the spatial nodes are located at $x = (j - 1)\Delta x$. The value of η_r^* at the time $t^* = n\Delta t$ is calculated using Eq. (13) together with β_1^* calculated by Eq. (34). Eq. (12) yields the value of h_1^*, while $u_1^* = [2(h_1^*)^{1/2} - \beta_1^*]$ using the definition of β given by Eq. (11). Thus, the values of h_1^*, u_1^* and $m_1^* = u_1^* h_1^*$ at $x = 0$ and $t^* = n\Delta t$ are obtained.

2.5 Wave Runup on Vertical Wall

For the case of no wave overtopping, the no flux boundary condition given by Eq. (14) is used in the following to find the unknown value of the vector $\mathbf{U}_s^*$ at $x = x_w$

at the time $t^* = n\Delta t$ using Eq. (17) together with the estimated value of the vector $\mathbf{U}_{s+1}$ at the hypothetical node located at $j = (s+1)$.

The second-order finite difference approximation corresponding to Eq. (14) may be expressed as

$$u_{s+1} = -u_{s-1} \tag{35}$$

The horizontal momentum equation (7) by use of Eq. (14) yields

$$\frac{\partial h}{\partial x} = -\theta \quad \text{at } x = x_w \tag{36}$$

To the second-order accuracy, Eq. (36) yields

$$h_{s+1} = h_{s-1} - 2\Delta x \theta_s \tag{37}$$

Since $m_{s+1} = u_{s+1} h_{s+1}$, the vector $\mathbf{U}_{s+1}$ can be estimated using Eqs. (35) and (37). The value of $\mathbf{U}_s^*$ is then computed using Eq. (17) together with the known values of $\mathbf{U}_{s-1}$, $\mathbf{U}_s$ and $\mathbf{U}_{s+1}$ where $c_{s+1} = h_{s+1}^{1/2}$ is used and $\theta_{s+1} = \theta_s$ is assumed.

Wave runup on the vertical wall can easily be found from the computed water depth h at the wall. For the one-dimensional numerical model, the computation of wave runup on the vertical wall does not involve the moving shoreline unlike the computation of wave runup on smooth slopes (e.g., Kobayashi and Karjadi, 1993).

3. COMPARISON WITH BENCHMARK PROBLEM NO. 3

For Benchmark Problem No. 3, the seaward boundary $x' = 0$ of the computation domain is taken at the toe of the 1:53 slope where the water depth below SWL was $d'_t = 21.8$ cm as shown in Fig. 2. The bottom profile consisted of three different slopes with $\cot \theta' = 53$, 150 and 13. The landward boundary of the computation domain is taken at the vertical wall located at $x' = x'_w = 819$ cm where the water depth below SWL was $d'_w = 4.697$ cm. Wave Gages 7 and 9 were located at the points of the slope changes where the water depth below SWL was 13.574 and 11.620 cm, respectively.

The perfect solitary wave given by Eq. (28) is assumed to have been generated by the wavemaker and arrived at the seaward boundary $x' = 0$. The initial time $t' = 0$ of the following computations is taken at the time when the edge of the incident solitary wave arrived at $x' = 0$. The incident solitary wave height H'_r is estimated using the theory of Goring (1978) together with the stroke S' of the wavemaker specified to generate the solitary wave

$$H'_r = \frac{3S'^2}{16d'_t} \tag{38}$$

The solitary wave profile measured by Wave Gage 5 located at $x' = 0$, which was given to us after the computations in this paper were made, is found to be in good agreement with the solitary wave profile based on Eq. (28). The solitary wave height

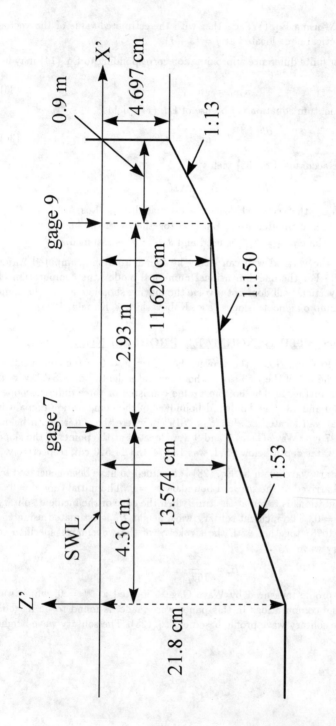

Figure 2: Bottom Profile in Computation Domain for Benchmark Problem No. 3.

Table 1: Incident Solitary Wave Characteristics for Three Cases of Benchmark Problem
No. 3

Case	S' (cm)	H'_r (cm)	T'_r (s)	d_t	σ	ξ
A	10.6	1.0	3.42	21.8	107	0.893
B	25.7	5.7	1.31	3.83	17.1	0.143
C	39.5	13.4	0.753	1.63	6.44	0.054

Table 2: Computational Parameters for Three Cases

Case	Δx	Δt	t_{max}	CPU time (min)	Memory (bytes)	Storage (bytes)
A	0.0191	0.002	10	2.2	1.5×10^6	3.1×10^5
B	0.0210	0.004	16	1.9	1.5×10^6	4.8×10^5
C	0.0237	0.005	20	1.8	1.5×10^6	6.0×10^5

H'_r estimated by Eq. (38) is slightly larger that the measured height for Cases A and
B but is slightly smaller than the measured height for Case C. The minor difference
between the measured and assumed incident solitary wave profile at $x' = 0$ does not
alter the essential aspects of the compared results presented in the following.

Table 1 list the values of S' estimated from the wave paddle trajectories for Cases
A, B and C of Benchmark Problem No. 3. The corresponding solitary wave heights
H'_r are estimated using Eq. (38) with $d'_t = 21.8$ cm. The normalized water depth,
$d_t = d'_t/H'_r$, listed in Table 1 indicates that the incident solitary waves were non-
breaking at the seaward boundary of the computation. The dimensionless parameter
σ calculated from Eq. (32) is used to find the representative wave period T'_r for each
case. The assumption of $\sigma^2 \gg 1$ made in the numerical model is satisfied for the
three cases. The surf similarity parameter ξ listed in Table 1 is based on the average
bottom slope, $\tan \theta'_\xi = 0.02088$, in the computation domain shown in Fig. 2. The
computations of breaking solitary wave runup on a 1:19.85 slope made by Kobayashi
and Karjadi (1994) were limited to the range $\xi = 0.125$–1.757. For Case A with $\xi =$
0.893, the presence of the vertical wall prevents wave breaking as will be shown later.

For the computations of Cases A, B and C, 401 nodes are placed between $x' = 0$
and $x' = x'_w = 819$ cm and the dimensional nodal spacing is $\Delta x' = 2.0475$ cm, which
may be small enough to resolve breaking solitary waves for Cases B and C as will be
shown later. The vertical wall is located at the node $j = s = 401$. The normalized

nodal spacing Δx, the normalized time step size Δt and the normalized computation duration t_{max} for each case are listed in Table 2. The value of Δt is selected in view of the numerical stability criterion given by Eq. (27). The value of t_{max} is chosen such that the reflected solitary wave propagates out of the computation domain before $t = t_{max}$. The computations presented in this paper are made using a Sun Sparc2 Workstation. The CPU time and the memory and storage requirements for each case are listed in Table 2. These computations are found to be much less demanding than the previous computations for irregular waves of longer durations.

The computed results for Cases A, B and C have been plotted in the same manner as in the paper of Karjadi and Kobayashi (1994) to interpret the computed results physically. Figs. 3, 4 and 5 show the essential features of the computed results for Cases A, B and C, respectively. In each figure, the free surface elevation, $\eta = \eta'/H'_r$, normalized by the incident solitary wave height H'_r at Wave Gages 7 and 9 and at the vertical wall is plotted as a function of the normalized time, $t = t'/T'_r$, together with the normalized incident solitary wave profile $\eta_i(t)$ given by Eq. (28) and the computed reflected wave profile $\eta_r(t)$ at $x = 0$. The computed maximum free surface elevation, η_{max}, at the vertical wall indicated in each figure may be compared with the visually measured runup in the absence of entrained air bubbles where the density of water is assumed to be constant in this numerical model.

Fig. 3 for Case A shows that the incident solitary wave propagates landward without any breaking and that the reflected solitary wave propagates seaward with its wave front steepening as it propagates seaward. The steepening front of the reflected solitary wave is likely to be caused by the shortcoming of this numerical model for nondispersive waves (Kobayashi et al., 1989). A dispersive Boussinesq wave model (e.g., Zelt, 1991) is expected to give a better prediction for non-breaking solitary waves. The computed maximum free surface elevation η_{max} at the wall is approximately twice the incident solitary wave height in front of the wall because of the essentially perfect reflection from the vertical wall. It is noted that the computed absolute values of u at the wall are found to be less than about 0.001 and essentially satisfy the no flux boundary condition given by Eq. (14) where the landward boundary algorithm adopted in Section 2.5 does not impose $u = 0$ at $x = x_w$, explicitly.

Figs. 4 and 5 for Cases B and C are similar except that the normalized free surface elevation η is larger for Case B with $d_t = 3.83$ than Case C with $d_t = 1.63$. Correspondingly, the normalized reflected solitary wave height is larger for Case B than Case C. The computed incident and reflected solitary waves at Wave Gages 7 and 9 in Figs. 4 and 5 exhibit the steep wave fronts which are regarded as wave breaking in this one-dimensional numerical model. Comparison of the computed spatial variations of the maximum crest elevation for Cases B and C suggest that the point of wave breaking is located further landward for Case B as expected. The computed normalized maximum runup η_{max} on the wall is larger for Case B.

Table 3: Measured and Computed Runup on Vertical Wall

Case	η'_{max} (cm) Measured	η'_{max} (cm) Computed
A	2.7	2.7
B	45.7	10.0
C	27.4	14.7

The measured time series of the dimensional free surface elevation η' at Wave Gages 7 and 9 were provided to us after the computed results shown in Figs. 3, 4 and 5 were obtained. The dimensional time t' used in the experiment is adjusted in the following comparisons such that the arrival time of the crest of the measured incident solitary wave at $x' = 0$ coincides with that of the solitary wave profile given by Eq. (28) and employed in these computations dealing with the solitary wave transformation landward of the toe of the 1:53 slope. Fig. 6 shows the comparisons of the measured and computed time series of the free surface elevation η' at Wave Gages 7 and 9 for Cases A, B and C. For Case A, the agreement is fairly good except for the steepening front of the reflected solitary wave. For Case B, the present numerical model for nondispersive waves can not predict the sharp crest of the shoaling solitary wave because of the premature steepening and breaking of the wave front. For Case C, the agreement improves at Wave Gage 9 where the incident solitary wave appears to have broken between Wave Gages 7 and 9.

Table 3 compares the measured and computed values of the dimensional runup, $\eta'_{max} = H'_r \eta_{max}$, on the vertical wall for Cases A, B and C. The agreement for Case A is excellent because the incident solitary wave with its small height $H'_r = 1$ cm did not break in front of the vertical wall where the water depth was $d'_w = 4.697$ cm as shown in Fig. 2. For Case B, the incident solitary wave with its height $H'_r = 5.7$ cm appears to have broken immediately in front of the vertical wall. For Case C, the incident solitary wave with its height $H'_r = 13.4$ cm broke between Wave Gages 7 and 9 in Fig. 2. The numerical model based on the assumptions of hydrostatic pressure and constant density significantly underpredicts the runup of breaking and broken waves on the vertical wall. The pressures computed by Cooker and Peregrine (1992) using potential flow theory and pressure impulse theory indicated very high upward fluid accelerations in front of the vertical wall.

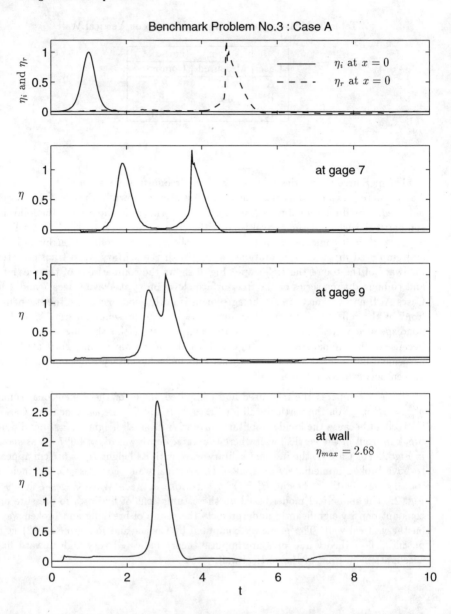

Figure 3: Specified Incident Wave Profile η_i and Computed Reflected Wave Profile η_r at $x = 0$ as well as Computed Free Surface Elevation η at Wave Gage 7 and 9 and at Vertical Wall as a Function of Normalized Time t for Case A.

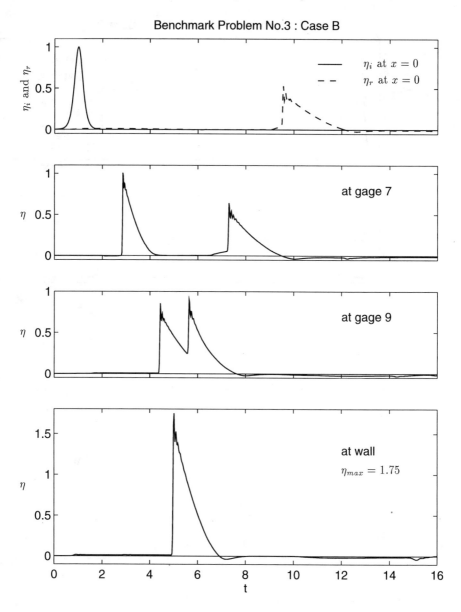

Figure 4: Specified Incident Wave Profile η_i and Computed Reflected Wave Profile η_r at $x = 0$ as well as Computed Free Surface Elevation η at Wave Gage 7 and 9 and at Vertical Wall as a Function of Normalized Time t for Case B.

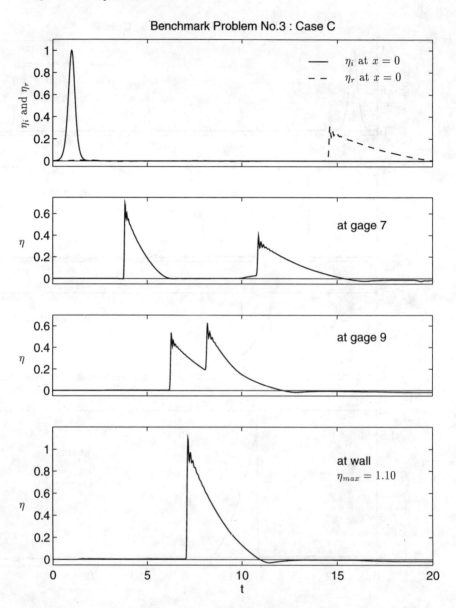

Figure 5: Specified Incident Wave Profile η_i and Computed Reflected Wave Profile η_r at $x = 0$ as well as Computed Free Surface Elevation η at Wave Gage 7 and 9 and at Vertical Wall as a Function of Normalized Time t for Case C.

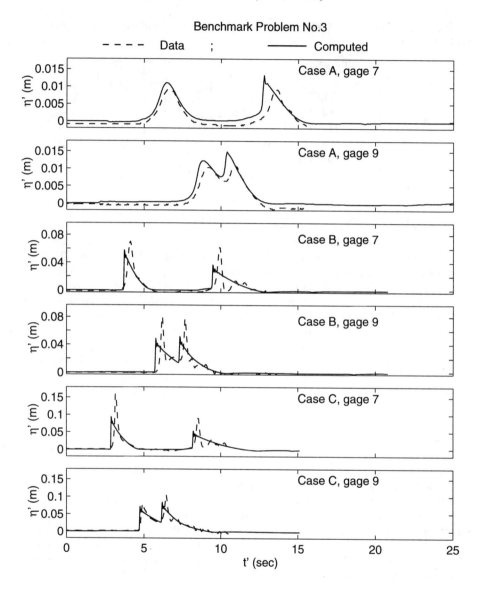

Figure 6: Comparison of Measured and Computed Free Surface Elevations η' at Wave Gages 7 and 9 for Cases A, B and C.

Table 4: Incident Solitary Wave Characteristics for Six Runs of Ramsden and Raichlen (1990).

Run	d'_t (cm)	H'_r (cm)	T'_r (s)	d_t	σ	H'_1 (cm)	c' (cm/s)
1	3.5	4.7	0.17	0.74	2.45	2.4	75
2	7.4	9.0	0.27	0.82	2.78	2.6	92
3	11.1	13.5	0.33	0.82	2.78	3.5	110
4	12.5	15.7	0.34	0.80	2.67	4.0	115
5	15.3	18.2	0.39	0.84	2.86	4.5	121
6	16.7	20.2	0.40	0.83	2.80	4.9	126

4. <u>COMPARISON WITH AVAILABLE DATA</u>

The numerical model is compared with the experiment on the impact forces on a vertical wall due to a broken solitary wave conducted by Ramsden and Raichlen (1990) in a tilting flume that was 40 m long, 110 cm wide, and 61 cm deep. The uniform bottom slope was $\tan \theta' = 0.02$. The water depth below SWL at the wall was $d'_w = 0.5$ cm. The solitary wave heights near the wavemaker and at breaking were measured. The present numerical model will not simulate solitary wave shoaling over a long distance without premature breaking (Kobayashi et al., 1989). Furthermore, the finite difference grid spacing needs to be small enough to resolve the steep wave front after wave breaking. Consequently, we are forced to take the seaward boundary $x' = 0$ of the computation domain at the breaker point. In the following computations, H'_r is the breaker height and d'_t is the breaker depth below SWL.

Table 4 lists the values of H'_r and d'_t for the six runs of Ramsden and Raichlen (1990). The value of T'_r, d_t and σ for each run are calculated in the same way as in Table 1. Their measured results were normalized using $H'_1 =$ maximum bore height at the time of bore impact on the vertical wall and $c' =$ bore celerity when the bore front was about 20 cm in front of the wall. These values for each run are also listed in Table 4. The surf similarity parameter ξ for the slope $\tan \theta' = 0.02$ is approximately 0.02 for the six runs and is less than those listed in Table 1. The problems associated with the selection of the breaker point as the seaward boundary $x = 0$ are that the values of σ listed in Table 4 may be too small to justify the assumption of $\sigma^2 \gg 1$ made in the numerical model and that the incident solitary wave profile at $x = 0$ may not be expressed by Eq. (28) for non-breaking solitary waves. The computed results presented in the following should be interpreted in light of these limitations.

Table 5 lists the computational parameters adopted for the six runs in a manner similar to Table 2. The horizontal distance of the computation domain is given by

Table 5: Computational Parameters for Six Runs

Run	x'_w (cm)	s	Δx	Δt	t_{max}
1	150	151	0.0869	0.02	33
2	345	346	0.0400	0.01	25
3	530	401	0.0353	0.01	25
4	600	401	0.0358	0.01	25
5	740	401	0.0356	0.01	25
6	810	401	0.0358	0.01	25

$x'_w = (d'_t - d'_w)/\tan\theta'$. The integer s indicates the node number at the vertical wall and is equal to the number of nodes used in the computation domain. The dimensional nodal spacing is given by $\Delta x' = x'_w/(s-1)$. The normalization of the dimensional variables is based on Eqs. (3), (4) and (5) with the values of H'_r and T'_r listed in Table 4. The selected value of Δt satisfies the numerical stability criterion given by Eq. (27). The computation duration t_{max} is set to be long enough to simulate the impact of the broken solitary wave on the vertical wall.

Fig. 7 compares the measured and computed relative incident bore profiles for the six runs at the instant the tip of the bore impacted the wall. The measured profiles are read from Fig. 8 in Ramsden and Raichlen (1990). The measured free surface elevation η' is normalized by H'_1 listed in Table 4. The horizontal distance, $(x'_w - x')$, from the wall is also normalized by H'_1. For the numerical model based on the normalized variables given by Eqs. (3), (4) and (5), the instant of the bore tip impacting the wall is taken as the time when the normalized free elevation η_w at the wall exceeds the small value δ where use is made of $\delta = 0.001$. The impact time based on $\eta_w > \delta$ should be insensitive to δ because η_w increases rapidly at the instant of the bore impact. The computed free surface elevation η is renormalized as $\eta'/H'_1 = \eta H'_r/H'_1$ where the measured value of H'_1 is used so as to check whether the numerical model predicts the height of the bore impacting the wall.

Fig. 7 shows that the numerical model predicts the incident bore profile fairly well except for Run 1 for which the bore height is underpredicted significantly. It is not certain whether this underprediction of the bore height is caused by the shortcoming of the numerical model or the approximation of the breaking solitary wave profile at $x = 0$ by Eq. (28). For Runs 2–6, the predicted bore front is too steep probably because the numerical model does not include any physical mechanism for stabilizing the steep bore front.

Figs. 8 and 9 compare the measured and computed spatial variations of η'/H'_1

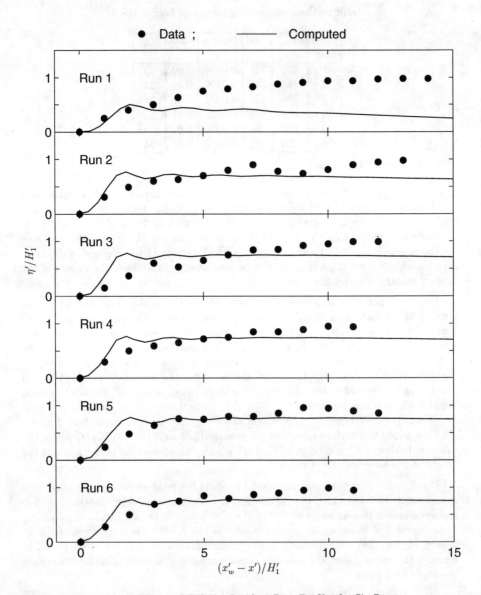

Figure 7: Measured and Computed Relative Incident Bore Profiles for Six Runs.

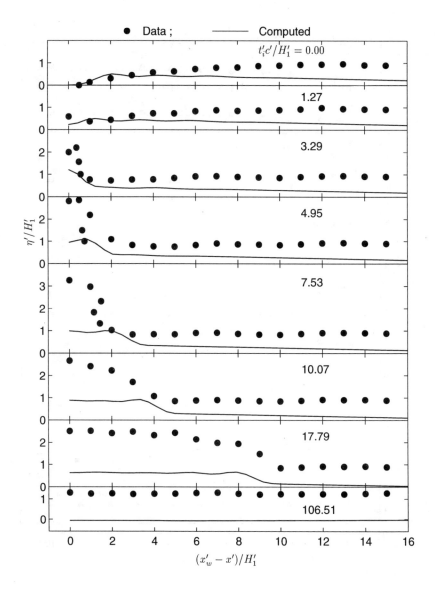

Figure 8: Measured and Computed Relative Free Surface Profiles at Selected Nondimensional Times for Run 1.

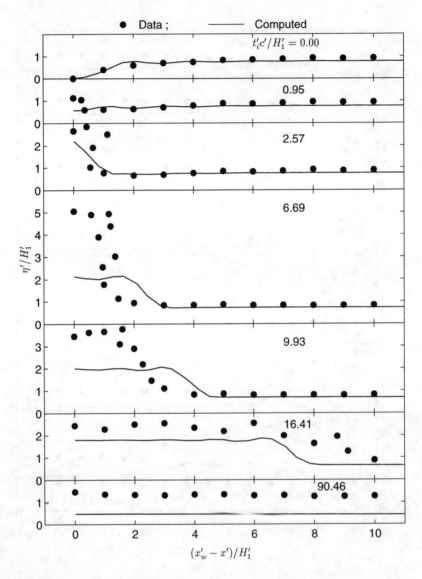

Figure 9: Measured and Computed Relative Free Surface Profiles at Selected Nondimensional Times for Run 6.

at the selected nondimensional times for Runs 1 and 6, respectively, where t_i' is the dimensional time after the bore tip impacting the wall and t_i' is normalized by the values of H_1' and c' listed in Table 4. The measured profiles in Figs. 8 and 9 are read from Figs. 11 and 12 in Ramsden and Raichlen (1990). The agreement for Run 1 shown in Fig. 8 is poor partly because of the significant underprediction of the incident bore height impacting the wall. The agreement for Run 6 shown in Fig. 9 is better but the numerical model underpredicts runup on the wall considerably. The measured free surface profiles were obtained from a frame-by-frame analysis of movies but their photographs for Run 1 indicated some air entrainment during the bore impact, runup and reflection from the wall. Consequently, the underprediction of bore runup may be caused partly by the increase of the free surface elevation due to the air entrainment. Admittedly, the assumption of essentially hydrostatic pressure below the instantaneous free surface employed in the numerical model is not really valid in the vicinity of the wall during the bore impact. The underprediction of bore runup on the wall may also be caused partly by the large upward fluid acceleration that is not accounted for in the numerical model. The computed reflected bore propagates away from the wall too quickly probably because of this underprediction of bore runup.

Fig. 10 compares the measured and computed temporal variations of the relative free surface elevation, η_w'/H_1', at the wall for the six runs. The measured variations in Fig. 10 are read from Fig. 14 in Ramsden and Raichlen (1990). Fig. 10 confirms that underprediction of bore runup on the vertical wall for Runs 2, 3, 4 and 5 as well. It is noted that the initial runup is predicted fairly well perhaps because of small vertical fluid acceleration and little air entrainment in the initial runup.

5. <u>CONCLUSIONS</u>

The one-dimensional time-dependent numerical model SBREAK (Kobayashi and Karjadi, 1994) based on the finite-amplitude shallow-water equations is modified to satisfy the no flux boundary condition at the vertical wall to the same second-order accuracy as the explicit dissipative Lax-Wendroff finite difference method used to solve the governing equations. No wave overtopping is assumed to occur but vertical walls are likely to be overtopped for most practical applications. The landward boundary algorithm developed herein may be expanded further to deal with overtopping flow over a vertical wall using the numerical procedure developed by Greenspan and Young (1978).

The modified numerical model is applied to Cases A, B and C of Benchmark Problem No. 3. Since the numerical model based on the finite-amplitude shallow-water equations for nondispersive waves can not simulate solitary wave propagation and shoaling over a long distance without premature wave breaking (Kobayashi et al., 1989), the wavemaker is assumed to have produced perfect solitary waves on the basis

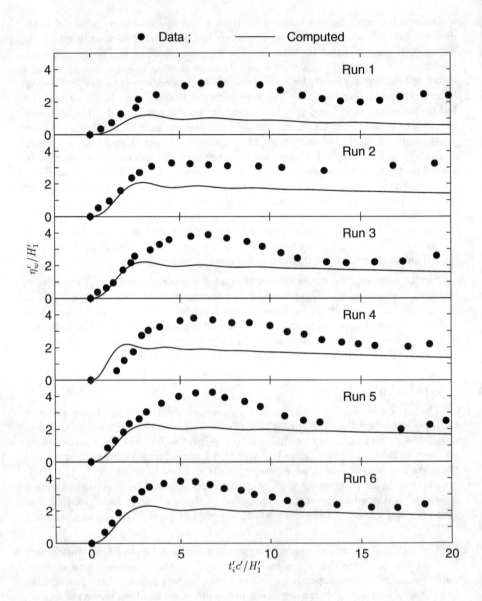

Figure 10: Measured and Computed Relative Free Surface Elevations on Vertical Wall for Six Runs.

of the theory of Goring (1978) at the toe of the compound slope in the experiment for Benchmark Problem No. 3. The measured and computed time series of the free surface elevation at Wave Gages 7 and 9 and the measured and computed runup on the vertical wall are compared to assess the limitations and capabilities of the numerical model. The computed results for Case A are physically realistic except for the steepening front of the reflected solitary wave due to the nondispersive nature of the numerical model. The computed results for Cases B and C reveal the premature wave breaking without the sharp crest of the shoaling solitary wave. The agreement improves landward of the actual wave breaking point.

The numerical model is also compared with the six runs of the experiment by Ramsden and Raichlen (1990) in which the spatial and temporal variations of the free surface elevation were measured for a solitary bore impacting on a vertical wall. The transformation from a breaking solitary wave to an incident bore on the gentle slope in front of the vertical wall is predicted as accurately as in the previous comparisons of the numerical model with solitary, regular and irregular wave data (Kobayashi, 1995). The numerical model underpredicts the free surface elevation during the solitary bore impact on the vertical wall partly because of the water depth increase due to air entrainment and partly because of the limitations of the one-dimensional model based on the assumption of hydrostatic pressure. A two-dimensional numerical model with air entrainment such as a volume of fluid method (e.g., Van der Meer et al., 1992) will be required to simulate the bore impact on the vertical wall in a more realistic manner but with considerably more numerical effort.

ACKNOWLEDGEMENT

The documentation of the computer program SBREAK modified herein was supported by the National Science Foundation under Grant No. BCS-9111827.

REFERENCES

Anderson, D.A., Tannehill, J.C. and Pletcher, R.H. (1984). *Computational Fluid Mechanics and Heat Transfer*, Hemisphere, New York.

Battjes, J.A. (1974). "Surf similarity." *Proc. 14th Coast. Engrg. Conf.*, ASCE, 466–480.

Cooker, M.J. and Peregrine, D.H. (1992). "Wave impact pressure and its effect upon bodies lying on the sea bed." *Coast. Engrg.*, 18, 205–229.

Cox, D.T., Kobayashi, N. and Okayasu, A. (1995). "Experimental and numerical modeling of surf zone hydrodynamics." *Res. Rept. CACR-95-07*, Ctr. for Applied Coast. Res., Univ. of Delaware, Newark, Del.

Dean, R.G. and Dalrymple, R.A. (1984). *Water Wave Mechanics for Engineers and Scientists*, Prentice-Hall, Englewood Cliffs, N.J.

Goring, D.G. (1978). "Tsunamis—the propagation of long waves onto a shelf." *Rep. No. KH-R-38*, W. M. Keck Lab. of Hydraulics and Water Resources, Cal. Inst. Tech., Pasadena, Cal.

Greenspan, H.P.and Young, R.E. (1978). "Flow over a containment dyke." *J. Fluid Mech.*, 87(1), 179–192.

Hibberd, S. and Peregrine, D.H. (1979). "Surf and run-up on a beach: A uniform bore." *J. Fluid Mech.*, 95(2), 323–345.

Karjadi, E.A. and Kobayashi, N. (1994). "Numerical modelling of solitary wave breaking, runup and reflection." *Proc. Intl. Symp. on Waves—Physical and Numerical Modelling*, IAHR, 426–435.

Kobayashi, N., Otta, A.K. and Roy, I. (1987). "Wave reflection and runup on rough slopes." *J. Wtrway. Port Coast. and Oc. Engrg.*, ASCE, 113(3), 282–298.

Kobayashi, N., DeSilva, G.S. and Watson, K.D. (1989). "Wave transformation and swash oscillation on gentle and steep slopes." *J. Geophys. Res.*, 94(C1), 951–966.

Kobayashi, N. and Wurjanto, A. (1992). "Irregular wave setup and run-up on beaches." *J. Wtrway. Port Coast. and Oc. Engrg.*, ASCE, 118(4), 368–386.

Kobayashi, N. and Karjadi, E.A. (1993). "Documentation of computer program for predicting long wave runup." *Res. Rep. CACR-93-03*, Ctr. for Applied Coast. Res., Univ. of Delaware, Newark, Del.

Kobayashi, N. and Karjadi, E.A. (1994). "Surf-similarity parameter for breaking solitary-wave runup." *J. Wtrway. Port Coast. and Oc. Engrg.*, ASCE, 120(6), 645–650.

Kobayashi, N. and Raichle, A.W. (1994). "Irregular wave overtopping of revetments in surf zones." *J. Wtrway. Port Coast. and Oc. Engrg.*, ASCE, 120(1), 56–73.

Kobayashi, N. (1995). "Numerical models for design of inclined structures." In *Wave Forces on Inclined and Vertical Wall Structures*, N. Kobayashi and Z. Demirbilek (eds.), ASCE, New York, 118–139.

Kobayashi, N. and Johnson, B.D. (1995). "Numerical model VBREAK for vertically two-dimensional breaking waves on impermeable slopes." *Res. Rept. No. CACR-95-06*, Ctr. for Applied Coast. Res., Univ. of Delaware, Newark, Del.

Kobayashi, N. and Karjadi, E.A. (1995). "Swash dynamics under obliquely incident waves." *Proc. 24th Coast. Engrg. Conf.*, ASCE, 2155–2169.

Kobayashi, N. and Karjadi, E.A. (1996). "Obliquely incident irregular waves in surf and swash zones." *J. Geophys. Res.*, (in press).

Packwood, A.R. (1980). "Surf and run-up on beaches." Ph.D. Thesis, School of Math., Univ. of Bristol, U.K.

Ramsden, J.D. and Raichlen, F. (1990). "Forces on vertical wall caused by incident bores." *J. Wtrway. Port Coast. and Oc. Engrg.*, ASCE, 116(5), 592–613.

Richtmyer, R.D. and Morton, K.W. (1967). *Difference Methods for Initial-Value Problems.* 2nd ed., Interscience, New York, NY.

Synolakis, C.E. (1987). "The runup of solitary waves." *J. Fluid Mech.*, 185, 523–545.

Van der Meer, J.W., Petit, H.A.H., Van den Bosch, P., Klopman, G. and Broekens, R.D. (1992). "Numerical simulation of wave motion on and in coastal structures." *Proc. 23rd Coast. Engrg. Conf.*, ASCE, 1772–1784.

Yeh, H.H., Ghazali, A. and Marton, I. (1989). "Experimental study of bore run-up." *J. Fluid Mech.*, 206, 563–578.

Zelt, J.A. (1991). "The run-up of nonbreaking and breaking solitary waves." *Coast. Engrg.*, 15, 205–246.

Fully Nonlinear Potential Flow Models Used For Long Wave Runup Prediction

S. Grilli
Department of Ocean Engineering
University of Rhode Island
Kingston 02881, RI

Abstract

A review of Boundary Integral Equation methods used for long wave runup prediction is presented in this chapter.

In Section 1, a brief literature review is given of methods used for modeling long wave propagation and of generic methods and models used for modeling highly nonlinear waves. In Section 2, fully nonlinear potential flow equations are given for the Boundary Element Model developed by the author, including boundary conditions for both wave generation and absorption in the model. In Section 3, details are given for the generation of waves in the model using various methods (wavemakers, free surface potential, internal sources). In Section 4, the numerical implementation of the author's model based on a higher-order Boundary Element Method is briefly presented. In Section 5, many applications of the model are given for the computation of wave propagation, shoaling, breaking or runup on slopes, and interaction with submerged and emerged structures. The last application presented in this Section is the Benchmark #3 problem for the runup of solitary waves on a vertical wall that was proposed as part of the "International Workshop on Long-wave Runup Models (San Juan Island, WA, USA, 09/95). Finally, Appendices A to F give more details about various aspects of the numerical model.

1. Introduction

1.1. Modeling of long wave propagation, shoaling, breaking and runup

Over the past forty years, ocean wave propagation, shoaling, breaking or runup over a slope, have been the object of numerous theoretical and numerical studies, particularly for the case of—essentially two-dimensional—long waves or swells generated by wind (wind waves) or earthquakes (tsunamis).

Main approaches pursued were based on using : (i) linear or nonlinear Shallow Water Wave equations (Carrier and Greenspan [8] 1958, Carrier [7] 1966, Camfield and Street [6] 1969, Hibberd and Peregrine [50] 1979, Kobayashi et al. [57] 1989, and Synolakis [89] 1990); (ii) Boussinesq or parabolic approximations of Boussinesq equations (Peregrine [72] 1967, Pedersen and Gjevik [71] 1983, Freilich and Guza [24] 1984, Zelt and Raichlen [97] 1990, and

Kirby [55] 1991) [a]. Most of the methods used in these works, however, are based on first- or low-order theories whose assumptions—for instance small amplitude, mildly nonlinear waves, or mild bottom slope—may no longer be valid for waves that, due to shoaling, may be close to breaking at the top of a slope (i.e., strongly nonlinear) before they run-up or break on the slope.

Until recently, state-of-the-art methods used for predicting characteristics of highly nonlinear waves shoaling over a sloping bottom up to impending breaking (e.g., shoaling coefficients, breaker height and kinematics), were based on higher-order expansion methods originally developed for waves of permanent form over constant depth (Stiassine and Peregrine [84] 1980, Peregrine [73] 1983, Sobey and Bando [82] 1991). These methods, however, by nature cannot include effects of finite bottom slope or changes of wave form during shoaling. Long waves, in particular, are known to become strongly asymmetric when shoaling over a gentle slope and approaching breaking (e.g., experiments by Skjelbreia [78] 1987, Grilli et al.[41] 1994), an effect that is not included in the above approaches. Griffiths et al. [28] 1992, compared measurements of internal kinematics of periodic waves shoaling up a 1:30 slope with predictions of the 5th-order Stokes theory, the 9th- and higher-order streamfunction theory, and the full nonlinear model by New et al. [65]. They found that horizontal velocities were correctly predicted by most theories below still water level but [b], in the high crest region, low-order theories underpredicted velocities by as much as 50% whereas predictions of the fully nonlinear theory were quite good up to the crest [c]. Grilli et al.[41] 1994 showed that computations with a fully nonlinear potential model quite well predicted the shape of solitary waves during shoaling over a 1:35 slope, as measured in well-controlled laboratory experiments. The agreement was within 2%, both in time and space, up to the breaking point. The same computations also showed that, even for long waves, horizontal velocities under a shoaling wave crest eventually become significantly non-uniform over depth (in some cases by more than 200%), an effect which is neglected in (first-order) nonlinear shallow water wave theories.

Grilli et al.[40] 1994 and Wei et al.[93] 1995 recently compared predictions of classical (i.e., weakly nonlinear and weakly dispersive) and modified (i.e., with improved dispersion characteristics and/or full nonlinearity) Boussinesq models (BM) to the full nonlinear potential flow solution—used as a reference—for the shoaling of solitary waves over slopes 1:100 to 1:8, up to the breaking point. They found that, in the region of large nonlinearity where the ratio wave height over depth is larger than 0.5, the classical BM significantly overpredicts crest height and particle velocity. This model also predicts spurious secondary troughs behind the main crest. The fully nonlinear BM, however, was found much more accurate in predicting both wave shape and horizontal velocity under the crests, from bottom to surface. Similar conclusions were reached for the propagation of highly nonlinear undular bores over constant depth.

---

[a]The reader can find details on various wave theories and summaries of some of the above referenced works in Mei [63] 1983, and Dean and Dalrymple [17] 1984.

[b]see Ref.[17] for definitions of these wave theories.

[c]Note that these comparisons were only done for a mild slope (i.e, with limited bottom effect) and for cases in which breaking occurred by spilling. The authors pointed out that "all theories are grossly in error when compared to severe plunging breakers".

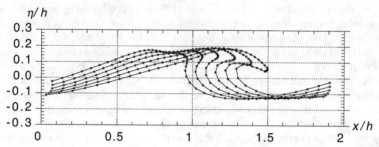

Fig. 1. Instability by plunging breaking of a large sine wave over constant depth h, as computed with the model by Grilli *et al.* [36], 1989. Initial wave height is $H/h = 0.333$, length $L/h = 1.85$, and period $T\sqrt{g/L} = 2.50$. A periodicity condition is used in the model on lateral boundaries, to create a situation similar to that examined by Longuet-Higgins and Cokelet [62]. Symbols (o) denote BEM discretization nodes, identical to individual fluid particles whose motion is calculated in time.

In the above studies, it is thus seen that a correct representation of both the shape and kinematics of strongly nonlinear long waves can only be achieved when using highly or fully nonlinear models, i.e., models in which no approximation are introduced for the free surface boundary conditions. Even for long waves with very small nonlinearity when approaching the deep water end of a slope, it is also seen that long distances of propagation over a gentle slope can make such waves both strongly asymmetric and nonlinear towards the top of the slope, whether they subsequently break or simply run up the slope.

These conclusions justify using fully nonlinear models for studying shoaling, runup or breaking, of large long waves close to the shore.

1.2. Modeling of highly nonlinear waves

Over the past twenty years, considerable efforts have been devoted to developing increasingly accurate and efficient models for fully nonlinear water waves at sea. Starting with the key work by Longuet-Higgins and Cokelet[62] 1976, the most successful approaches so far have been based on describing the physical problem based on potential flow theory (i.e., neglecting both viscous and rotational effects on the wave flow) while keeping full nonlinearity in the free surface boundary conditions (i.e., a "Fully Nonlinear Potential Flow" (FNPF) model). Most methods have also used a representation of the flow that allows for multi-valued free surface elevations appearing during breaking (i.e., a mixed Eulerian-Lagrangian representation; see Fig. 1). Despite its intrinsic limitations, potential flow theory has been shown in many applications to model the physics of wave propagation and overturning in deep water, and wave shoaling up to breaking or runup over slopes, with a surprising degree of accuracy (e.g., Dommermuth *et al.*[20], Grilli[30], and Grilli *et al.*[41,47,48]; see below for a discussion).

Many quite exhaustive reviews of the relevant literature have been published to date and can be consulted for more information (e.g., Grilli[30], Grilli *et al.*[36,38], Peregrine[73,74], Yeung[95]). For the purpose of introducing the present numerical model and its applications to long wave propagation and runup, the following is a brief description of the main steps in

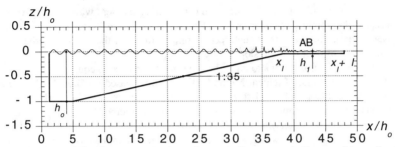

Fig. 2. Generation and shoaling over a 1:35 slope of numerically exact periodic waves (streamfunction waves) with initial height $H_o/h_o = 0.1$ and period $T\sqrt{g/h_o} = 3.55$, as computed by Grilli and Horrillo [31]. To achieve zero-mass-flux and thus constant volume in the computational domain, waves are generated on top of an opposite current equal to the mean mass transport velocity. An absorbing beach (AB) of length l, with counteracting free surface pressure, is specified for $x'_l > 38$ over a shelf of depth $h_1 = 0.05$.

the development of FNPF models that will identify key elements of the problem. Starting with Longuet-Higgins and Cokelet [62], the problem was first formulated in deep water by assuming that waves were two-dimensional in the vertical plane—i.e., long crested—and periodic in space, thus making it possible to use conformal mapping techniques which wrap the computational domain on itself and eliminate the need for lateral boundaries in the model. Doing so, deep water plunging breakers could be calculated up to touch down of the jet on the free surface (Fig. 1). Along this line, various increasingly accurate and stable numerical formulations were proposed for both deep and constant water depth, and applications sometimes also included periodic structures (Dold and Peregrine [18], New *et al.*[65], and Vinje and Brevig [92]) [d]. Results of such computations were compared to laboratory measurements and found to agree with them up to the latest stages of wave breaking, thus confirming the validity of the FNPF approach to model the physics of wave breaking far from the shore (e.g., Dommermuth *et al.* [20]).

Most of the previous and similar models—except perhaps the improvement of Dold and Peregrine's model by Cooker[13] and Cooker *et al.*[15]—due to their intrinsic nature, were unable or had the greatest difficulties generating and propagating waves over complex bottom topography. This, however, is required for solving problems of wave shoaling and breaking in shallow water and over beaches, and problems of wave interaction with coastal structures and runup on slopes. To solve such problems, models working in the so-called *physical space* must be used which brings additional problems of wave generation and/or wave absorption in the computational domain and of treatment and representation of corners in the modeled boundary (these aspects are discussed in individual Sections below). Early works that addressed problems of wave generation by a wavemaker in the physical space are the model by Kim *et al.*[54] which, however, was limited to non-breaking (single-value elevation) waves, and the model by Lin *et al.*[60], who used and improved Vinje and Brevig's formulation but somewhat restricted their scope of application. More recent models working

---

[d]Also note the somewhat different method introduced by Zaroodny and Greenberg[96], and Baker *et al.*[5], based on a vortex sheet approach.

in the physical space can accommodate almost arbitrary incident waves, complex bottom topography, and moving boundaries (e.g., Cointe[10], Grilli et al.[36,39], Klopman[56], Ohyama and Nadaoka[67]). An example of such recent computations for the shoaling and absorption of periodic waves over a gentle slope is given in Fig. 2.

In most FNPF applications to date, the governing (Laplace's) equation is solved using a higher-order Boundary Element Method (BEM), either based on Green's identity or on Cauchy integral theorem formulations, and on time integrating the free surface boundary condition using either a time marching predictor-corrector method[62,92] (Runge-Kutta and/or Adams-Bashforth-Moulton schemes) or a Taylor series expansion method[18,36]. The FNPF model by Grilli et al.[36,39,46], which will be used in the present applications, was developed in the physical space following the strategy of deep water and constant depth nonlinear wave models mentioned above (e.g., Dold and Peregrine [18]). It is based on a mixed Eulerian-Lagrangian representation with full nonlinearity in the free surface boundary conditions. FNPF equations are solved by a BEM based on Green's identity, which easily accounts for arbitrary bottom topography and almost arbitrary incident wave conditions. Development of this model was carried out under a 2D formulation, which makes the model directly applicable to shoaling and breaking and/or runup over arbitrary slopes of normally incident long crested waves, without any approximation on the wave shape or on the free surface boundary conditions [e]. Many validations (both analytical and experimental) of Grilli et al.'s model and of its more recent improved versions were carried out, mostly using solitary waves, for : (i) shallow water wave generation, propagation, and reflection, by Grilli and Svendsen[45,47]; (ii) wave runup over a steep slope, by Svendsen and Grilli[87]; (iii) shoaling and breaking over both gentle and steep slopes, by Grilli et al.[41,48]; (iv) wave impact on a mixed breakwater, by Grilli et al.[33,35]; and (v) wave propagation over a submerged obstacle, by Grilli et al.[32,34].

For completeness, other fully nonlinear wave models used for calculating wave propagation and runup on slopes (most of them based on boundary integral formulations) will be mentioned. These models have either inherently been limited to non-breaking waves (Fenton and Rienecker [23] 1982, Nakayama [64] 1983, Liu et al.[61] 1992) or have represented extensions (e.g., to axisymmetric problems) or variant of existing methods—mostly by[18,62,92]—(Isaacson [51] 1982, Jansen [53] 1986, Dommermuth and Yue [19] 1987, Gravert [26] 1987, Greenhow [27] 1987, Tanaka et al. [91] 1987, Romate [76] 1990, Seo and Dalrymple [77] 1990).

Detailed equations and numerical procedures for Grilli et al.'s wave model are presented in Sections 2,3, and 4, and applications of the model to cases of long wave propagation in shallow water and runup on slopes are presented in Section 5.

---

[e]Note that all elements in Grilli et al.'s model were selected to allow implementation of a three-dimensional model as a direct extension of the 2D formulation. This is unlike 2D FNPF models based on complex variable formulations. Such extensions of FNPF models in the physical space to three-dimensional problems have already been proposed by Romate [75,76] 1990, Yue[94] 1992, and Broeze[3] 1993, but still face challenges posed by the formidable size of the computational problem as well as problems of both representation of the free surface and boundary conditions at intersections between side walls and the free surface.

2. Mathematical model

Governing equations for the two-dimensional FNPF model by Grilli *et al.* [36,46] and its most recent extensions are presented in the next subsections. Full nonlinearity is maintained in the free surface boundary conditions, and time integration of these conditions is based on higher-order Taylor expansions, for both the free surface position and the potential. No-flow boundary conditions are prescribed along solid boundaries of the domain (bottom, coastal structures) and arbitrary waves are generated in the model, either by specifying an initial wave on the free surface, either by simulating a wavemaker at the open-sea boundary of the computational domain (as in laboratory experiments), or by using a line of internal sources. Finally, wave energy absorption can be specified in the model using an absorbing beach.

2.1. Governing equations and solid boundary conditions

The velocity potential $\phi(\mathbf{x}, t)$ is used to describe inviscid irrotational 2D flows in the vertical plane (x, z), where the velocity is given by $\mathbf{u} = \nabla \phi = (u, w)$. Continuity equation in the fluid domain $\Omega(t)$, with boundary $\Gamma(t)$, is a Laplace's equation for the potential (see Fig. 3 for definitions),

$$\nabla^2 \phi = 0 \qquad\qquad \text{in } \Omega(t) \qquad\qquad (1)$$

On the free surface $\Gamma_f(t)$, ϕ satisfies the nonlinear kinematic and dynamic boundary conditions,

$$\frac{D\mathbf{r}}{Dt} = \mathbf{u} = \nabla \phi \qquad\qquad \text{on } \Gamma_f(t) \qquad\qquad (2)$$

$$\frac{D\phi}{Dt} = -gz + \frac{1}{2} \nabla \phi \cdot \nabla \phi - \frac{p_a}{\rho} \qquad\qquad \text{on } \Gamma_f(t) \qquad\qquad (3)$$

respectively, with $\mathbf{r}$ the position vector of a free surface fluid particle, g the acceleration due to gravity, z the vertical coordinate (positive upwards, and $z = 0$ at the undisturbed free surface), p_a the atmospheric pressure, and ρ the fluid density. The material derivative is defined as,

$$\frac{D}{Dt} \equiv \frac{\partial}{\partial t} + \mathbf{u} \cdot \nabla \qquad\qquad (4)$$

Along the stationary bottom Γ_b and other fixed boundaries denoted as Γ_{r2}, a no-flow condition is prescribed as,

$$\nabla \phi \cdot \mathbf{n} \equiv \frac{\overline{\partial \phi}}{\partial n} = 0 \qquad\qquad \text{on } \Gamma_b \text{ and } \Gamma_{r2} \qquad\qquad (5)$$

in which $\mathbf{n}$ is the unit outward normal vector.

2.2. Boundary conditions for wave generation

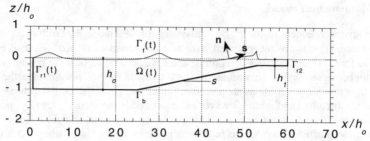

Fig. 3. Typical computational domain for wave shoaling over a slope, with definition of various boundaries. The domain has a slope s, terminated by a shelf of depth h_1 at its upper part (case where waves break before reaching the top of the slope). The sketched free surface profile corresponds to a cnoidal wave of initial height $H_o/h_o = 0.2$ and period $T\sqrt{g/h_o} = 25$, generated by a piston wavemaker on boundary $\Gamma_{r1}(t)$.

In models developed in the physical space, it is necessary to generate waves at one extremity of the computational domain. Kim *et al.*[54], Lin *et al.*[60], Cointe[10], Dommermuth *et al.*[19,20], and Grilli *et al.*[36] generated waves in their FNPF models using surface-piercing numerical wavemakers. Brorsen and Larsen[4] proposed a different approach for generating waves using internal sources, which was also used by Grilli and Svendsen[43] and Ohyama and Nadaoka[66] in their models.

When using a wavemaker to generate waves, there is a corner in the model, at the intersection between the wavemaker and the free surface, separating boundary segments with both different boundary conditions and normal directions. Possible singularity of the flow near such an intersection has given rise to substantial concern in the literature. Grilli and Svendsen[46] reviewed such singularity problems and showed that, in the particular context of wavemakers starting from a state of rest ("cold start"), provided the initial acceleration of the wavemaker is small with respect to gravity and corner boundary conditions are well-posed in the model, numerically speaking, there will be no strong singularity at the free surface corner (at least in the FNPF regime). Thus, in the applications, they used an initial damping function for the wavemaker motion in such a way that the acceleration remain small during the first few time steps of the computations. They also ensured well-posedness of governing equations and boundary conditions on both sides of corners using a double-node representation combined with (continuity and compatibility) conditions expressing that potential and velocity are unique at corners (see Grilli and Svendsen[46] for more details). Many validations of Grilli and Svendsen's model were conducted for numerical piston or flap type wavemakers, particularly with solitary waves (e.g., Grilli *et al.*[33,34,36,41,45,46,47]). Extensions of compatibility relationships that further improve numerical accuracy at corners were proposed by Otta *et al.*[70] and Svendsen *et al.*[88], and their application and validation for the wavemaking problem was further discussed by Grilli and Subramanya[39].

Assuming "numerically well-posed" wavemaker boundary conditions, it is well known, however, that a clean finite amplitude wave cannot be generated using solid wavemakers, whether in laboratory tanks or in nonlinear wave models (see, e.g., Mei[63] p. 578). Essentially, due to wave nonlinearity, higher-order harmonics are being generated that modulate the shape of the wave one intends to generate. This is because sinusoidal waves or other first-order solutions like Boussinesq solitary waves are not exact solutions of the fully non-linear problem. To overcome this difficulty and generate "clean" finite amplitude waves in their model, Grilli and Svendsen[46] used the numerically exact method by Tanaka[90] to generate solitary waves, and Klopman[56], Subramanya and Grilli[85], and Grilli and Horrillo[31] used the exact periodic wave solution of the FNPF problem (i.e., a *streamfunction wave* (SFW) solution; Dean and Dalrymple[17] p. 305) to generate periodic waves ⨍.

In the present model, waves are thus generated either by prescribing a wavemaker motion on the "open sea" boundary $\Gamma_{r1}(t)$ of the computational domain, either by prescribing the elevation and potential on the free surface of a known "exact" wave solution of flow equations, or by using an internal line of sources.

General boundary conditions for these three types of wave generation are given in the following. Generation of specific waves is discussed in Section 3.

2.2.1. Plane wavemaker

A plane wavemaker motion $\bar{x} = x_p(z, t)$ can be specified on the moving boundary $\Gamma_{r1}(t)$ to generate waves as in laboratory experiments. In this case, normal velocity is specified over the surface of the paddle as,

$$\frac{\overline{\partial \phi}}{\partial n} = \mathbf{u}_p \cdot \mathbf{n} = \frac{\frac{\partial x_p}{\partial t}}{\sqrt{1 + (\frac{\partial x_p}{\partial z})^2}} \qquad \text{on } \Gamma_{r1}(t) \qquad (6)$$

in which the right hand side represents the normal paddle velocity. Eq. (6) is developed in Section 3 for the case of piston or flap wavemakers.

2.2.2. Exact wave solutions

"Numerically exact" permanent form solutions of the FNPF boundary value problem over constant depth (eqs. (1)-(5); i.e., solitary or streamfunction waves) can be generated either by specifying their potential $\phi(x, t_o)$ and elevation $\eta(x, t_o)$ on the free surface $\Gamma_f(t_o)$ at initial time t_o (solitary waves), or by specifying their horizontal velocity and acceleration $(u(z), \partial u/\partial t)$ along a vertical wavemaker boundary (streamfunction waves).

For *exact solitary waves*, normal velocity is also prescribed to $U(t)$ over the fixed vertical lateral boundaries Γ_{r1}, Γ_{r2}. We thus get,

$$\bar{\phi} = \phi(x, t_o) , \ \bar{z} = \eta(x, t_o) \qquad \text{on } \Gamma_f(t_o)$$

$$\frac{\overline{\partial \phi}}{\partial n} = U(t) \qquad \text{on } \Gamma_{r1}, \Gamma_{r2} \qquad (7)$$

in which overbars denote prescribed values.

⨍Note that SFW's were also used in periodic FNPF models by Skourup *et al.*[80] and Grilli *et al.*[36].

Streamfunction waves, unlike linear periodic waves, have a non-zero horizontal mass flux. When specified at one extremity of the model, such waves thus lead to a continuous accumulation of water in the computational domain. In Klopman's[56] computations, only steep slopes were modeled in fairly short computational domains and waves were computed over a few periods only. Hence, water accumulation was small and did not cause any apparent problem. In Subramanya and Grilli's[85] shoaling computations, however, with a longer computational domain and for a larger number of wave periods, water accumulation resulted in a significant increase in the mean water level that clearly affected wave shape. For the corresponding coastal problem, one would expect an offshore return flow to occur under wave troughs (undertow) and cancel the incoming wave mass flux at some distance from the shore, thereby ensuring constant water volume in the nearshore region. Hence, water accumulation in the computational domain is non-physical and should be prevented. Grilli and Horrillo[31] 1995 proposed a method for achieving zero-average-mass flux in a SFW generation which they implemented and tested in their model. In this method, a depth uniform current, equal and opposite to the wave mean mass transport velocity is superimposed to the SFW [g].

For generating SFW's, a vertical wavemaker boundary is horizontally moved at one extremity of the model, following the motion, $x_p(t) = x_1(t)$, of the first node on the free surface, and wave kinematics is specified along the vertical boundary according to the SFW solution. Wave phase at time t is thus calculated along the wavemaker as,

$$\theta(t) = k\left(x_p(t) - ct\right) - \theta_o \tag{8}$$

where θ_o is an initial shift to the location of "zero-up-crossing" towards the wave crest, for which both wave elevation and horizontal velocity are zero. To avoid problems due to the "cold start" of the wave generation, the SFW velocity field is multiplied by a ("tanh-like") damping function $\mathcal{D}(t)$ smoothly varying between 0 to 1 over a specified number of wave periods. Boundary conditions on the wavemaker boundary thus read,

$$\frac{\overline{\partial \phi}}{\partial n} = -u(\theta, z)\,\mathcal{D}(t) \qquad \text{on } \Gamma_{r1}(t) \equiv \{x = x_p(t); z \in [-h_o, \eta(x_p(t))]\}$$

$$\frac{\overline{\partial^2 \phi}}{\partial t \partial n} = -u(\theta, z)\,\dot{\mathcal{D}}(t) - \frac{\partial u}{\partial t}(\theta, z)\,\mathcal{D}(t) \tag{9}$$

where u and $\partial u/\partial t$ are calculated using both the coefficients and the wave characteristics obtained from Dean's[17] streamfunction solution [h].

2.2.3. Internal sources

---

[g]Note that since a current affects wave characteristics due to Doppler effects, SFW's have to be calculated by iteration so as to satisfy the zero-mass-flux condition as well as the streamfunction wave equations (see details in Ref.[31]).

[h]Acceleration terms have been mentioned in Eq. (9) since they will be needed in the model, as described below in Section 2.4.

The traditional way of generating waves by specifying a velocity distribution or the motion along part of the boundary has the disadvantage that this boundary also reflects waves propagating towards the boundary, from inside the computational domain (such as the scattered wave field from a structure). This is a major problem in any physical model. In a computational model, this can be avoided to a large degree by generating waves by internal sources (an idea first suggested by Brorsen and Larsen [4], for a linear wave model). If oscillating sources are distributed along a vertical, say, line placed a short distance inside the fluid domain, waves will be generated and will propagate away from the sources in both directions. The waves moving into the computational domain are the ones we are interested in. On the other hand waves scattered from structures inside the computational domain will essentially pass through the sourceline. Those scattered waves, along with waves generated away from the domain, should be leaving the domain through its open sea boundary. Hence, a radiation condition or an absorbing beach should also be specified with this type of wave generation (see, Grilli and Svendsen[43], Otta *et al.* [70], Ohyama and Nadaoka[66], and Grilli and Horrillo[31], for detail).

When sources (or sinks which are negative sources) are introduced in the fluid domain, Laplace's equation (1) becomes the Poisson equation,

$$\nabla^2 \phi = b(\mathbf{x}, t) \qquad \qquad \text{in } \Omega(t) \qquad \qquad (10)$$

where $b(\mathbf{x}, t)$ is the density of a known distribution of sources inside the domain $\Omega(t)$. Values of $b(\mathbf{x}, t)$ are discussed in Section 3 for the generation of specific waves in the model.

2.3. Boundary condition for wave absorption

Energy absorption may be necessary in a FNPF model to calculate shoaling of a train of waves for sufficiently long time over a slope, whether these waves break and/or runup on the slope. As discussed above, absorption may also be necessary for the generation of waves by an internal line of sources.

Within the frame of potential flow theory, no purely dissipative process can be used to absorb the energy of incident waves. To overcome this difficulty, two main approaches were proposed in the literature mostly for the absorption of linear waves or weakly nonlinear long waves : (i) wave radiation through an open boundary (e.g., Engquist and Majda[22], Israeli and Orszag[52], Orlanski[68], Sommerfeld[83]); and/or (ii) wave damping directly on the free surface or within a so-called "sponge layer" (e.g., LeMehaute[59], Larsen and Dancy[58]). No general method has yet been proposed for the absorption/radiation of fully nonlinear transient waves. Instead, some rather heuristic boundary conditions were proposed.

Along the line (i), Lin *et al.*[60] matched exterior linear solutions to the nonlinear interior solution at finite distance. Dommermuth and Yue[19] used the same method as Lin *et al.*'s to compute the forced heaving motion of an axisymmetric cylinder. Some of the radiation conditions developed for linear waves are also applicable to fully nonlinear waves of permanent form like solitary or streamfunction waves. Grilli *et al.*[36], for instance, developed an implicit iterative radiation condition based on Sommerfeld's[83] condition. A more accurate explicit approach was proposed by Otta *et al.*[70] who combined Orlanski's[68] radiation condition with the incident wave field kinematics calculated at internal nodes in the model, close to the radiation boundary. The method worked well for periodic waves but only showed limited success when applied to irregular waves.

Along the line (ii), Larsen and Dancy[58] developed a sponge layer method based on the idea of an "absorbing beach" (AB), first suggested by LeMehaute[59]. They only implemented the method in a (weakly nonlinear) Boussinesq model but their method was later used by Ohyama and Nadaoka[66] in a FNPF model. Similar methods were successfully used by Baker *et al.*[5] and Cointe[10] in their FNPF models, and by Subramanya and Grilli[85] and Grilli and Horrillo[31], who implemented an AB with active control of the beach parameter in their FNPF model. Boundary conditions for the latter AB are briefly presented in the following. In this case, the AB is always located at the top of a slope but the same principle can be (and has been) used to generate waves at an open ocean boundary in combination with a distribution of internal sources (Ohyama and Nadaoka[66]) or to simulate bottom discontinuities—like shelf-breaks or reefs—inducing local energy loss in incident waves.

The principle of the AB is similar to the ideas developed in Refs.[5,10] : a negative work is created against incident waves over a given section of the free surface by specifying an external counteracting pressure, $p_a = P$, in the dynamic free surface condition (3) (with $z = \eta$), which effectively extracts energy from the incident wave train. For shoaling problems, the AB is located in the model over a shallow shelf region of maximum depth h_1 in the upper part of the slope (Fig. 2). In most earlier approaches, P was specified proportional to the free surface potential ϕ but this could result in creating a positive work in the AB in some cases and, hence, lead to increased wave energy in the beach. In order for the AB to always produce a negative work against the wave motion and thus to always remove energy from the wave train, as suggested by Cao *et al.*[9], the external pressure is defined here as proportional to the normal particle velocity, $\partial\phi/\partial n$, along the free surface. The modified dynamic free surface condition thus reads,

$$\frac{D\phi}{Dt} - \frac{1}{2}\nabla\phi \cdot \nabla\phi + g\eta + \frac{P}{\rho} = 0 \tag{11}$$

with,

$$P(x, \eta, t) = \nu(x, t)\frac{\partial\phi}{\partial n}(\eta(x, t)) \tag{12}$$

in which ν, the beach absorption function, varies smoothly along the AB as,

$$\nu(x, t) = \nu_o(t)\,\rho\sqrt{gh_1}\,\left(\frac{x - x_l}{l}\right)^\alpha \tag{13}$$

where $\alpha = 2$ to 3 and ν_o is a non-dimensional beach absorption coefficient. In earlier approaches, ν_o was specified as constant (e.g., Refs.[5,9,10]). To optimize absorption of incident wave energy in the AB and make it easier, at a later stage, to deal with irregular waves, Grilli and Horrillo[31] adaptively calculated ν_o in the model as a function of time (i.e., for each time step in the model) for the AB to exactly absorb the period-averaged wave energy entering the beach over time step Δt. Details and validation of adaptive energy absorption in the AB can be found in Ref.[31].

2.4. The time integration

Free surface boundary conditions (2) and (3) are integrated at time t, to establish both the new position and the relevant boundary conditions on the free surface, at a subsequent time $t + \Delta t$ (with Δt being a small time step). In the model, this is done following the approach introduced by Dold and Peregrine [18], using Taylor expansions for both the position $r(t)$ and the potential $\phi(r(t))$ on $\Gamma_f(t)$. Series, truncated to Nth-order, are expressed in terms of the material derivative (4) and of time step Δt, as,

$$\bar{r}(t + \Delta t) = r(t) + \sum_{k=1}^{N} \frac{(\Delta t)^k}{k!} \frac{D^k r(t)}{Dt^k} + \mathcal{O}[(\Delta t)^{N+1}] \tag{14}$$

for the free surface position, and,

$$\bar{\phi}(r(t + \Delta t)) = \phi(r(t)) + \sum_{k=1}^{N} \frac{(\Delta t)^k}{k!} \frac{D^k \phi(r(t))}{Dt^k} + \mathcal{O}[(\Delta t)^{N+1}] \tag{15}$$

for the potential. The last terms in Eqs. (14) and (15) represent truncation errors. The time updating of the free surface geometry described by Eq. (14) actually corresponds to following the motion of fluid particles in time. This procedure is often referred to as a "Mixed Eulerian-Lagrangian" formulation.

Second-order series are used in the present case ($N=2$). Higher-order Taylor series, however, have successfully been used by others to provide highly accurate solutions for periodic problems (e.g., Dold and Peregrine [18] ($N=3$), and Seo and Dalrymple [77] 1990 ($N=4$)).

First-order coefficients in Eqs. (14) and (15) are obtained, based on Eqs. (2) and (3), using ϕ and $\frac{\partial \phi}{\partial n}$ as provided by the solution of Laplace's equation (1) at time t. Second-order coefficients are expressed as $\frac{D}{Dt}$ of (2) and (3), and are calculated using the solution of a second elliptic problem of the form (1) for ($\frac{\partial \phi}{\partial t}, \frac{\partial^2 \phi}{\partial t \partial n}$). This is because all time derivatives of the potential satisfy Laplace's equation. Higher-order series would simply require that more Laplace's equations are solved for higher-order time derivatives of ϕ. Detailed expressions of the coefficients of Taylor series (14) and (15) are given in Appendix A, in a curvilinear coordinate system (s, n) defined along the boundary (Fig. 3).

No-flow boundary conditions for a second Laplace's equation for $\frac{\partial \phi}{\partial t}$ are readily obtained along solid boundaries, as,

$$\overline{\frac{\partial^2 \phi}{\partial t \partial n}} = 0 \qquad \qquad \text{on } \Gamma_b \text{ and } \Gamma_{r2} \tag{16}$$

The boundary condition at the free surface is obtained from Eqs. (3) and (4) as,

$$\overline{\frac{\partial \phi}{\partial t}} = -\frac{1}{2} \nabla \phi \cdot \nabla \phi - \frac{p_a}{\rho} - gz \qquad \qquad \text{on } \Gamma_f(t) \tag{17}$$

which indicates that $\overline{\frac{\partial \phi}{\partial t}}$ can be specified on the free surface as a function of known geometry and potential at time t.

When $\Gamma_{r1}(t)$ represents a wavemaker boundary moving at velocity $\mathbf{u}_p(\mathbf{x}_p(t), t)$, we have by (6),

$$\overline{\frac{\partial^2 \phi}{\partial t \partial n}} = \frac{\partial}{\partial t}(\mathbf{u}_p \cdot \mathbf{n})$$

or,

$$\overline{\frac{\partial^2 \phi}{\partial t \partial n}} = [\frac{d(\mathbf{u}_p \cdot \mathbf{n})}{dt} - \mathbf{u}_p \cdot \nabla(\mathbf{u}_p \cdot \mathbf{n})] \qquad \text{on } \Gamma_{r1}(t) \tag{18}$$

in which, $\frac{d}{dt} = \frac{\partial}{\partial t} + \mathbf{u}_p \cdot \nabla$, denotes time derivative following the motion of the boundary $\mathbf{x}_p(t)$. This boundary condition is further developed in Section 3.

When waves are generated by a line of internal sources, the time derivative of the source strength $\frac{\partial b}{\partial t}(\mathbf{x}, t)$ is introduced in a Poisson equation of the form (10), for $\frac{\partial \phi}{\partial t}$.

2.5. Discussion of model assumptions and limitations

No approximations other than potential flow theory have been made in the model. In particular, unlike analytical or numerical expansion wave theories (see, Dean and Dalrymple [17]), no small parameter, periodicity, or permanent form wave conditions, have been assumed. This makes the present model valid from deep to shallow water and for arbitrary length waves.

The main limitations—inherent to potential flow theory—of this type of model are that bottom friction and flow separation cannot be modeled, and that computations have to be interrupted shortly after breaking of a wave first occurs. These limitations are discussed in the following :

- Long wave theory shows that bottom friction should attenuate long waves in shallow water, whereas short waves should be relatively unaffected.
  For solitary waves shoaling over gentle bottom slopes, however, experiments by Camfield and Street [6] showed that "bottom roughness has no measurable effect". This was later confirmed in other experiments by Grilli *et al.* [41] (see Fig. 14 and applications in Section 5.4). The likely reason for this is that bottom friction only becomes significant when wave height is large and this only occurs in a small region over the slope, just before the wave starts breaking.
  For large solitary waves running up a steep slope, Grilli and Svendsen[44,45,47] and Svendsen and Grilli [87] compared their nonlinear computations to experiments and found that frictional effects were also negligible. In this case, the distance of propagation over steep slopes was likely too small for friction to significantly affect waves, despite their large amplitude.
  Hence, bottom friction is not an important factor when wave height and/or distance of propagation are small.
- Flow separation over obstacles on the bottom is significant for steep obstacles (like steps or rectangular bars) of large height to depth ratios, and for high waves (Grilli *et al.* [32,33] 1992; see Figs. 5-8 in Section 5.2).

Flow separation leads to an energy loss at the obstacle that reduces wave height downstream of the obstacle. As mentioned before, although not yet tested in the model, localized energy loss could be specified to model dissipation at steps and obstacles on the bottom based on the energy absorption method used by Grilli and Horrillo[31] for their absorbing beach.

- When a wave starts overturning, a small horizontal jet forms in the highest region of the wave crest (Figs. 1 and 4). The jet curls up on itself and falls towards the free surface. Breaking occurs when the tip of the falling jet impinges on the free surface, leading to a local violation of continuity equation manifesting itself by strongly unstable numerical results. Hence, computations with the model are in essence limited to prior to the time such an impact of a wave on the free surface first occurs. Because of potential flow theory hypotheses, however, computationally accurate results may not be physically realistic up to that stage. This is discussed below.

Dommermuth *et al.* [20] compared wave profiles calculated using a FNPF model to experimental results for deep water overturning breakers. They concluded that potential theory is valid up to the moment the tip of the breaker jet hits the free surface (i.e., slightly further in time than in the situation illustrated in Fig. 1).

Skyner *et al.* [81] confirmed this conclusion and compared computed and measured velocities inside plunging breakers. The good agreement they found for the velocities further confirmed the validity of potential flow theory.

- For a train of solitary or periodic waves shoaling over a sloping beach, the front wave of the train is also the steepest wave that first breaks in the shallower water. Hence, the model can be used to calculate detailed shoaling coefficients over the length of the beach, up to the point the front wave breaks (breaker line). In this case, computations are not greatly affected by the limitation of the model to the first breaking wave, discussed above.

For periodic waves, computations can be pursued for a longer time by using an absorbing beach in the upper slope region of the model (Subramanya and Grilli[85], Grilli and Horrillo[31]). Doing so, waves can shoal the slope up to a very large fraction of their breaking height and then be absorbed in the beach. A quasi-steady state can thus be reached in the model for which characteristics of fully nonlinear waves shoaling over a slope (or more complex bottom geometry) can be calculated.

For irregular wave trains and/or complex bottom geometry, breaking is likely to occur almost anywhere in the shoaling region, due to nonlinear interactions between wave components and between waves and bottom geometry. Hence, computations may have to be stopped when breaking first occurs, and this limitation may reduce the utility of the model in its present form for addressing these situations.

- Finally, runup of non-breaking waves on steep or gentle slope can be accurately calculated in a FNPF model (e.g., Grilli *et al.*[41], Grilli and Svendsen[44,45,47], Svendsen and Grilli[87]), again, provided wave reflection does not make another incoming wave break, or a thin jet of water is not expelled at runup (like, e.g., in the computations with a vertical wall in Cooker[12], Cooker and Peregrine[14], and Grilli *et al.*[33,35]), or breaking does not occur during the backwash (like in Grilli and Svendsen[47], Svendsen and Grilli[87], Otta et al.[69]).

3. Wave generation in the model

3.1. Exact solitary waves

Tanaka [90] proposed a method to calculate numerically exact solitary wave solutions of the FNPF problem in constant depth h_o. This method has been implemented in the model to specify initial *exact* solitary waves for which surface elevation and potential are directly prescribed on the free surface [i], using (7).

In the applications with solitary waves, standard dimensionless variables, x', z', t' and c', will be used with definitions,

$$x' = \frac{x}{h_o}\,,\ z' = \frac{z}{h_o}\,,\ t' = t\sqrt{\frac{g}{h_o}}\,,\ c' = \frac{c}{\sqrt{gh_o}} = F \tag{19}$$

in which c denotes wave celerity and F is the wave Froude number. For solitary waves, initial wave height H_o is identical to the maximum elevation above $z = 0$, and we further denote by, $H' = H/h_o$, the nondimensional wave height. Details of Tanaka's method are given in Appendix B.

3.2. Exact periodic waves

The streamfunction wave (SFW) theory was introduced by Dean[16] (see also Dean and Dalrymple[17]) to calculate numerically exact periodic solutions of the FNPF problem in constant depth h_o. The original method worked in a coordinate system moving with the wave celerity, $c = L/T$ (with L the wavelength and T the wave period), and accounted for the presence of a depth-uniform current U.

A streamfunction wave solution is thus defined as,

$$\psi(\theta, z) = \sum_{j=1}^{N} X(j)\,\sinh jk(h_o + z)\,\cos j\theta\ - (U - c)z \tag{20}$$

where, $\theta = k\,(x - c\,t)$, is the wave phase and $X(j)$ is a set of N coefficients that are numerically calculated, along with $L = 2\pi/k$, to satisfy free surface boundary conditions (2) and (3), and specified wave height and period (H, T).

Horizontal velocity is easily obtained from Eq. (20) in the original coordinate system as a function of depth as,

$$u(\theta, z) = -\frac{\partial \psi}{\partial z} + c = -\sum_{j=1}^{N} (jk)\,X(j)\,\cosh jk(h_o + z)\,\cos j\theta\ + U \tag{21}$$

---

[i] In applications, initial *exact* solitary waves are specified far enough from lateral boundaries of the model for $U(t) \simeq 0$ to be assumed with sufficient accuracy.

Noting that, $\partial \theta / \partial t = -c\,k$, local horizontal acceleration is obtained as,

$$\frac{\partial u}{\partial t}(\theta, z) = -c \sum_{j=1}^{N} (jk)^2 \, X(j) \, \cosh jk(h_o + z) \, \sin j\theta \tag{22}$$

Equations (21) and (22) are used to specify the kinematics of an incident SFW over a vertical wavemaker boundary located at, $x = x_p$, in the model (Eq. (9)).

Following the method by Grilli and Horrillo[31], current U can be specified as opposite to the direction of wave propagation, with a magnitude such as to generate zero-mass-flux SFW's in the model.

3.3. Wave generation by a plane wavemaker

3.3.1. Introduction

An oscillating paddle wavemaker can be specified on boundary $\Gamma_{r1}(t)$ to generate waves the same way as in laboratory wave tanks. The wavemaker motion $x_p(t)$ and velocity $u_p(x_p(t), t)$ required to generate specific incident waves can be obtained from first-order wave theory (i.e., Boussinesq theory for long waves and first-order Stokes theory for periodic short waves) [j].

Waves generated with a first-order method propagate without change of form only in a model solving first-order theory equations. In the present fully nonlinear model—or for this respect in a laboratory wavetank—such waves are not expected to correspond to permanent form solutions (for this matter, a SFW solution would be needed). Goring[25], for instance, found that solitary waves of small amplitude ($H' < 0.2$) generated by a piston wavemaker in a wave flume kept their shape constant within a very small margin. For such small waves, the first-order wave profile is quite close to an *exact* solitary wave. For steeper waves ($H' \geq 0.2$), however, Goring found that solitary waves shed a tail of oscillations behind them as they propagated down the flume. Similarly, in computations with their model, Grilli and Svendsen[47] observed that waves of significantly large height generated by a wavemaker adjusted their shape as they propagated down a numerical tank. Such results were reproduced in many different numerical set-ups and found to agree quite well with corresponding laboratory experiments (Grilli and Svendsen[47], Grilli *et al.*[32,33,34,41], Svendsen and Grilli[87]).

3.3.2. General wavemaker boundary condition

General boundary conditions for $\frac{\partial \phi}{\partial n}$ and $\frac{\partial^2 \phi}{\partial t \partial n}$ can be derived for any specified wavemaker motion and velocity, based on Eqs. (6) and (18). The latter equation for $\frac{\partial^2 \phi}{\partial t \partial n}$ includes a time derivative with respect to the rigid body motion that needs to be carefully derived.

---

[j]Note that second-order corrections can also be applied to wavemaker motion in the model as done in laboratory flumes (e.g., Skourup[79] 1995).

This was done by Cointe [11] for the motion of a rigid body of arbitrary shape. In the case of a plane rigid body like a wavemaker, Cointe's expression reads,

$$\overline{\frac{\partial^2 \phi}{\partial t \partial n}} = (\ddot{\boldsymbol{\alpha}} \cdot \mathbf{n}) + \dot{\theta}\,[(\dot{\boldsymbol{\alpha}} \cdot \mathbf{s}) - \frac{\partial \phi}{\partial s}] - \frac{\partial^2 \phi}{\partial n \partial s}(\dot{\boldsymbol{\alpha}} \cdot \mathbf{s}) + \frac{\partial^2 \phi}{\partial s^2}(\dot{\boldsymbol{\alpha}} \cdot \mathbf{n}) \tag{23}$$

in which $\boldsymbol{\alpha}$ denotes the position vector for points on the wavemaker surface, θ is the angle of rotation around point $\mathbf{x}_g$, and dots denote absolute time derivatives with respect to the body motion, d/dt, defined as in Eq. (18).

Expressions for the velocity and the acceleration of boundary points $(\dot{\boldsymbol{\alpha}}, \ddot{\boldsymbol{\alpha}})$ can be derived for various types of wavemakers as a function of wavemaker stroke x_p and used in Eq. (23) to specify boundary conditions in the model. This is done in Appendix C for both piston and flap type wavemakers.

In the next two sections, expressions of wavemaker stroke used for generating first-order waves in the model are discussed.

3.3.3. Generation of long waves by a piston wavemaker

In a long wave of permanent form over constant depth h_o, due to mass conservation, we have at any instant,

$$\int_{-h_o}^{\eta} u \; \mathrm{d}z = c_a \eta + Q_s + u_c h_o \tag{24}$$

in which c_a is the propagation speed of the wave in a fixed frame of reference, $\eta(x,t)$ is the wave elevation above still water level, Q_s is the nonlinear mass flux averaged over a wave period, and u_c is the speed of the current defined as the averaged particle velocity below wave trough level.

For a first-order long wave, the right hand side of Eq. (24) simply reduces to $c\eta$, where c is the speed of the wave relative to the water, so that Eq. (24) becomes the simpler expression used, e.g. by Goring [25], for determining the motion required by a piston wavemaker to generate a specified water surface elevation immediately in front of the wavemaker. Since the piston motion creates a depth uniform horizontal velocity $u_p(x_p(t), t)$, Eq. (24) reduces to,

$$u_p(t) = \frac{c\eta}{h_o + \eta} \tag{25}$$

which means that a surface elevation η can be generated by specifying the piston velocity u_p as defined above. In this case, horizontal piston motion $x_p(t)$ is given by,

$$x_p(t) = \int_0^t \frac{c\eta(x,\tau)}{h_o + \eta(x,\tau)} \; \mathrm{d}\tau \tag{26}$$

Developments of this equation for generating first-order solitary or cnoidal waves are given in Appendix D.

As mentioned before, this method will only generate accurate permanent form long waves for sufficiently small initial wave height (i.e., smaller than $\simeq 0.2 h_o$). This is illustrated in the applications in Section 5.

3.3.4. Generation of a sum of periodic sine waves by a flap wavemaker

As commonly done in laboratory experiments, a sum of sine waves can be generated in the model using a flap wavemaker in water of depth h_o and specifying boundary conditions based on *first-order* Stokes theory.

To do so, the paddle stroke $x_p(t)$ is specified as the sum $S(t)$ of n sine functions of frequency $2\pi\omega_i$, phase φ_i, and amplitudes A_i. The latter are related (in a linear sense) to corresponding wave component amplitudes a_i to be generated, by a linear transfer function, $T(\omega_i, h_o)$, which can be obtained from wavemaker theory (e.g., Dean and Dalrymple [17]). Furthermore, a smooth start of the wavemaker, with small initial acceleration, is ensured by multiplying $x_p(t)$ by a damping function $\mathcal{D}(t)$ varying from 0 to $(1 - \varepsilon_z)$ over a given time $2\,t_{\varepsilon_z}$. For $\varepsilon_z \ll 1$, the damping function gives a smooth transition from 0 to $\sim S(t)$ over a time $2\,t_{\varepsilon_z}$.

We thus get,

$$x_p(t) = S(t)\mathcal{D}(t) \qquad \text{with} \qquad S(t) = \sum_{i=1}^{n} A_i\,[1 - \cos(\omega_i\,t + \varphi_i)]/2$$

$$a_i = A_i\,T(\omega_i, h_o) \qquad \text{with} \qquad T(k_i(\omega_i, h_o), h_o) = \frac{4\sinh^2 k_i h_o}{2k_i d + \sinh 2k_i h_o} \qquad (27)$$

with, $H_i = 2a_i$, the wave height (predicted by linear wave theory) and $k_i(\omega_i, h_o)$, the wavenumber of a given sine wave component to be generated obtained using the linear dispersion relation as,

$$k_i \tanh k_i h_o = \frac{\omega_i^2}{g} \qquad\qquad (28)$$

Detailed expressions for $\mathcal{D}(t)$ and resulting wavemaker boundary conditions are given in Appendix E.

As discussed before, due to nonlinearities, it is well known that free second and higher-order harmonics will be created when monochromatic waves of finite amplitude propagate down a tank (see, e.g., Mei [63]). This is illustrated in the applications in Section 5.

3.4. Wave generation by an internal line of sources

Using a BIE representation based on free space Green's function, Poisson equation (10) transforms into (see Section 4.2),

$$\alpha(\mathbf{x}_l)\phi(\mathbf{x}_l) = \int_{\Gamma(\mathbf{x})} [\frac{\partial\phi}{\partial n}(\mathbf{x})G(\mathbf{x}, \mathbf{x}_l) - \phi(\mathbf{x})\frac{\partial G(\mathbf{x}, \mathbf{x}_l)}{\partial n}]\,d\Gamma(\mathbf{x})$$
$$+ \int_{\Omega(\mathbf{x})} b(\mathbf{x}, t)G(\mathbf{x}, \mathbf{x}_l)\,d\Omega(\mathbf{x}) \qquad (29)$$

where $b(\mathbf{x}, t)$ denotes the source field contribution. Eq. (29) can be solved by a Boundary Element Method (BEM) (see Section 4.3.) but, in the present case, besides boundary integrals, domain integrals must be calculated to account for the source field contribution.

For a vertical line of sources with linear density $q(s(\mathbf{x}), t)$ (with $s(\mathbf{x})$ measured along the line Γ_ℓ), the source contribution in Eq. (29) reduces to,

$$\int_\Omega b(\mathbf{x}, t) G(\mathbf{x}, \mathbf{x}_l) \, d\Omega = \int_{\Gamma_\ell} q(s(\mathbf{x}), t) G(\mathbf{x}, \mathbf{x}_l) \, d\Gamma_\ell \tag{30}$$

In two dimensions, a line of sources with continuously varying strength creates a velocity normal to the line equal to $q/2$. Thus, specification of the strength of the source distribution q is straightforward if particle velocities are known along the line for the waves to be generated. In most cases, it is sufficient to specify the source strength only at N_s points along the line Γ_ℓ. In this case, only point sources of strength $B_s(t)$ are specified along a vertical line from bottom to surface at say, $x = x_s$, thus defining N_s segments of constant strength,

$$B_s(t) = 2 \frac{\eta(x_s, t) + h_o}{N_s} \, \overline{u_w}(x_s, z_s, t) \qquad ;s = 1, \dots, N_s \tag{31}$$

where $\overline{u_w}(x_s, z_s, t)$ denotes the mean horizontal velocity of the wave within the s-th segment and $\eta = \eta(x_s, t)$ is the wave elevation above the source line (a stretching is applied to the line to account for changes in wave elevation above the line). Hence, in Eq. (30), we have,

$$q(s(\mathbf{x}), t) = \sum_{s=1}^{N_s} B_s(t) \delta(\mathbf{x} - \mathbf{x_s}) \qquad ;s = 1, \dots, N_s \tag{32}$$

where $\delta(\mathbf{x} - \mathbf{x_s})$ denotes a Dirac function at point $\mathbf{x_s}$ and, due to the sifting property of the Dirac function, Eq. (30) simplifies into,

$$\int_\Omega b(\mathbf{x}, t) G(\mathbf{x}, \mathbf{x}_l) \, d\Omega = \sum_{s=1}^{N_s} B_s(t) G(\mathbf{x}_s, \mathbf{x}_l) \tag{33}$$

This method of wave generation makes it possible to model any wave motion for which particle velocity distribution is given along a chosen bottom-to-surface line. Two such cases are detailed in Appendix F.

4. Numerical Model

4.1. Time stepping method

If initial conditions are known at time t on the free surface boundary $\Gamma_f(t)$, i.e., the position $\mathbf{r}(t)$ and the potential $\phi(t)$, together with relevant boundary conditions on the rest of the boundary, one can calculate $\frac{\partial \phi}{\partial n}$ and the time derivatives $\frac{\partial \phi}{\partial t}$ and $\frac{\partial^2 \phi}{\partial t \partial n}$ along $\Gamma_f(t)$ by solving two Laplace's equations of the type (1) for ϕ and $\frac{\partial \phi}{\partial t}$, expressed in the same geometry ^{k}. At this stage, both the free surface position and potential can be updated to subsequent time, $t + \Delta t$, using Taylor series expansions (14) and (15), truncated to second-order in Δt ($N = 2$). Lateral boundary conditions (e.g., wavemakers) are then updated, if needed, to

---

^{k}This is done in the model using a Boundary Element Method (BEM), as detailed in the following Sections.

complete a full time stepping loop. The whole process is repeated to carry computations further in time.

Coefficients in the Taylor series are expressed as function of $\{\phi, \frac{\partial\phi}{\partial n}, \frac{\partial\phi}{\partial s}, \frac{\partial^2\phi}{\partial n\partial s}, \frac{\partial^2\phi}{\partial s^2}, \frac{\partial\phi}{\partial t},$ $\frac{\partial^2\phi}{\partial t\partial n}, \frac{\partial^2\phi}{\partial t\partial s}, \beta, \frac{\partial\beta}{\partial s}, p_a, \frac{Dp_a}{Dt}\}$ along the free surface, using equations (A.7), (A.14), (A.15) and (A.21) developed in Appendix A, with s and n given by (A.1),(A.2) as a function of β, the angle between s and the x-axis. Tangential s-derivatives of field variables that appear in some of these coefficients are computed within a 4th-order "sliding" polynomial on the boundary. At the intersection between the free surface and a moving wavemaker boundary, the accuracy of the s-derivatives is in general not sufficient and special relationships developed by Grilli and Svendsen [46] ("compatibility conditions") are used for calculating derivatives [l].

More specifically, for any given time t, values of $\frac{\partial\phi}{\partial n}$ and the geometry are specified along lateral boundaries depending on the specific problem under consideration [m]. These boundary conditions, together with the specification of ϕ on the free surface at time t, define a first Laplace problem which is solved to calculate ϕ or $\frac{\partial\phi}{\partial n}$ along Γ (whichever is unknown). Following this, $\frac{\partial\phi}{\partial t}$ is specified on the free surface using Bernoulli equation (17) as,

$$\frac{\overline{\partial\phi}}{\partial t} = -\frac{1}{2}[(\frac{\partial\phi}{\partial s})^2 + (\frac{\partial\phi}{\partial n})^2] - \frac{1}{\rho}p_a - gz(\mathbf{r}) \qquad \text{on } \Gamma_f(t) \qquad (34)$$

in which all right hand side variables and the geometry are known at time t. Depending on the type of conditions along the rest of the boundary, $\frac{\partial^2\phi}{\partial t\partial n}$ is similarly specified and a second Laplace problem is solved to calculate $\frac{\partial\phi}{\partial t}$ or $\frac{\partial^2\phi}{\partial t\partial n}$ whichever is unknown [n]. At this stage, both the geometry and values of $\{\phi, \frac{\partial\phi}{\partial n}, \frac{\partial\phi}{\partial t}, \frac{\partial^2\phi}{\partial t\partial n}\}$ are known at time t along the boundary and the free surface updating to subsequent time $t + \Delta t$ can proceed as described above.

These operations are globally referred to as "time stepping" at time t, with time step value being Δt.

4.2. Transformation of Laplace's equations into BIE's

In the model, Laplace's equations for ϕ and $\frac{\partial\phi}{\partial t}$ are transformed into Boundary Integral Equations (BIE) using third Green's identity and free space Green's function G defined as,

$$\nabla^2 G(\mathbf{x}, \mathbf{x}_l) + \delta(\mathbf{x}, \mathbf{x}_l) = 0 \qquad (35)$$

in which $\delta(\mathbf{x}, \mathbf{x}_l)$ is a Dirac function at point $\mathbf{x}_l$ of domain Ω. With definition (35), third

---

[l] These relationships were later extended by Otta *et al.*[70], Svendsen *et al.*[88], and Grilli and Subramanya[39], and the extended expressions are used in the applications of the model in Section 5.

[m] Boundary motion and $\frac{\partial\phi}{\partial n}$ can for instance be calculated using Eq. (C.5) for a piston wavemaker and $\frac{\partial\phi}{\partial n}$ is invariably zero along solid boundaries.

[n] Since both Laplace problems are expressed for the same boundary geometry $\Gamma(t)$, the additional computational effort required to solve the second problem is quite small.

Green's identity for the potential ϕ reads,

$$\phi(\mathbf{x}_l) = \int_{\Gamma(\mathbf{x})} [G(\mathbf{x}, \mathbf{x}_l) \frac{\partial \phi}{\partial n}(\mathbf{x}) - \phi(\mathbf{x}) \frac{\partial G}{\partial n}(\mathbf{x}, \mathbf{x}_l)] \, d\Gamma(\mathbf{x}) \tag{36}$$

in which the "sifting" property of the Dirac function has been used to eliminate the domain integral. In two-dimensions, the solution of Eq. (35) yields (e.g., Brebbia[2])

$$G(\mathbf{x}, \mathbf{x}_l) = -\frac{1}{2\pi} \log | \mathbf{x} - \mathbf{x}_l | \tag{37}$$

Thus, Green's function $G(\mathbf{x}, \mathbf{x}_l)$, also referred to as *fundamental solution* of Laplace's equation, has a logarithmic singularity when point $\mathbf{x}$ approaches point $\mathbf{x}_l$.

A system of BIE's for values of $\phi(\mathbf{x}_l)$ is obtained by selecting a set of points $\mathbf{x}_l$ on the boundary. Doing so, some of the integrals in Eq. (36) become strongly singular and the "extraction" of such singularities (in a Cauchy Principal Value sense) creates so-called jumps in the potential value when moving from inside the domain to the boundary (e.g., Brebbia[2]). After some transformations, weakly singular BIE's corresponding to Laplace problems for ϕ and $\frac{\partial \phi}{\partial t}$ are derived as,

$$\alpha(\mathbf{x}_l)\phi(\mathbf{x}_l) = \int_{\Gamma(\mathbf{x})} [\frac{\partial \phi}{\partial n}(\mathbf{x})G(\mathbf{x}, \mathbf{x}_l) - \phi(\mathbf{x})\frac{\partial G}{\partial n}(\mathbf{x}, \mathbf{x}_l)] \, d\Gamma(\mathbf{x})$$

$$\alpha(\mathbf{x}_l)(\mathbf{x}_l) = \int_{\Gamma(\mathbf{x})} [\frac{\partial^2 \phi}{\partial t \partial n}(\mathbf{x})G(\mathbf{x}, \mathbf{x}_l) - (\mathbf{x})\frac{\partial G}{\partial n}(\mathbf{x}, \mathbf{x}_l)] \, d\Gamma(\mathbf{x}) \tag{38}$$

in which $\mathbf{x} = (x, z)$ and $\mathbf{x}_l = (x_l, z_l)$ are points on boundary Γ and $\alpha(\mathbf{x}_l)$ is a geometric coefficient function of the angle of the boundary at point $\mathbf{x}_l$ which contains the jumps in potential value mentioned above.

Other integral equations approaches can be (and have been) proposed for solving potential flow equations in FNPF models. Cauchy Integral theorem can be used to derive BIE's for the complex velocity potential (e.g. Dold and Peregrine[18], Vinje and Brevig[92]). A vortex sheet method can also be used to derive BIE's for the vorticity density (Biot-Savart equations; Zaroodny and Greenberg[96] and Baker et al.[5]).

In all cases singular BIE's are obtained which are discretized into algebraic equations and numerically evaluated (see next Section).

4.3. Discretization and solution of Boundary Integral Equations

The numerical solution of the two BIE's (38) requires both the selection of N *collocation nodes* $\mathbf{x}_l$ along the entire boundary (discretization), to describe the variation of boundary geometry as well as boundary conditions and unknown functions of the problem, and *interpolation functions* to describe this variation in between the collocation nodes. In the present model, this is done using a Boundary Element method (BEM) (Brebbia[2]) in which the variation of all quantities is represented by means of shape functions or splines and the boundary is divided into M elements, each of which contains two or more nodes. In the applications in Section 5, quadratic isoparametric elements (Grilli, et al.[36]) are used on lateral and bottom boundaries, and cubic elements ensuring continuity of the boundary slope are used on the free surface. In these elements, geometry is modeled by a cubic spline approximation and field variables are interpolated between each pair of nodes on the free surface either using linear shape functions (Quasi-spline elements (QS); Grilli and Svendsen[46]) or the mid-section of a four-node "sliding" isoparametric element (Mixed Cubic Interpolation (MCI); Grilli and Subramanya[39]).

Using a set of boundary elements, each boundary integral is transformed into a sum of M integrals over each element. Non-singular integrals are computed by a standard Gauss quadrature rule. A kernel transformation is applied to weakly singular integrals which are then integrated by a numerical quadrature which is exact for the logarithmic singularity (Grilli *et al.*[36]). Adaptive integration methods based on subdividing the integrals are used to improve the accuracy of regular integrations near corners and in other areas of the domain where elements on different parts of the boundary may get close to each other and create almost singular situations (Grilli and Subramanya[37]).

Corners are represented by double nodes and compatibility relationships are specified for boundary velocity components on each side of corners, to ensure both uniqueness and regularity of the solution (Grilli and Subramanya[39]; Grilli and Svendsen[46]). Double nodes represent two nodes of identical coordinates with different nodal values of the field variables. Hence, two algebraic BIE's are obtained for each double node, which, however, are not independent. Continuity conditions express uniqueness of ϕ or $\frac{\partial \phi}{\partial t}$ for both nodes of a double node and compatibility conditions express uniqueness of the velocity or the acceleration vectors, based on values of ($\frac{\partial \phi}{\partial s}$, $\frac{\partial \phi}{\partial n}$) or ($\frac{\partial^2 \phi}{\partial t \partial s}$, $\frac{\partial^2 \phi}{\partial t \partial n}$), respectively, on both intersecting boundaries at the corner.

Discretization and numerical integrations transform the BIE's into a system of N linear algebraic equations in which boundary conditions are directly specified. The system is then solved for the unknowns at collocations nodes using a direct elimination method. After solution, Eq. (38) can be expressed for known boundary values to explicitly calculate the solution (and its gradient : the velocity and acceleration) for any location inside the domain, without further numerical approximation. This, in fact, represents one of the major advantages of a BEM approach versus domain discretization type methods (e.g., finite differences or finite elements) : the representation of the solution over the computational domain is *exact*. The only approximation in the method resides in the discretization of the boundary and the numerical evaluation of integrals in the BIE's. Other more obvious advantages result from the limitation of the discretization to boundaries which makes the generation of discretization data and analysis of results much easier than when using domain discretization methods, and usually allows for a higher-order representation of the boundary solution and thus of the internal solution.

In an Eulerian-Lagrangian modeling approach, free surface discretization nodes represent fluid particles which, for nonlinear wave flows, slowly drift away in the direction of the mean mass transport. With time, particularly for periodic wave problems, such a node drift leads either to a concentration of nodes in flow convergence regions of the free surface (like wave crests and breakers) which creates quasi-singular situations due to node proximity, or to a poor resolution of the discretization in regions of flow divergence, like close to a wavemaker °, which may induce instability of computations. To either add and

---

°Note that, in applications with a SFW generation, the vertical wavemaker boundary Γ_{r1} is horizontally

redistribute nodes in regions of poor resolution of the free surface or to remove and redistribute nodes in regions of flow convergence, a *node regridding* technique was introduced by Grilli and Subramanya[39] (see also Subramanya and Grilli[86]) and implemented in the model in combination with the MCI interpolation method.

4.4. Global accuracy of the solution

In the applications, accuracy of computations is checked for each time step by computing errors in total volume m and energy e of the generated wave train. As a general rule, results are deemed inaccurate and computations are stopped when—usually due to impending breaking—these errors become larger than 0.05% or so.

Based on results of computations made in various spatio-temporal discretizations, for a large solitary wave propagating over constant depth h_o, Grilli and Svendsen [46] showed that numerical errors in the model are function of both the size (i.e., the initial distance between nodes Δx_o) and the degree (i.e., quadratic, cubic,...) of boundary elements used in the spatial discretization, and of the size of the selected time step Δt_o. Based on these computations, they developed a criterion for selecting the optimum time step in the model. Using QS elements on the free surface and quadratic isoparametric elements elsewhere, they showed that, for a constant time step, errors in m and e are minimum when the mesh Courant number is approximately 0.5 or,

$$C_o = \sqrt{gh_o}\,\frac{\Delta t_o}{\Delta x_o} \simeq 0.5 \tag{39}$$

Based on these results, they developed an adaptive time step procedure, applicable to highly transient waves like breakers, in which the time step is calculated as a function of time based on the optimum mesh Courant number C_o and on the minimum distance between nodes on the free surface, $\Delta \mid r(t) \mid^{min}$, for the given time t as,

$$\Delta t = C_o \frac{\Delta \mid r(t) \mid^{min}}{\sqrt{gh_o}} \tag{40}$$

Similar calculations were carried out by Grilli and Subramanya [39] using the more accurate MCI elements for the interpolation on the free surface. These showed that the optimum value of C_o is around 0.35-0.40 for the MCI elements.

5. Applications

Many applications of the FNPF model described in Sections 2-4 were performed over the past few years for various types of wave propagation, shoaling and runup, and for wave interaction with emerged and submerged coastal structures or obstacles in the bottom. A brief review of these applications is given in Section 5.1, along with references to selected publications with more details on both computational and physical aspects of the problems.

---

moved in time with the Lagrangian motion of the first free surface node/particle, which eliminates resolution problems mentioned above, close to the wavemaker boundary.

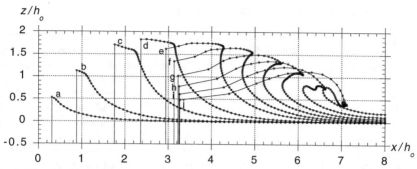

Fig. 4. Wave breaking induced by a piston wavemaker (solitary wave motion with $H'_o = 2.0$) with : $\Delta x'_o = 0.10$ and $C_o = 0.4$; (o) denote discretization nodes and vertical lines mark successive locations of the piston wavemaker. Time of plotted wave profiles is $t' = $ a : 1.442, b : 2.064, c : 2.843, d : 3.381, e : 4.011, f : 4.353, g : 4.681, h : 4.936, i : 5.191, j : 5.438.

In Sections 5.2 to 5.5, details of data and results are given for specific applications of the model to problems of long wave shoaling, runup, and/or breaking over plane slopes. Although the model can address more general problems, new applications presented here have been limited to academic cases both for sake of simplicity and because of the focus of the present work on long wave runup. In this line, Section 5.5 reports on the Benchmark Problem #3 for solitary wave runup on a vertical wall that was proposed as part of the "International Workshop on Long-wave Runup Models (San Juan Island, WA, USA, 09/95).

5.1. Review of past applications of the model to long wave propagation, runup, and interaction with coastal structures

5.1.1. Wave generation by a moving vertical boundary

Grilli and Svendsen [46] studied the generation of breaking waves by horizontally moving vertical boundaries. They analyzed the accuracy of computed results as a function of both discretization and time step and evaluated the performance of corner compatibility relationships in the very demanding case where both lateral and free surface boundaries take large displacements. Similar computations were performed by Grilli and Subramanya[39], using improved free surface discretization methods (MCI elements), extended corner compatibility conditions, and node regridding methods. Fig. 4 gives an example of such computations.

Grilli [29] extended the model to the calculation of breaking bow waves and wave resistance coefficients for forward moving slender ships. This application is implemented in the present model but has not been described in this chapter.

5.1.2. Wave runup over and reflection from steep slopes

Grilli and Svendsen [42,44,45,47] and Svendsen and Grilli[87], through careful numerical experiments, extensively studied the runup on, and reflection of solitary waves from steep slopes, and from vertical walls. They compared model results to laboratory experiments and, in general, found surprisingly good agreement between both of these.

As an illustration of such computations, two applications are presented in Section 5.2 for the runup of a solitary wave of incident height $H_o/h_o = 0.12$ over slopes of angle $\theta = 20°$ and $45°$, and one application is presented in Section 5.3 for the runup of a cnoidal wave of incident height $H_o/h_o = 0.10$ over a slope of angle $\theta = 20°$.

These applications were selected for sake of comparison with results earlier obtained by Liu et al.[61] with their nonlinear model, and experiments by Hall and Watts[49].

5.1.3. Wave shoaling and breaking over a gentle slope

Grilli et al.[45], Otta et al.[69], and Grilli et al.[41] used the model to calculate shoaling of solitary waves over a gentle slope up to initiation of breaking.

Grilli et al.[41] compared their results to classical Green's and Boussinesq's shoaling laws and to careful laboratory experiments. They concluded that none of the theoretical laws could accurately predict observed shoaling and breaking behaviors but that the present FNPF model agreed quite well with experiments up to the breaking point.

Otta et al. [69], based on their calculations with the model, developed a criterion for breaking of solitary waves over slopes and analyzed the kinematics of waves at breaking. Using improved numerical methods by Grilli and Subramanya[39] (particularly node regridding), Grilli et al.[48] performed a more detailed analysis of breaking types and characteristics of breaking jets for solitary wave shoaling over slopes 1:4 to 1:100. Based on their computations, they proposed an improved breaking criterion for solitary waves on plane slopes that was shown to agree quite well with experimental results. In particular, no solitary wave that can propagate stably over constant depth was found to break on a slope steeper than 12°. In Section 5.4, a similar application is presented for the shoaling and breaking of an incident solitary wave of initial height $H_o/h_o = 0.20$ over a slope $s = 1:35$.

More recently, cases with periodic waves shoaling up to breaking over a slope were calculated by Subramanya and Grilli[85] and Grilli and Horrillo[31], using a combination of zero-mass-flux SFW's and an absorbing beach, to study the kinematics and integral properties of waves on beaches (Fig. 2). Such results are of importance to surf-zone dynamics modelers.

5.1.4. Wave interactions with submerged obstacles

Accurate prediction of water wave interaction with submerged obstacles is of prime importance in coastal engineering. Submerged breakwaters are becoming increasingly used as both aesthetic and economical means of shoreline protection against extreme storms and even tsunamis. Natural reefs and sandbars are frequent coastal features that function as natural submerged breakwaters. In addition, the study of waves close to the shoreline and in the surf zone requires that the offshore wave climate be accurately "propagated" over

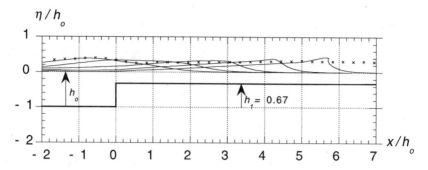

Fig. 5. Computed profiles at successive times, $t' = t\sqrt{g/h_o} = 19.45, 20.78, 22.11, 23.44$ and 24.77 (left to right), for a solitary wave of height $H_o/h_o = 0.33$ propagating and breaking over a step $h_1 = 0.67h_o$ in the bottom. Initial discretization is with $\Delta x'_o = 0.1875$ and $C_o = 0.43$. Symbols ($\times$) represent the free surface envelope measured by Grilli *et al.* [32].

any existing submerged obstacle, man-made or natural.

Propagation of waves was calculated with the present nonlinear model over three different types of submerged obstacles of various engineering implications. Cases with both large incident waves or shallow submerged obstacles led to stronger nonlinear interactions between incident waves and the obstacles and to various instabilities and breaking of incident waves on or downstream of the obstacles. It is worth pointing out that most of these phenomena cannot be accurately modeled by standard wave theories but require fully (or highly) nonlinear theories to be accurately described,

- **Step in the bottom :** The simplest possible steep obstacle on the bottom is the step discontinuity between two constant depth regions (Fig. 5). Numerous studies of the interaction of a long wave with a step have been carried out using various wave theories, from linear to mildly nonlinear, and numerical models. The main motivation for these studies has been to answer the question : How do long waves behave when they propagate from deep water into shallow water over the continental shelf ? More specific questions have also been addressed, by assuming that the step represents a first approximation for a wide crested obstacle in shallow water—like a bar or a reef—or even a submerged breakwater.
  In this line, Grilli *et al.*[32] used the present model to study strong nonlinear interactions— leading to breaking—of large solitary waves over steps in the bottom. They compared numerical results to laboratory experiments and found fairly good agreement between both of these for wave shape and wave envelope. An illustration of such computations is given in Fig. 5.
- **Rectangular bar :** After the step in the bottom, the rectangular obstacle has the simplest possible geometry for representing submerged bars or breakwaters (Fig. 6). One may expect, in fact, that most of the phenomena observed or computed for rectangular bars also occur, at least qualitatively, for obstacles of more complex geometry.

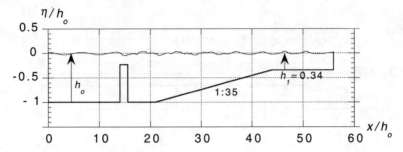

Fig. 6. Propagation of a cnoidal wave of height $H_o/h_o = 0.05$, and period $T' = T\sqrt{g/h_o} = 7.52$ over a submerged rectangular bar of height 0.768 and width 1.576. Free surface profile is plotted at $t' = 85.57$ (or $11.38T'$). Initial free surface discretization has $\Delta x'_o = 0.25$ and $C_o = 0.50$. Vertical exaggeration is 10.

Driscoll et al. [21] studied the propagation of small amplitude cnoidal waves over a submerged shallow bar with rectangular cross-section. They compared laboratory experiments to first and second-order analytic models and to the present fully nonlinear BEM model. They found that the BEM model could accurately predict the generation of higher-order harmonics observed in laboratory in the wave train, downstream of the obstacle. An illustration of these computations is given in Fig. 6.

A similar, more extensive, numerical study was recently presented by Ohyama and Nadaoka[66].

• **Submerged trapezoidal breakwaters :**   Submerged breakwaters used for shoreline protection are usually built by dropping rocks from barges at selected offshore locations and, hence, take an approximate trapezoidal shape (Fig. 7). The protection offered by submerged breakwaters consists in inducing breaking and partial reflection-transmission of large incident waves, while small wave propagation and, in some cases, local navigation can still take place over the structure during normal conditions.

Cooker et al. [15] used an extension of Dold and Peregrine's [18] nonlinear model to calculate solitary wave interactions with a submerged semicircular cylinder of radius R in water of depth h_o. Results showed that a variety of behaviors occur depending on wave height and cylinder radius. In short, for small cylinders ($R/h_o < 0.5$), waves essentially transmit and exhibit a tail of oscillations. This is a regime of weak interactions. For larger cylinders ($R/h_o \geq 0.5$), interactions are much stronger : small waves partially transmit and reflect (crest exchange); medium waves undergo a stronger crest exchange over the cylinder, and the first oscillation in their tail may break backward onto the cylinder (direction opposite to propagation); and large waves break forward (plunging), slightly after passing over the cylinder. A limited number of experiments confirmed these theoretical predictions.

Grilli et al. [34] repeated the above study for submerged breakwaters with a more realistic trapezoidal cross-section. Computations using the present model were compared

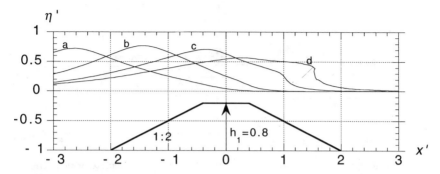

Fig. 7. Propagation of a solitary wave of height $H_o/h_o = 0.70$, (generated using Tanaka's[90] method) over a submerged trapezoidal breakwater of height $h_1 = 0.8$. Free surface profiles are given at successive times, $t' =$ a: 2.72; b: 3.69; c: 4.65; and (d) 5.20. Initial free surface discretization has $\Delta x'_o = 0.125$ and $C_o = 0.50$.

to laboratory experiments for a large number of solitary waves of various heights H and for a breakwater geometry defined by : a height $h_1 = 0.8h_o$, a width at the crest $b = h_1$, and two (seaward and landward) 1:2 slopes. Results qualitatively agreed with earlier observations by Cooker *et al.* [15] as far as crest exchange and breaking behaviors are concerned. In all cases, a reflected wave formed at the breakwater seaward face and propagated backward into the tank. An illustration of such computations is given in Fig. 7.

Despite the renewed interest for underwater breakwaters mentioned above, the general conclusion of these studies is that underwater breakwaters only offer limited protection against long waves, since they only create large reflection (i.e., low energy transmission) for very low depth of their crest.

5.1.5. Wave impact on coastal structures

Two cases with more realistic coastal structures were studied in earlier applications with the model that illustrated its ability to predict shoaling of incident waves from deep to shallow water over a mild slope and interaction with a structure in the shallow water region.

In the latter application, the model was able to predict peak impact pressures from breaking waves on the vertical wall of mixed breakwater. Such numerical simulations are helpful for designing coastal structures,

- **Mixed berm breakwaters :** Most classical breakwaters used for shoreline or harbor protection are made of a main trapezoidal breakwater, with a small submerged berm at the toe of the emerged structure. Part of the incident wave energy dissipates by breaking over the berm which, hence, offers some protection to the main structure. Such a case was studied by Grilli and Svendsen [45], for which, unlike with traditional berm breakwaters, a small detached submerged structure was simply located slightly

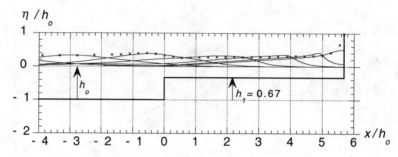

Fig. 8. Computation of solitary wave impact over a mixed vertical breakwater with a berm height $h_1 = 0.67h_o$, and length $5.7h_o$, and an incident wave height $H_o/h_o = 0.33$. Symbols ($\times$) mark free surface envelope measured by Grilli *et al.* [33].

in front of the main structure. The combination of the two structures was called a "mixed berm breakwater". This configuration, while offering the same degree of protection as classical berm breakwaters, may be more economical and simpler to build. It was found, in particular, that the toe structure could substantially reduce maximum runup of solitary waves on the main steep slope.

- **Mixed vertical breakwaters :**   Mixed vertical breakwaters are composed of a vertical concrete caisson sitting on a wide berm made of rocks. They function as vertical walls during high tide and as mound breakwaters during low tide (Fig. 8). Their upper section is designed to be safe against sliding and overturning induced by wave impact force. Laboratory and field experiments show that impacts of normally incident breaking waves are the most severe. In this case, the maximum impact force on the wall may rise up to 10 times the hydrostatic force based on wave elevation at the wall.

Cooker [12] and Cooker and Peregrine [14] confirmed these observations by solving 2D fully nonlinear potential flows. Their model, however, although very accurate, was limited to a simple vertical wall and used a large incident long wave with characteristics selected to create a large scale breaker in the model.

Grilli *et al.* [33,35] computed violent impacts of breaking waves on mixed vertical breakwaters with the present nonlinear model, i.e., using both more realistic incident waves and a breakwater geometry closely reproducing the experimental set-up. An example of such computations is given in Fig. 8. Good qualitative agreement was found between laboratory experiments and computations but the model overpredicted peak pressures by up to 50%. This is believed to be due in large part to geometric irregularities in the experimental set-up that limited both the jet formation and the pressure build up at the wall. In fact, a poor repeatability was found for measured peak pressures whereas wave shape and kinematics could be reproduced to within a few percent for 9 repetitions of the same experiment.

5.2. Runup of solitary waves on a steep slope

The computational domain for this problem is similar to the case sketched in Fig. 3, except that, due to the steep slope used in the present case, there is no need for a shallow shelf at the rightward extremity of the computational domain (Γ_{r2}). The domain length is 30 times the depth. The runup of a solitary wave of incident height $H_o/h_o = 0.12$ is calculated over two slopes of angle $\theta = 20°$ and $45°$. The incident wave is generated by simulating a piston wavemaker on the leftward boundary (Γ_{r1}).

For the first slope, a discretization with 120 two-node QS elements is used on the free surface. Three-node isoparametric elements are used on the leftward boundary, on the rightward boundary (i.e., the slope in the present case), and on the bottom. The discretization thus has 254 nodes and 185 elements with an initial distance between nodes $\Delta x_o = 0.25$ on the free surface, 0.167 on boundary Γ_{r1}, 0.20 on boundary Γ_{r2}, and 0.25 on the bottom Γ_b. The initial time step is $\Delta t_o = 0.09$ and the Courant number is thus $C_o = 0.36$ (with $g = h_o = 1$). The average CPU time per time step is 3.3 s (IBM9000/3) or about 66 min for the whole run of 1200 time loops. Similar data are generated for the 45° slope.

Results for the free surface elevation at selected times are presented in Fig. 9 (20° slope) and Fig. 10 (45° slope). One can see that waves propagate from left to right up to about $t' = 43$ (curves a-f) and 41 (curves a-e), respectively. The maximum runup calculated for both slopes is $R_u = 2.351 H_o$ (at $t' = 43.07$), and $R_u = 2.275 H_o$ (at $t' = 41.16$), respectively, which agrees quite well with both computations by Liu *et al.*[61] and experiments by Hall and Watts[49]. After runup, waves rundown, reflect on the slopes, and propagate backward into the numerical tank (curves f-l and e-l), trailing a (well resolved) tail of oscillations behind them, slightly more pronounced for the smaller slope. For time $t' \geq 60$, the leading oscillations in the reflected waves re-reflect on the wavemaker.

Fig. 11 gives indicators of global accuracy of computations for each case. These are the relative errors on total (dimensionless) wave energy $\Delta e/e$ and volume $\Delta v/v$, in which $e = 0.06762$ and $v = 0.83227$ for the generated solitary wave. Both numerical errors are small quantities for the initial stages of wave propagation ($\mathcal{O}(10^{-7})$) and then gradually increase. Errors temporarily decrease during runup and rundown of the waves on the slopes and then increase to stabilize at about $\mathcal{O}(10^{-3})$ or smaller [p].

5.3. Runup of cnoidal waves on a steep slope

The runup of a cnoidal wave of incident height $H_o/h_o = 0.10$ and period $T\sqrt{g/h_o} = 20$ (i.e., $\omega = 0.31416$, for which $L/h_o \simeq 20$) is calculated over a slope of angle $\theta = 20°$, using the same discretization and initial data as for the solitary wave in the previous Section. The wave is generated by a piston wavemaker with motion $x_p(t)$ as in Fig. 13a.

Results for the free surface elevation at selected times are presented in Fig. 12. One can see that waves propagate from left to right and a first crest runs-up the slope at about $t' = 36$ (curve d), reflects, and then propagates back into the tank. This crest then interacts with the second crest to produce a slightly larger runup for the second crest at about $t' = 56$

---

[p]Accuracy of these results can be greatly improved by using MCI elements on the free surface instead of QS elements. This will be illustrated in the applications in Sections 5.4 and 5.5.

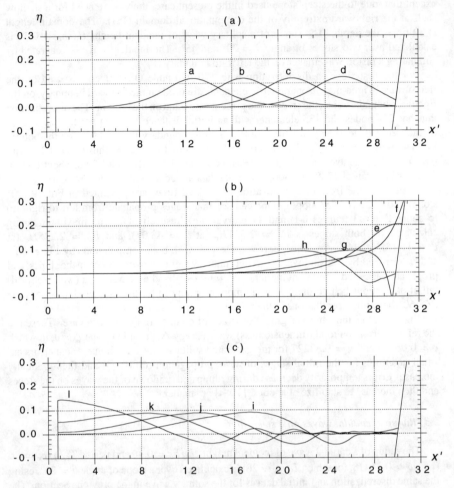

Fig. 9. Runup of a solitary wave of height $H_o/h_o = 0.12$, on a 20° slope. Axes are non-dimensional with respect to depth h_o and curves correspond to successive dimensionless time $t' =$ a: 24; b: 28; c: 32; d: 36; e: 40; f: 44; g: 48; h: 52; i: 56; j: 60; k: 64; l: 68.

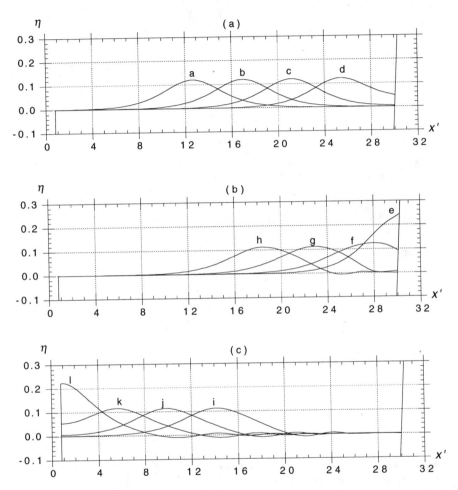

Fig. 10. Runup of a solitary wave of height $H_o/h_o = 0.12$, on a 45° slope. Axes are non-dimensional with respect to depth h_o and curves correspond to successive dimensionless time $t' =$ a: 24; b: 28; c: 32; d: 36; e: 40; f: 44; g: 48; h: 52; i: 56; j: 60; k: 64; l: 68.

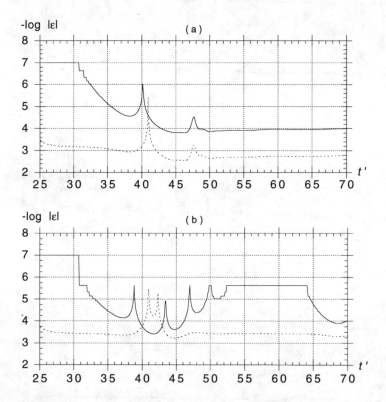

Fig. 11. Relative numerical error on total wave energy (- - - - -) $\varepsilon = \Delta e/e$, and volume (———) $\varepsilon = \Delta v/v$, for the computations reported in : (a) Fig. 9; and (b) Fig. 10.

(curve i), and so forth. The time history of runup on the slope in Fig. 13b confirms that the second and later runups are reinforced by successive incoming crests. The first crest runs-up to about twice the incident wave height, and the second and third crests run-up to about 2.4 times the incident wave height, while keeping the same rundown value. These results also fairly well agree with results by Liu *et al.* [61], as far as one can tell from their figures.

Fig. 13c shows the relative error, $\Delta V/V$, on total volume of the computational domain ($V = 28.626$), as a function of time. One sees that this error stays quite small during all computations.

5.4. Solitary wave shoaling and breaking over a gentle slope

A case similar to those calculated by Grilli *et al.*[41] and Otta *et al.* [69] (also analyzed by Wei *et al.*[93]) is presented in the following, for the shoaling and breaking of an incident solitary wave of initial height $H'_o = 0.20$ over a 1:35 slope. The computational domain is as sketched in Fig. 3 with the toe of the slope at $x' = 5$. The incident solitary wave is generated on the leftward lateral boundary of the domain using the numerical piston wavemaker. To improve accuracy of regular integrations in the upper part of the slope where the domain geometry becomes very narrow, a small shelf is specified to the right of the domain in depth $h_1 = 0.1 h_o$ (unlike in computations with steeper slopes reported in the previous Sections) [q].

The free surface discretization has 226 two-node MCI elements, with $\Delta x'_o = 0.20$, and there are 100 quadratic elements on the bottom and lateral boundaries. The total number of nodes is 429. The distance between nodes on the bottom is 0.5 in the constant depth region, and reduces to 0.40, 0.25, 0.20, 0.15, and 0.10 on the slope, in order to get increased resolution where depth decreases. The distance between nodes is 0.15 on the shelf bottom. Adaptive integration with up to 2^6 subdivisions (as function of the geometry) is specified on the free surface and on the bottom, for the elements located between $x' = 31$ and 40. The mesh Courant number is $C_o = 0.40$ and, hence, $\Delta t'_o = 0.08$. With these data, the CPU time is 12.9 sec per time step (IBM9000).

Fig. 14a shows computed stages of wave shoaling and breaking. During propagation, time step reduces down to $\Delta t' = 0.020$ at the time of breaking defined by a vertical tangent occurring on the front face (curve d in Fig. 14a, $t' = 44.52$, $x'_b = 36.3$). The total number of time steps up to this stage is 950 and the average time step is 0.047. The wave height at breaking is $H_b = 0.364$ and the ratio wave height over depth at breaking is $H_b/h_b = 1.402$. This is much larger than the usually accepted value for gentle slopes (~ 0.80) and agrees to within 5% with measurements by Grilli *et al.*[41]. A detailed comparison of free surface elevations measured at several locations over the slope (gages) to computed results is given in Fig. 14c. One can see that the agreement between both of these is very good up to the last gage which is virtually at the computed breaking point.

---

[q]This is to avoid that elements on different parts of the boundary get too close to each other, leading to a loss of accuracy of numerical integrations of the Green's function kernels. This change in geometry—as compared to a plain slope—does not affect shoaling and breaking of a solitary wave, provided these occur, as observed in the present case, before reaching the shelf, i.e., for $x' \le 41$.

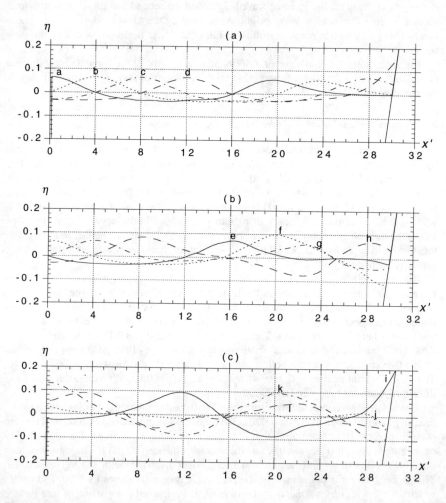

Fig. 12. Runup of a cnoidal wave of height $H_o/h_o = 0.10$ and period $T' = T\sqrt{g/h_o} = 20$, on a $20°$ slope. Axes are non-dimensional with respect to depth h_o and curves correspond to successive dimensionless time $t' =$ a: 24; b: 28; c: 32; d: 36; e: 40; f: 44; g: 48; h: 52; i: 56; j: 60; k: 64; l: 68.

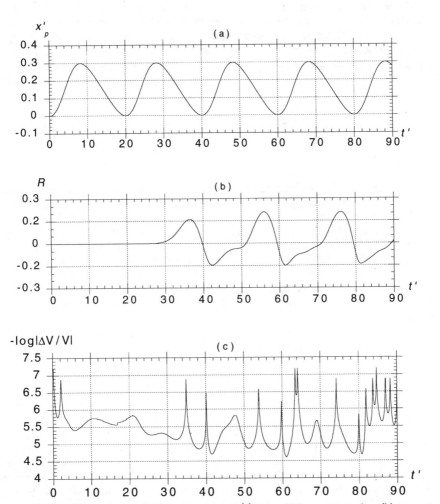

Fig. 13. (a) Horizontal motion of piston wavemaker $x_p(t)$ for cnoidal wave generation; (b) runup at the slope $R(t)$; and (c) relative numerical error on computational domain volume, $\Delta V/V$, for the computations reported in Fig. 12.

To be able to accurately pursue computations beyond the breaking point, the regridding method by Subramanya and Grilli[86] and Grilli and Subramanya[39] is used to add 40 nodes in the crest region, between $x' = 33.26$ and 37.80, at the time of curve d (Fig. 14). Computations are restarted and Fig. 14b shows blow-ups of the region over the slope where breaking occurs. Discretization nodes are marked on the figure and one sees that the breaker jet is well resolved up to touch down on the free surface. The wave breaks as a large scale plunging breaker. Details and accuracy for such computations beyond the breaking point are discussed in Grilli and Subramanya[39] and a further analysis of results is done in Grilli et al.[48].

5.5. Solitary wave runup on a vertical wall : Benchmark #3

5.5.1. Wave generation

In Benchmark #3 application, three solitary waves were generated in a laboratory tank at the U.S. Army Engineering Waterways Experiment Station (Vicksburg) using a piston wavemaker with motion $x_p(t)$ provided to the workshop participants as a set of digital data.

Wavemaker motion Eq. (D.6) was first used to best fit the digitized paddle trajectories and find corresponding incident waves $H'_o = H_o/h_o$ to be used in Eq. (D.7) to calculate boundary conditions (C.5), for the paddle velocity and acceleration as a function of time on boundary Γ_{r1}, needed to generate solitary waves in the model (Fig. 15). We thus obtained $H'_o =$ (case A) 0.0440; (case B) 0.2602; and (case C) 0.6087. As can be seen, with these wave heights, only small differences are observed between experimental and calculated curves.

5.5.2. Wave propagation over constant depth

The wave tank geometry provided to the workshop participants is sketched in Fig. 16 with a region with constant depth, $h_o = 0.218$ m, length $68.991h_o$, and three successive slopes 1:53 (length $20.00h_o$), 1:150 (length $13.44h_o$), and 1:13 (length $4.13h_o$). A vertical wall is located at the tank far end.

In all three cases, due to the large region of constant depth in front of the slopes, to save computational time, waves were first generated and propagated in a (shorter) computational domain of constant depth h_o and length $69h_o$. Waves were then introduced in a second computational domain containing part of the constant depth region and the rightward region of the tank with varying water depth. Doing so, parts of the oscillatory tails shed behind the generated waves were cut out of the second computational domain whereas the main leading waves were kept for further propagation over the slopes.

The same discretization was used in all three cases for the constant depth domain, with 141 nodes on the free surface and 140 MCI elements ($\Delta x'_o = 0.5$) and three-nodes quadratic elements elsewhere, with a node spacing 0.5 on the bottom. Initial time step was $\Delta t'_o = 0.15$ for a Courant number 0.3. For the propagations over constant depth, relative errors on wave volume and energy were typically small in all three cases, around 0.001%

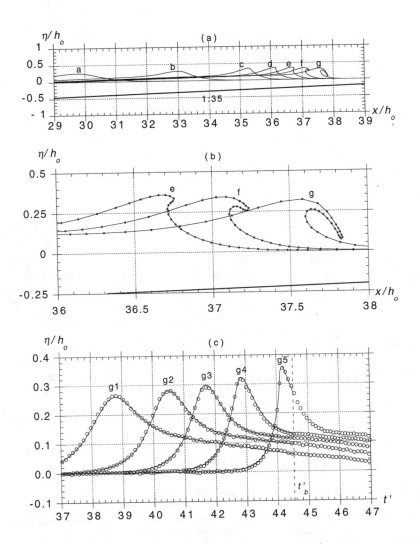

Fig. 14. Shoaling and breaking of a solitary wave with initial height $H'_o = 0.20$, over a 1:35 slope. The wave is generated by a piston wavemaker at $x' = 0$. The slope starts at $x' = 5$ and plots in Fig. (a) correspond to profiles at time $t' =$ a: 37.17; b : 40.73; c: 43.48; d: 44.53; f: 44.94; f: 45.40; g: 46.00. Fig. (b) is a blow-up of last three profiles in (a) with (o) denoting BEM discretization nodes. Fig. (c) shows a comparison of computed (—) and measured (o) free surface elevation at gages at $x' =$ g1: 30.96; g2: 32.55; g3: 33.68; g4: 34.68; and g5: 36.91 (from Grilli *et al.*[41]).

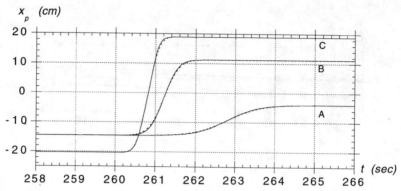

Fig. 15. Piston wavemaker paddle trajectories $x_p(t)$ for Benchmark #3 application : (——) experimental; (- - - - -) numerical with $H'_o =$ (case A) 0.0440; (case B) 0.2602; and (case C) 0.6087.

or less.

Time histories of surface elevation computed at gages 1-3 are given in Fig. 17 for all three cases. One can see that, as expected, the larger the wave the larger and the longer the tail of oscillations it sheds behind its main crest. This is due, as discussed before, to the piston wavemaker generation of approximate (first-order) solitary waves in the tank. When shedding tails of oscillations, as pointed at by Grilli and Svendsen[47], the main crest height of waves also gradually decreases over constant depth until the wave shape stabilizes and the main crest adjusts to the (more peaky) shape of exact solitary waves (those that Tanaka's[90] method would have generated). Thus at gage 2, for instance, the computed maximum wave height is, 0.0440, 0.254, and 0.589, for cases A, B, and C, respectively, i.e., 0.0, 2.4, and 3.2% smaller than the incident wave generated at $x' = 0$, respectively. It is seen that these maximum values do not significantly change from gage 2 to 3, indicating that incident waves have reached their permanent form. Finally, one also sees from the results that the wave for case A has the theoretical height specified at the wavemaker and no noticeable oscillatory tail, confirming that such a small wave (with $H'_o < 0.2$ according to Goring[25]) has a shape very close to a first-order solitary wave.

5.5.3. Wave propagation over the sloping bottom and runup on the wall

Computations are pursued in the second computational domain. For case A, the domain extends from $x' = 18.72$ onward, with a total of 491 nodes, 150 of which on the free surface ($\Delta x'_o = 0.5$) and 370 elements. The distance between nodes on the bottom is progressively reduced from 0.5 to 0.15. For case B, the domain extends from $x' = 46.1$ onward (the wave is higher and much narrower and travels faster than in case A), with a total of 471 nodes, 240 of which on the free surface ($\Delta x'_o = 0.25$) and 378 elements. The distance between nodes on the bottom is progressively reduced from 0.5 to 0.15. For case C, the domain is chosen identical to case B.

Fig. 18 gives a summary of results computed for case A. In Fig. 18a, the time history of

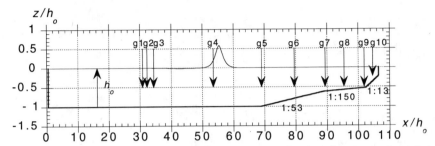

Fig. 16. Sketch of computational domain for Benchmark #3 application with 10 gages located at $x' =$ g1: 31.00; g2: 32.00; g3:34.00; g4: 54.00; g5: 69.00; g6: 79.00; g7: 89.00; g8: 95.69; g9: 102.43; g10: 104.50. Sketched wave corresponds to case C at $t' = 49.00$.

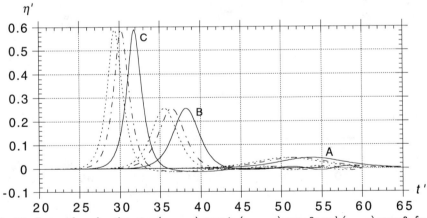

Fig. 17. Free surface elevation at : (- - - - -) gage 1; (— . —) gage 2; and (———) gage 3, for computations with cases A, B, C.

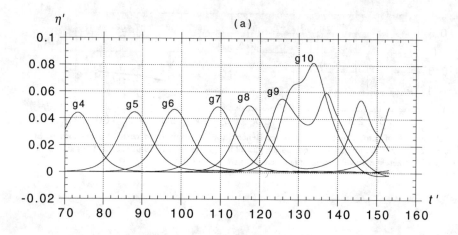

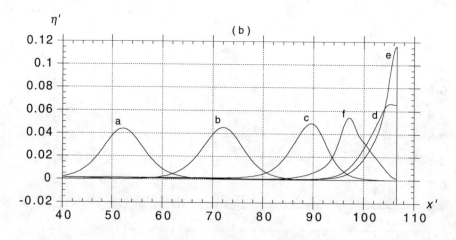

Fig. 18. Case A. (a) Free surface elevation as a function of time at gages 4-10; (b) Free surface profile at $t' =$ a: 71.55; b: 91.04; c: 109.99; d: 128.25; e: 131.25 (runup); and f: 153.12 (reflection).

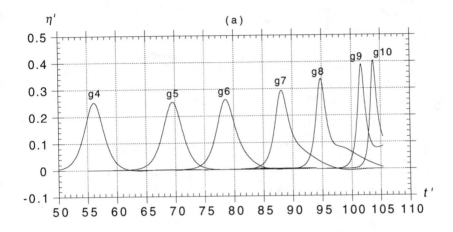

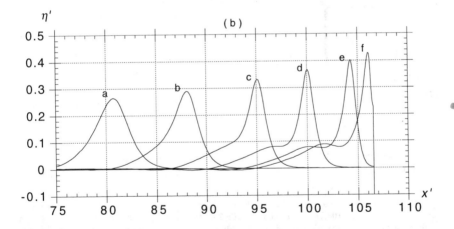

Fig. 19. Case B. (a) Free surface elevation as a function of time at gages 4-10; (b) Free surface profile at $t' =$ a: 80.24; b: 87.20; c: 94.108; d: 99.18; e: 103.92 ; and f: 105.32 (impending runup).

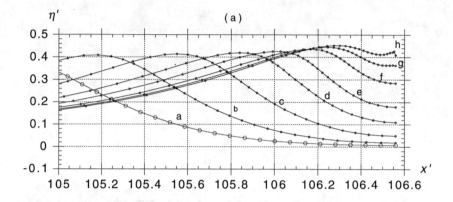

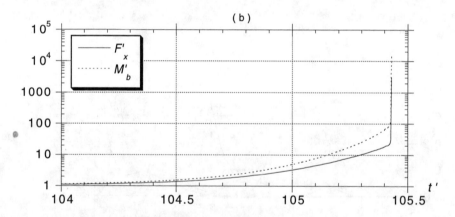

Fig. 20. Case B. after node regridding (a) Free surface profile at $t' =$ a: 104.00 (first regridded profile); b: 104.42; c: 104.78; d: 105.05; e: 105.20 ; f: 105.33; g: 105.39 ; and h: 105.43 (impending upward jet expulsion) ((○) indicate regridded discretization nodes; no scale distorsion); (b) Dimensionless horizontal pressure force, $F_x' = F_x/(\frac{1}{2}\rho g d^2)$ and moment with respect to the bottom, $M_b' = M_b/(\frac{1}{6}\rho g d^3)$, on the vertical wall ($d = 0.22$ is the depth at the wall, and denominators denote still water values).

water elevation at gages 4-10 shows wave transformation up to runup and reflection from the wall and, in Fig. 18b, selected free surface profiles are given for 6 times (curves a-f), with curve e representing the maximum runup occurring at $t' = 131.25$ (i.e., $t = 19.57$ sec) from the start of wave generation in the model. Maximum runup computed on the wall is $\mathcal{R}' = 0.1163 = 2.643H_o'$. During these computations, relative errors on wave volume ($v = 0.4842$) and energy ($e = 0.2564$) were less than 0.032 and 0.027%, respectively.

Fig. 19 gives a summary of results for case B. In Fig. 19a, the time history of water elevation at gages 4-10 shows wave transformation up to impending runup on the wall and in Fig. 19b, selected free surface profiles are given for 6 times (curves a-f) with curve f, at $t' = 105.32$, representing the closest time to maximum runup that can be accurately computed in this discretization. Unlike with case A, the much larger wave generated in case B significantly feels the bottom and becomes more and more asymmetric as it propagates towards the wall. This is similar to shoaling computations in Fig. 14. Curves e and f in Fig. 19b even show the formation of a secondary crest behind the main wave crest indicating the initiation of reflection from the steep 1:13 slope and from the wall. During these computations, relative errors on wave volume ($v = 1.202$) and energy ($e = 1.164$) were less than 0.017 and 0.0017%, respectively, up to $t' = 95$. For later times, these errors increased up to 0.39% and 0.085%, respectively, at the time of curve f. These increased errors clearly indicate the need for a finer discretization of the free surface close to the wall, to be able to compute further in time and resolve the "flip-through" motion of the free surface which is about to happen in front of the main crest.

This can be achieved through regridding of the free surface to a finer discretization : 30 nodes are added to the free surface at $t' = 104.00$ (Fig. 20a, curve a), from $x' = 103.13$ to the wall. The new spacing between nodes in the regridded region is approximately 0.07. Computations are accurately pursued in the new discretization up to the time a small jet of water is about to be vertically expelled at the wall (Fig. 20a, curve h, $t' = 105.43$). At this stage, numerical errors reach (an acceptable) 0.25% and one can see in Fig. 20a that small features in free surface shape are quite well resolved. Beyond this time, the few nodes close to the wall are vertically expelled with very large upward velocity and acceleration (thousands of g's) and computations break down. This is consistent with computations by Cooker and Peregrine[14] and Grilli *et al.*[33] and is also supported by experiments in the latter study. For the last computed profile, the runup at the wall is $\mathcal{R}_u = 0.435 = 1.67H_o'$. After jet expulsion, however, this value is likely to become much larger. As in the previous studies, we also see in Fig. 20b that the computed horizontal pressure force on the wall and moment with respect to the toe of the wall also reach very large (impact) values shortly after the time of curve h in Fig. 20a (0.0033 time unit later).

Fig. 21 gives a summary of results for case C. In Fig. 21a, selected free surface profiles are given for 6 times (curves a-f) with curve f, at $t' = 74.86$ (or $t = 11.16$ sec from the start of wave generation), representing the time at which the wave breaks, with $H_b' = 0.7536$, $h_b' = 0.644$, $x_b' = 87.85$, and a breaking index $H_b/h_b = 1.169$. Fig. 21b gives a blow up without scale distorsion of the region of Fig. 21a where breaking occurs and Fig. 21c gives a blow-up of the breaking crest in curve f, with indication of computational points. The breaker shape is fairly well resolved. During these computations, relative errors on wave volume ($v = 1.803$) and energy ($e = 1.663$) were less than 0.05% for most

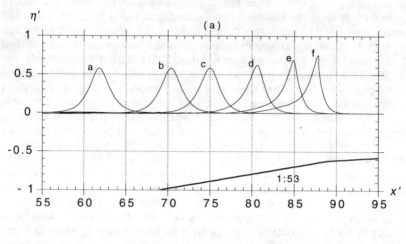

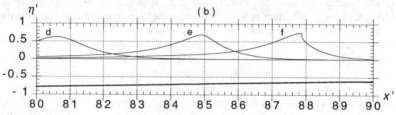

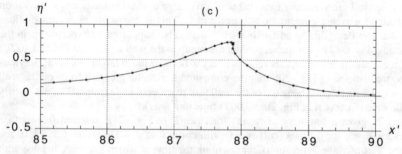

Fig. 21. Case C. (a) Free surface profile at $t' =$ a: 54.21; b: 61.08; c: 64.76; d: 69.23; e: 72.65 ; and f: 74.86 (breaking at $x'_b = 87.85$); (b) blow-up of (a) without distorsion of scales; (c) further blow-up of (b) showing the breaking crest with discretization nodes ($\circ$).

of the propagation but increased to 0.12 and 0.55%, respectively at the time of breaking, when nodes in the crest move very close to each other (Fig. 21c). To accurately compute results further than this time would require using regridding techniques similar to those in Grilli and Subramanya[39] illustrated by results in Fig. 14b.

It is of interest to note that, according to the (numerical/experimental) studies of breakers by Grilli *et al.*[48], the wave in case C has a slope parameter $S_o = 1.521s/\sqrt{H'_o} = 0.0246 <$ 0.025 (with $s = 1 : 53$) and should thus (barely) break as a spilling breaker. This seems to be supported by results in Fig. 21c when considering the fairly small size of the breaker jet. Empirical relationships based on parameter S_o (best fit through both experimental and numerical results) would further give, for the breaking characteristics at the time the wave front face reaches vertical tangent (i.e., slightly before the time of curve f), $H_b/h_b = 0.985$ and $h'_b = 0.729$, $H'_b = 0.72$. This is also in fairly good agreement with the present results.

In conclusion, for case C, breaking interrupts computations with the model and wave runup cannot be further calculated.

5.5.4. Comparison of numerical results with experiments

During the workshop, participants were provided with surface elevations measured at gages 7 and 9 in the experiments carried out in Vicksburg. Since the wave for case C broke before reaching those gages, a comparison with numerical results has only been made in Fig. 22 for cases A and B and numerical and experimental results have been synchronized in time at the location of gage 7.

For case A (Fig. 22a), the incident profiles at gage 7 are in quite good agreement, and similarly at gage 9. The experimental incident wave just seems a little smaller, which might be due to frictional effects during propagation in the tank. The reflected wave at gage 9 is also slightly smaller in the experiments and propagates through the gage slightly later in time than the computed wave, due to its smaller speed. Overall, the agreement is quite good.

For case B (Fig. 22b), the incident profiles at gage 7 are in quite good agreement, and similarly in gage 9, except that, in this case, the experimental wave is higher and, in fact, has more volume than the numerical wave. The reason for this is unknown. Computations unfortunately had to be interrupted before the reflected wave came back, due to the violent flip-through motion at the wall discussed before.

Acknowledgments

The author wishes to acknowledge support for this research from the NRL-SSC grant N00014-94-1-G607, from the US Department of the Navy Office of the Chief of Naval Research. The information reported in this work does not necessarily reflect the position of the US Government. Frederic Estadieu, a visiting student from ECN Nantes in France is acknowledged for his help in running the computations for Benchmark #3 applications.

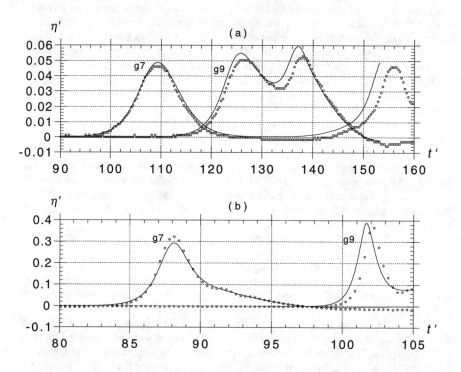

Fig. 22. Comparison of computed (——) and measured (○) free surface elevations $\eta' = \eta/h_o$ at gages 7 (g7) and 9 (g9) as a function of time t', for : (a) case A; (b) case B.

Appendix A Detailed expressions of coefficients in time updating

Detailed expressions of coefficients in Taylor series expansions (14) and (15) are given in the following.

Appendix A.1. Curvilinear coordinates

Derivations are carried out in a curvilinear coordinate system $(\mathbf{s}, \mathbf{n})$ defined along the boundary as (Fig. 3),

$$\mathbf{s} = [\cos\beta , \sin\beta] , \ \mathbf{n} = [-\sin\beta , \cos\beta] \tag{A.1}$$

$$\cos\beta = \frac{\partial x}{\partial s} , \ \sin\beta = \frac{\partial z}{\partial s} \tag{A.2}$$

where β denotes the angle between the horizontal axis x and the tangential vector $\mathbf{s}$ at the free surface.

Derivatives of vectors $(\mathbf{s}, \mathbf{n})$ with respect to their directions are obtained from Eq. (A.1) as,

$$\frac{\partial \mathbf{s}}{\partial s} = \frac{\partial \beta}{\partial s}\mathbf{n} , \ \frac{\partial \mathbf{n}}{\partial s} = -\frac{\partial \beta}{\partial s}\mathbf{s} \tag{A.3}$$

$$\frac{\partial \mathbf{s}}{\partial n} = \frac{\partial \beta}{\partial n}\mathbf{n} , \ \frac{\partial \mathbf{n}}{\partial n} = -\frac{\partial \beta}{\partial n}\mathbf{s} \tag{A.4}$$

Now, in a family of curves $n = \mathrm{cst}$ and of straight lines $s = \mathrm{cst}$ along the free surface, the derivative $\frac{\partial \beta}{\partial n}$ vanishes in Eq. (A.4).

With definitions in Eqs. (A.1) to (A.4), the curvilinear gradient operator reads,

$$\nabla \equiv \frac{1}{h_s}\frac{\partial}{\partial s}\mathbf{s} + \frac{\partial}{\partial n}\mathbf{n} \tag{A.5}$$

where h_s is a scale factor associated with curves $n = \mathrm{cst}$, defined along the free surface as,

$$-\frac{1}{h_s}\frac{\partial h_s}{\partial n} = \frac{1}{R} = \frac{\partial \beta}{\partial s} \quad \text{with,} \quad h_s = 1 \tag{A.6}$$

where $R(\mathbf{x})$ is the radius of curvature of the free surface. Thus, h_s is independent of s and only depends on n.

Appendix A.2. Taylor series coefficients

The kinematic free surface boundary condition Eq. (2) provides the first-order coefficient in Eq. (14) for the updating of free surface position vector $\mathbf{r}$,

$$\frac{D\mathbf{r}}{Dt} = \frac{\partial \phi}{\partial s}\mathbf{s} + \frac{\partial \phi}{\partial n}\mathbf{n} \tag{A.7}$$

Applying the material derivative Eq. (4) to Eq. (2), we get the general expression for the second-order coefficient in Eq. (14) as,

$$\frac{D^2\mathbf{r}}{Dt^2} = \frac{D\mathbf{u}}{Dt} = \frac{\partial \mathbf{u}}{\partial t} + \mathbf{u} \cdot \nabla\mathbf{u} \tag{A.8}$$

By definition of potential theory, the first term in the right hand side of Eq. (A.8) reads,

$$\frac{\partial \mathbf{u}}{\partial t} = \nabla \frac{\partial \phi}{\partial t} = \frac{\partial^2 \phi}{\partial t \partial s} \mathbf{s} + \frac{\partial^2 \phi}{\partial t \partial n} \mathbf{n} \tag{A.9}$$

Using the curvilinear system defined above, the second term in the right hand side of Eq. (A.8) becomes,

$$\mathbf{u} \cdot \nabla \mathbf{u} = \nabla \phi \cdot [\frac{1}{h_s} \frac{\partial \nabla \phi}{\partial s} \mathbf{s} + \frac{\partial \nabla \phi}{\partial n} \mathbf{n}]$$

which, using the orthogonality of $\mathbf{s}$ and $\mathbf{n}$, can be expressed as,

$$\mathbf{u} \cdot \nabla \mathbf{u} = \frac{1}{h_s^2} \frac{\partial \phi}{\partial s} \frac{\partial \nabla \phi}{\partial s} + \frac{\partial \phi}{\partial n} \frac{\partial \nabla \phi}{\partial n}$$

or,

$$\begin{aligned}
\mathbf{u} \cdot \nabla \mathbf{u} &= \frac{1}{h_s^2} \frac{\partial \phi}{\partial s} [\frac{1}{h_s}(\frac{\partial^2 \phi}{\partial s^2} \mathbf{s} + \frac{\partial \phi}{\partial s} \frac{\partial \mathbf{s}}{\partial s}) + \frac{\partial^2 \phi}{\partial s \partial n} \mathbf{n} + \frac{\partial \phi}{\partial n} \frac{\partial \mathbf{n}}{\partial s}] \\
&\quad + \frac{\partial \phi}{\partial n} [\frac{1}{h_s} \mathbf{s}(\frac{\partial^2 \phi}{\partial n \partial s} - \frac{1}{h_s} \frac{\partial h_s}{\partial n} \frac{\partial \phi}{\partial s}) + \frac{\partial^2 \phi}{\partial n^2} \mathbf{n}]
\end{aligned} \tag{A.10}$$

in which $\frac{\partial \beta}{\partial n} = 0$ and $\frac{\partial h_s}{\partial s} = 0$ were used.

Using Eqs. (A.1)-(A.6), it can be shown that continuity equation $\nabla \cdot \mathbf{u} = 0$ and irrotationality condition $\nabla \times \mathbf{u} = 0$ transform into,

$$\frac{\partial^2 \phi}{\partial s^2} + \frac{\partial^2 \phi}{\partial n^2} - \frac{\partial \beta}{\partial s} \frac{\partial \phi}{\partial n} = 0 \tag{A.11}$$

$$\frac{\partial^2 \phi}{\partial s \partial n} = \frac{\partial^2 \phi}{\partial n \partial s} \tag{A.12}$$

respectively, along the free surface.

Hence, with Eqs. (A.11) and (A.12), Eq. (A.10) can be expressed as,

$$\mathbf{u} \cdot \nabla \mathbf{u} = \{\frac{\partial \phi}{\partial s} \frac{\partial^2 \phi}{\partial s^2} + \frac{\partial \phi}{\partial n} \frac{\partial^2 \phi}{\partial n \partial s}\} \mathbf{s} + \{\frac{\partial \phi}{\partial s} \frac{\partial^2 \phi}{\partial n \partial s} - \frac{\partial \phi}{\partial n} \frac{\partial^2 \phi}{\partial s^2} + \frac{\partial \beta}{\partial s} [(\frac{\partial \phi}{\partial s})^2 + (\frac{\partial \phi}{\partial n})^2]\} \mathbf{n} \tag{A.13}$$

Combining Eqs. (A.8), (A.9), and (A.13), we get the final expression for the second-order coefficient in Eq. (14) as,

$$\begin{aligned}
\frac{D^2 \mathbf{r}}{Dt^2} &= \{\frac{\partial^2 \phi}{\partial t \partial s} + \frac{\partial \phi}{\partial s} \frac{\partial^2 \phi}{\partial s^2} + \frac{\partial \phi}{\partial n} \frac{\partial^2 \phi}{\partial n \partial s}\} \mathbf{s} + \\
&\quad \{\frac{\partial^2 \phi}{\partial t \partial n} - \frac{\partial \phi}{\partial n} \frac{\partial^2 \phi}{\partial s^2} + \frac{\partial \phi}{\partial s} \frac{\partial^2 \phi}{\partial n \partial s} + \frac{\partial \beta}{\partial s} [(\frac{\partial \phi}{\partial s})^2 + (\frac{\partial \phi}{\partial n})^2]\} \mathbf{n}
\end{aligned} \tag{A.14}$$

Similarly, dynamic free surface boundary condition Eq. (3) provides the first-order coefficient in the Taylor series (15) for the free surface potential updating. Using Eqs.

(A.5) and (A.6) we get,

$$\frac{D\phi}{Dt} = -gz + \frac{1}{2}[(\frac{\partial \phi}{\partial s})^2 + (\frac{\partial \phi}{\partial n})^2] - \frac{p_a}{\rho} \tag{A.15}$$

The second-order coefficient in Eq. (15) is obtained by material derivation of Eq. (3) as,

$$\frac{D^2\phi}{Dt^2} = -g\frac{Dz}{Dt} + \frac{1}{2}\frac{D}{Dt}(\nabla\phi \cdot \nabla\phi) - \frac{D}{Dt}(\frac{p_a}{\rho}) \tag{A.16}$$

with, using Eqs. (2), (A.1) and (A.5),

$$\frac{Dz}{Dt} = w = \frac{\partial \phi}{\partial n}\cos\beta + \frac{\partial \phi}{\partial s}\sin\beta \tag{A.17}$$

and, by definition of potential theory,

$$\frac{1}{2}\frac{D}{Dt}(\nabla\phi \cdot \nabla\phi) = \mathbf{u} \cdot \frac{D\mathbf{u}}{Dt} = \mathbf{u} \cdot \frac{\partial \mathbf{u}}{\partial t} + \mathbf{u} \cdot (\mathbf{u} \cdot \nabla\mathbf{u}) \tag{A.18}$$

Now, using orthogonality of $\mathbf{s}$ and $\mathbf{n}$, and Eqs. (2), (A.7), (A.9), and (A.13), we get the first term in the right hand side of Eq. (A.18) as,

$$\mathbf{u} \cdot \frac{\partial \mathbf{u}}{\partial t} = \frac{\partial \phi}{\partial s}\frac{\partial^2 \phi}{\partial t \partial s} + \frac{\partial \phi}{\partial n}\frac{\partial^2 \phi}{\partial t \partial n} \tag{A.19}$$

and the second term as,

$$\mathbf{u} \cdot (\mathbf{u} \cdot \nabla\mathbf{u}) = \frac{\partial \phi}{\partial s}[\frac{\partial \phi}{\partial s}\frac{\partial^2 \phi}{\partial s^2} + \frac{\partial \phi}{\partial n}\frac{\partial^2 \phi}{\partial n \partial s}] + \frac{\partial \phi}{\partial n}[\frac{\partial \phi}{\partial s}\frac{\partial^2 \phi}{\partial n \partial s} - \frac{\partial \phi}{\partial n}\frac{\partial^2 \phi}{\partial s^2} + \frac{\partial \beta}{\partial s}(\nabla\phi \cdot \nabla\phi)] \tag{A.20}$$

Finally, combining Eqs. (A.16)-(A.20) and Eq. (A.5), we get the final expression for the second-order coefficient in Eq. (15) as,

$$\begin{aligned} \frac{D^2\phi}{Dt^2} = & \frac{\partial \phi}{\partial s}\{\frac{\partial^2 \phi}{\partial t \partial s} + \frac{\partial \phi}{\partial s}\frac{\partial^2 \phi}{\partial s^2} + \frac{\partial \phi}{\partial n}\frac{\partial^2 \phi}{\partial n \partial s}\} + \\ & \frac{\partial \phi}{\partial n}\{\frac{\partial^2 \phi}{\partial t \partial n} - \frac{\partial \phi}{\partial n}\frac{\partial^2 \phi}{\partial s^2} + \frac{\partial \phi}{\partial s}\frac{\partial^2 \phi}{\partial n \partial s} + \frac{\partial \beta}{\partial s}[(\frac{\partial \phi}{\partial s})^2 + (\frac{\partial \phi}{\partial n})^2]\} - \\ & g\{\frac{\partial \phi}{\partial n}\cos\beta + \frac{\partial \phi}{\partial s}\sin\beta\} - \frac{1}{\rho}\frac{Dp_a}{Dt} \end{aligned} \tag{A.21}$$

where $\frac{Dp_a}{Dt}$ is the total rate of change of the free surface atmospheric pressure in time.

Appendix B   Generation of exact solitary waves by Tanaka's method

Tanaka's[90] method is based on using Cauchy's integral theorem for the complex velocity potential, in a frame of reference moving with the wave celerity c. In this frame, the crest velocity V_c fully defines the wave field and the dimensionless crest velocity $q_c = V_c/c$ is used as a parameter for the problem. The original method by Tanaka was modified by Cooker [12] to use wave height H' instead of q_c as a parameter.

Main steps in the calculations of *exact* solitary waves of height H' are as follows (superscripts denote iteration numbers),

- An approximate initial crest velocity $\tilde{q}_c^o$ is estimated for the specified H' by interpolation in a table of (H', q_c) values predetermined within the interval $(H' = 0.833197, q_c = 0)$ for the highest possible wave (like found, e.g., in Tanaka [90]) to $(H' = 0, q_c = 1)$ for a flat free surface.
- Velocity on the free surface is calculated for the approximate crest velocity $\tilde{q}_c^o$, using the original Tanaka's method.
- Wave celerity $\tilde{c}^o$ and Froude number [r]$(\tilde{F}^2)^o$ are calculated using the free surface velocities and the corresponding wave amplitude $\tilde{H}'^o$ is obtained from Bernoulli equation as,

$$\tilde{H}'^o = \frac{1}{2}[1 - (\tilde{q}_c^{\,2})^o](\tilde{F}^2)^o \tag{B.1}$$

- A better approximation for the crest velocity $\tilde{q}_c^1$ is re-estimated from $(H', \tilde{H}'^o)$ in the table of values (H', q_c).
- And so on, iteratively, until, $\Delta H' = |\,(H' - \tilde{H}'^n)/H'\,|$ is found sufficiently small [s].
- When convergence is reached for both F^2 and H', the wave shape and potential are calculated from free surface velocities. Normal velocity $\frac{\partial\phi}{\partial n}(x, t_o)$ is also calculated on the free surface at this stage (to be used as initial data for the first time step of computations with the BEM model), by noting that for a wave of constant shape,

$$\frac{\partial\phi}{\partial n}(x) = F\sin\beta(x) \tag{B.2}$$

- The wave area (or dimensionless mass) m above still water level and kinetic and potential energies (e_k, e_p) are calculated using standard integrals $(\rho' = g' = 1)$,

$$
\begin{aligned}
m &= \rho'\int_{\Gamma_f} z'dx' \\
e_k &= \frac{1}{2}\rho'\int_{\Gamma_f} \phi\frac{\partial\phi}{\partial n}d\Gamma \\
e_p &= \frac{1}{2}\rho'g'\int_{\Gamma_f} z'^2 dx'
\end{aligned} \tag{B.3}
$$

- The resulting *exact* solitary wave is finally truncated left and right to points for which free surface elevation, $\eta' = \varepsilon_z H'$ (with $\varepsilon_z \ll 1$, a pre-selected threshold), and wave elevation, potential, and normal velocity are re-interpolated within a constant step grid $\Delta x_o'$, with the crest being located at a specified x_o' value, to be used as initial data in the BEM model.

---

[r]Tanaka's method involves an iterative solution of Cauchy's integral theorem using the Froude number as the convergence parameter. The convergence criterion selected here is 10^{-10} in relative value of F^2. It was found that 70 to 75 iterations were necessary to achieve convergence within this accuracy.
[s]The convergence criterion selected here is $\Delta H' \le 10^{-5}$. Three to four iterations only are necessary to achieve convergence within this accuracy.

The overall method is found to be quite computationally efficient. Convergence on both F^2 and H' is reached and all wave data are calculated within less than 0.6s CPU time using 80 points on the free surface to describe the wave (for the author's program on an IBM3090/300).

Appendix C Boundary conditions for piston and flap wavemakers

Velocity and acceleration for points along a plane wavemaker boundary are derived in the following for both piston and flap type wavemakers, to be used as boundary conditions in Eq. (23).

Appendix C.1. Plane paddle wavemaker

If r_g denotes the distance between points $\boldsymbol{\alpha} = (\alpha, \beta)$ and $\mathbf{x}_g = (x_g, z_g)$ on the wavemaker (Fig. C.1), we get,

$$
\begin{aligned}
\alpha &= x_g + r_g \cos \theta \\
\beta &= z_g + r_g \sin \theta
\end{aligned}
\tag{C.1}
$$

Since r_g is constant with respect to any rigid body motion, we also have,

$$
\begin{aligned}
\dot{\alpha} &= \dot{x}_g - r_g \sin \theta\, \dot{\theta} = \dot{x}_g - (\beta - z_g)\dot{\theta} \\
\dot{\beta} &= \dot{z}_g + r_g \cos \theta\, \dot{\theta} = \dot{z}_g + (\alpha - x_g)\dot{\theta}
\end{aligned}
\tag{C.2}
$$

and,

$$
\begin{aligned}
\ddot{\alpha} &= \ddot{x}_g - r_g \cos \theta\, \dot{\theta}^2 - r_g \sin \theta\, \ddot{\theta} \\
\ddot{\beta} &= \ddot{z}_g - r_g \sin \theta\, \dot{\theta}^2 + r_g \cos \theta\, \ddot{\theta}
\end{aligned}
$$

or,

$$
\begin{aligned}
\ddot{\alpha} &= \ddot{x}_g - (\alpha - x_g)\dot{\theta}^2 - (\beta - z_g)\ddot{\theta} \\
\ddot{\beta} &= \ddot{z}_g - (\beta - z_g)\dot{\theta}^2 + (\alpha - x_g)\ddot{\theta}
\end{aligned}
\tag{C.3}
$$

Motion and boundary conditions are expressed in the following for two standard types of plane paddle wavemakers.

Appendix C.2. Piston wavemaker

This corresponds to a flat vertical plate with $\theta = \pi/2$, horizontally moving in depth h_o (Fig. 3). The specified horizontal piston motion (stroke) is $x_p(t)$ and $u_p(x_p(t), t) = \dot{x}_p(t)$ is the stroke velocity.

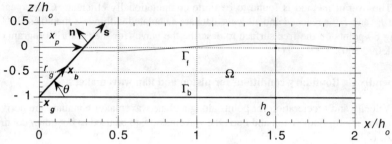

Fig. C.1. Sketch and definitions for a flap wavemaker motion on boundary Γ_{r1} of the computational domain.

Along the wavemaker paddle, by Eqs. (C.1),(C.2),(C.3), we have,

$$\mathbf{n} = [-1, 0], \qquad \mathbf{s} = [0, 1], \qquad \dot\theta = \ddot\theta = 0$$

$$\boldsymbol{\alpha} = \mathbf{x}_p = [x_p(t), z], \qquad \dot{\boldsymbol{\alpha}} = \mathbf{u}_p = [u_p(t), 0], \qquad \ddot{\boldsymbol{\alpha}} = \dot{\mathbf{u}}_p = [\dot{u}_p(t), 0]$$

$$\dot{\boldsymbol{\alpha}} \cdot \mathbf{n} = -u_p, \qquad \qquad \ddot{\boldsymbol{\alpha}} \cdot \mathbf{n} = -\dot{u}_p, \qquad \qquad \ddot{\boldsymbol{\alpha}} \cdot \mathbf{s} = 0 \qquad (\mathrm{C.4})$$

and from Eqs. (6),(23), and (C.4), boundary conditions on the piston wavemaker boundary read,

$$\frac{\overline{\partial \phi}}{\partial n} = -u_p(t)$$

$$\frac{\overline{\partial^2 \phi}}{\partial t \partial n} = -\dot{u}_p(t) - u_p(t) \frac{\partial^2 \phi}{\partial s^2} \qquad \text{on } \Gamma_{r1}(t) \qquad (\mathrm{C.5})$$

in which $\frac{\partial^2 \phi}{\partial s^2} = \frac{\partial^2 \phi}{\partial z^2}$ and $\dot{u}_p = \ddot{x}_p(t)$ denotes the specified wavemaker acceleration.

Appendix C.3. Flap wavemaker

This corresponds to a flat plate, hinged at $\mathbf{x}_g = (0, -h_o)$ on the bottom and oscillating with an angle $\theta(t) \in [\pi/2, 0]$ (defined trigonometrically with respect to the bottom; Fig. C.1). The specified horizontal piston motion (stroke) is $x_p(t)$ at $z = 0$ and $u_p(x_p(t), t) = \dot{x}_p(t)$ is the stroke velocity.

Along the wavemaker paddle, we have by Eq. (C.1),

$$\mathbf{n} = [-\sin\theta(t), \cos\theta(t)], \quad \mathbf{s} = [\cos\theta, \sin\theta], \quad \boldsymbol{\alpha} = \mathbf{x}_g + r_g \mathbf{s} = [\alpha(t), \beta(t)] \qquad (\mathrm{C.6})$$

in which r_g is given by,

$$r_g(t) = \alpha(t) \cos\theta(t) + [\beta(t) + h_o] \sin\theta(t) \qquad (\mathrm{C.7})$$

Now, by Eqs. (C.2) and (C.3), with $\dot{\mathbf{x}}_g = \ddot{\mathbf{x}}_g = 0$, we have,

$$\dot{\boldsymbol{\alpha}} = \mathbf{u}_p(t) = [-\beta(t) - h_o, \alpha(t)] \dot{\theta}$$

$$\ddot{\boldsymbol\alpha} = \ddot{\mathbf{u}}_p(t) = [-\beta(t) - h_o, \alpha(t)]\,\ddot{\theta}(t) - [\alpha(t), \beta(t) + h_o]\,\dot{\theta}^2(t) \tag{C.8}$$

Hence, by (C.6),(C.7),(C.8),

$$\dot{\boldsymbol\alpha}\cdot\mathbf{n} = [\alpha(t)\cos\theta(t) + (\beta(t) + h_o)\sin\theta(t)]\,\dot{\theta}(t)$$
$$= r_g(t)\,\dot{\theta}(t)$$
$$\ddot{\boldsymbol\alpha}\cdot\mathbf{n} = [\alpha(t)\cos\theta(t) + (\beta(t) + h_o)\sin\theta(t)]\,\ddot{\theta}(t)$$
$$+ [-(\beta(t) + h_o)\cos\theta(t) + \alpha(t)\sin\theta(t)]\,\dot{\theta}^2(t)$$
$$= r_g(t)\,\ddot{\theta}(t)$$
$$\dot{\boldsymbol\alpha}\cdot\mathbf{s} = [-(\beta(t) + h_o)\cos\theta(t) + \alpha(t)\sin\theta(t)]\,\dot{\theta}(t)$$
$$= 0 \tag{C.9}$$

since one can show, by simple geometric considerations, $[-(\beta + h_o)\cos\theta + \alpha\sin\theta] = 0$.

From Eqs. (6),(23), and (C.9), boundary conditions on the flap wavemaker boundary read,

$$\frac{\overline{\partial\phi}}{\partial n} = r_g(t)\,\dot{\theta}(t)$$
$$\frac{\partial^2\phi}{\partial t\partial n} = r_g(t)\,\ddot{\theta}(t) + \dot{\theta}(t)\,[r_g(t)\frac{\partial^2\phi}{\partial s^2} - \frac{\partial\phi}{\partial s}] \tag{C.10}$$

After some elementary developments, r_g and time derivatives of $\theta(t)$ in Eq. (C.10) can be expressed as a function of wavemaker stroke $x_p(t)$ and its time derivatives as,

$$\dot{\theta}(t) = -R(t)\,u_p(t)$$
$$\ddot{\theta}(t) = -R(t)\,[\dot{u}_p(t) - 2\,u_p^2(t)\frac{x_p(t)}{h_o}]$$
$$r_g(t) = R(t)\,\sqrt{h_o^2 + x_p^2(t)}\,[\alpha(t)\frac{x_p(t)}{h_o} + \beta(t) + h_o] \tag{C.11}$$

in which $[\alpha(t), \beta(t)]$ denote coordinates of points along the flap wavemaker and, $R(t) = h_o/(h_o^2 + x_p^2(t))$.

Appendix D Piston wavemaker motion for the generation of first-order solitary and cnoidal waves

Development of Eq. (26) is done in the following for the generation of first-order solitary or cnoidal waves by a piston wavemaker.

Appendix D.1. First-order solitary wave

The surface elevation for a *first-order* solitary wave of height H' in depth h_o is obtained as a permanent wave solution of Boussinesq equations as (e.g. Dean and Dalrymple [17]),

$$\eta'(x', t') = H' \mathrm{sech}^2[\kappa(x' - c't')] \tag{D.1}$$

where $\kappa = \sqrt{3H'}/2$ and the celerity $c' = \sqrt{1 + H'}$.

Substituting Eq. (D.1) into Eq. (26) while specifying $x' = x'_p(t)$ throughout the integration gives the piston stroke required for generating the wave.

Since the wave in Eq. (D.1) extends to infinity in both directions, however, before it is used in the model, it is necessary to truncate it at some distance from the origin. Goring [25] introduced the significant horizontal extension of the wave $2\lambda'$ corresponding to a reduction in wave elevation to $\eta' = \varepsilon_z H'$ (with $\varepsilon_z \ll 1$). Using this definition and Eq. (D.1), we get,

$$\varepsilon_z H' = H' \mathrm{sech}^2[\kappa\lambda']$$
$$\varepsilon_z^{-\frac{1}{2}} = \cosh \kappa\lambda' \tag{D.2}$$

and,

$$\ell = \mathrm{arcosh}[\varepsilon_z^{-\frac{1}{2}}] \qquad \text{with} \qquad \lambda' = \frac{\ell}{\kappa} \tag{D.3}$$

Now (Abramowitz and Stegun [1]),

$$\mathrm{arcosh}[\varepsilon_z^{-\frac{1}{2}}] = \log\left\{\varepsilon_z^{-\frac{1}{2}}[1 + (1 - \varepsilon_z)^{\frac{1}{2}}]\right\} \tag{D.4}$$

Hence, since $\varepsilon_z \ll 1$,

$$\ell \simeq \log \frac{4 - \varepsilon_z}{2\varepsilon_z^{\frac{1}{2}}} \tag{D.5}$$

In the numerical applications, we usually select $\varepsilon_z = 0.002$ to which it corresponds $\ell \simeq 3.80$.

Wave generation by the piston wavemaker thus starts at $t'_o = 0$ with $x' = x'_p + \lambda'$. Introducing this initial condition in the theoretical wave profile (D.1) and integrating (26) we get,

$$x'_p(t') = \frac{H'}{\kappa}[\tanh \chi(t') + \tanh \kappa\lambda'] \qquad \text{with} \qquad \chi(t') = \kappa(c't' - x'_p(t') - \lambda') \tag{D.6}$$

which is solved for x'_p for any given time t' using Newton iterations.

Wavemaker velocity, $u'_p(t')$ is then computed by Eq. (25) for $\eta'(x'_p(t'), t')$ and acceleration $\dot{u}'_p(t')$ is found by time derivation of the velocity,

$$u'_p(t) = H'(1 + H')^{\frac{1}{2}} \frac{1}{\cosh^2 \chi(t') + H'}$$
$$\dot{u}'_p(t) = \sqrt{3} H'^{\frac{3}{2}}(1 + H') \frac{\cosh^3 \chi(t') \sinh \chi(t')}{(\cosh^2 \chi(t') + H')^3} \tag{D.7}$$

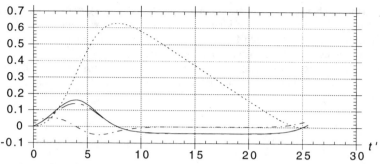

Fig. D.1. Generation of a first-order cnoidal wave by a piston wavemaker. Surface elevation and paddle motion as a function of time t', for $H' = 0.2$, $T' = 25$, with $\eta(\text{——})$, $x_p(\text{- - - -})$, $u_p(\text{— — -})$ and $\dot{u}_p(. \text{ — }.)$.

These values are introduced in Eq. (C.5) to define boundary conditions for the piston wavemaker.

Initial wavemaker velocity and acceleration at $t'_o = 0$ can be found as functions of H' and ε_z, by introducing Eqs. (D.3) and (D.6) into Eq. (D.7) as,

$$u'_p(t'_o) = H'(1 + H')^{\frac{1}{2}} \frac{\varepsilon_z}{1 + \varepsilon_z H'}$$

$$\dot{u}'_p(t'_o) = \sqrt{3}\, H'^{\frac{3}{2}}(1 + H')\varepsilon_z \frac{(1 - \varepsilon_z)^{\frac{1}{2}}}{(1 + \varepsilon_z H')^3} \tag{D.8}$$

which both are approximately proportional to ε_z, for a given H'. Hence, initial wavemaker acceleration, which should be kept small to avoid initial singularity problems (Section 2.2), is controlled by selecting a small enough truncation parameter ε_z. For $\varepsilon_z = 0.002$ and $H'=0.5$, for instance, $u_p(t_o) \simeq 0.00122\sqrt{gd}$ and $\dot{u}_p(t_o) \simeq 0.00184g$, which is quite small compared to gravity.

Appendix D.2. First-order cnoidal wave

First-order cnoidal waves are periodic wave solutions of Boussinesq equations. In water of constant depth h_o, a cnoidal wave elevation of height H', period T', and length $L' = c'T'$ is given by (e.g., Dean and Dalrymple [17]),

$$\eta'(x', t') = H'\{B + \text{cn}^2[\frac{2K}{L'}(x' - c't'), m]\} \tag{D.9}$$

in which, $L' = 4K\sqrt{m/(3H')}$, the celerity $c' = \sqrt{1 + AH'}$, with $A(m) = (2 - m - 3E/K)/m$, and the dimensionless trough $B(m) = (1 - m - E/K)/m$. Symbol "cn" denotes the Jacobian elliptic function of parameter m and $(K(m), E(m))$ denote complete elliptic integrals of the 1st and 2nd kind, respectively (Abramowitz and Stegun [1]).

Substituting Eq. (D.9) into Eq. (26) while specifying $x' = x'_p(t)$ throughout the integration gives the piston stroke required for generating the wave.

Wave generation starts for $x'_p = t' = 0$ at a given initial phase $x' = \lambda'$ of the wave. Setting $x' = x'_p + \lambda'$ in Eq. (D.9) and integrating (26), we get the following equation for the stroke $x'_p(t')$,

$$x'_p(\chi(t)) = \frac{L'}{2K} H' \{ \frac{E}{mK} (\chi(t) - \chi_o) - \frac{1}{m} [E(\chi(t), m) - E(\chi_o, m)] \}$$

$$\chi(t) = \frac{2K}{L'} [x'_p(t) + \lambda' - c't'] \tag{D.10}$$

which is solved by Newton iterations for any time t'. In Eq. (D.10), $\chi_o = \lambda'(2K/L')$ and $E(\chi(t), m)$ is the incomplete elliptic integral of the 1st kind.

Finally, $u'_p(t')$ and $\dot{u}'_p(t')$ are obtained by derivation of Eq. (D.10) and introduced into Eq. (C.5) to provide boundary conditions on the wavemaker. For a cnoidal wave of height $H' = 0.2$, and period $T' = 25$, which is close to the upper limit of long wave theory, for instance, using the above equations we get $L' = 25.99$, $c' = 1.040$ and $K = 5.035$, and Fig. D.1 shows the free surface elevation and paddle motion, velocity and acceleration calculated as a function of time for these data.

In the present case, initial acceleration of the wavemaker $\dot{u}'_p(t'_o)$ varies with the selected initial phase λ' and, hence, can be made sufficiently small by adjusting the phase. For $\lambda' = 0$, for instance, initial acceleration is zero. For cnoidal waves, however, this also corresponds to maximum crest elevation and velocity. The origin can be shifted to a point with zero water elevation and velocity by selecting,

$$\lambda' = \frac{L'}{2K} [2K - \mathrm{cn}^{-1} \sqrt{-B}] + x'_p(0) \tag{D.11}$$

where $x'_p(0)$ is obtained from Eq. (D.10) with $\chi = 0$. This is the situation plotted in Fig. D.1. For this case, however, the initial acceleration is no longer zero but, for long waves, it is still quite small compared to gravity ($\mathcal{O}(4c'KH'/L')$; in Fig. D.1 the initial acceleration is about $0.03g$).

Appendix E   Boundary conditions for the generation of a sum of sine waves by a flap wavemaker

Eqs. (27) and (28) used for generating sine waves with a flap wavemaker are further detailed in the following.

By analogy with the smooth initial motion obtained in Eq. (D.6) for the generation of solitary waves by a piston wavemaker, the initial damping function for sine waves is selected as,

$$\mathcal{D}(t) = \frac{1 + \varepsilon_z}{2} [\tanh \mu(t - t_{\varepsilon_z}) + \frac{1 - \varepsilon_z}{1 + \varepsilon_z}] \tag{E.1}$$

with μ, a damping coefficient obtained from the requirement that $\mathcal{D}(0) = 0$ as,

$$\mu = -\frac{1}{2\,t_{e_z}} \log \varepsilon_z \tag{E.2}$$

One can easily check that Eqs. (E.1) and (E.2) also satisfy $\mathcal{D}(2\,t_{e_z}) = 1 - \varepsilon_z$, which allows to select the rate of damping corresponding to given values of t_{e_z} and ε_z. For $\varepsilon_z = 0.001$, for instance, we get $\mu \simeq 3.454/t_{e_z}$.

In the applications, the time $2t_{e_z}$ is selected as an integer multiple N_n of the average wave period $\tilde{T}$ of the wave components to be generated,

$$t_{e_z} = \frac{N_n \tilde{T}}{2} \qquad \text{and} \qquad \tilde{T} = \frac{1}{n}\sum_{i=1}^{n}\frac{2\pi}{\omega_i} \tag{E.3}$$

By time derivation of Eq. (27), we get the paddle velocity and acceleration at $z = 0$ as,

$$
\begin{aligned}
u_p(t) &= \dot{\mathcal{S}}\mathcal{D} + \mathcal{S}\dot{\mathcal{D}} & \text{and} && \dot{u}_p(t) &= \ddot{\mathcal{S}}\mathcal{D} + 2\dot{\mathcal{D}}\dot{\mathcal{S}} + \mathcal{S}\ddot{\mathcal{D}} \\
\dot{\mathcal{S}}(t) &= \sum_{i=1}^{n}\frac{1}{2}A_i\,\omega_i \sin\left(\omega_i t + \varphi_i\right), && & \ddot{\mathcal{S}}(t) &= \sum_{i=1}^{n}\frac{1}{2}A_i\,\omega_i^2 \cos\left(\omega_i t + \varphi_i\right) \\
\dot{\mathcal{D}}(t) &= \frac{\mu}{2}\frac{1+\varepsilon_z}{\cosh^2 \mu(t - t_{e_z})}, && & \ddot{\mathcal{D}}(t) &= -\mu^2\left(1 + \varepsilon_z\right)\frac{\tanh \mu(t - t_{e_z})}{\cosh^2 \mu(t - t_{e_z})}
\end{aligned} \tag{E.4}
$$

Hence, boundary conditions (C.10) and (C.11) can be defined on the wavemaker.

The initial wavemaker velocity and acceleration at time $t_o = 0$ are obtained from (E.4) as,

$$
\begin{aligned}
\dot{\mathcal{D}}(t_o) &= 2\mu\varepsilon_z \frac{1}{1+\varepsilon_z} & \text{and} && \ddot{\mathcal{D}}(t_o) &= 4\mu^2 \varepsilon_z \frac{1-\varepsilon_z}{(1+\varepsilon_z)^2} \\
u_p(t_o) &= \mathcal{S}(t_o)\dot{\mathcal{D}}(t_o) & \text{and} && \dot{u}_p(t_o) &\simeq 2\dot{\mathcal{D}}(t_o)\left(\dot{\mathcal{S}}(t_o) + \mu\mathcal{S}(t_o)\right)
\end{aligned} \tag{E.5}
$$

Since for $\varepsilon_z \ll 1$, we have $\dot{\mathcal{D}}(t_o) \simeq 2\mu\varepsilon_z$ and $\ddot{\mathcal{D}}(t_o) \simeq 2\mu\dot{\mathcal{D}}(t_o)$. If we further require that $\mathcal{S}(t_o) = 0$ in Eq. (E.5), we get, $u_p(t_o) = 0$ and $\dot{u}_p(t_o) \simeq 4\mu\varepsilon_z\dot{\mathcal{S}}(t_o)$. For $\varepsilon_z = 0.001$, for instance and $\mu \simeq 1$ for $t_{e_z} = 3.454$, the initial acceleration is $\dot{u}_p(t_o) \simeq 0.004\dot{\mathcal{S}}(t_o)$, which is thus a rather small fraction of the initial paddle velocity.

Appendix F Generation of second-order solitary and periodic waves using internal source distributions

Wave velocity distributions $u_w(x_s, z, t)$ to be used in source distributions Eqs. (31) and (33) for the generation of waves by internal sources are given in the following for second-order waves. Note that higher-order solutions (even SFW) can be used and this was recently done by Ohyama and Nadaoka[66] for 5th-order Stokes waves.

Appendix F.1. Second-order solitary waves

For a solitary wave whose *first-order* profile is given by Eq. (D.1), the horizontal velocity can be deduced as a function of depth from Boussinesq's theory (see Mei [63]). The horizontal velocity is constant over depth, to the first order in H'. Identical developments can be made up to the 2nd-order accuracy and we get,

$$u_w(x_s, z, t) = \frac{Hg}{c}\mathrm{sech}^2\chi(t)\left[1 + (\frac{\kappa}{h_o})^2(z + h_o)^2(2\tanh^2\chi(t) + \sinh^2\chi(t))\right] \quad (\text{F.1})$$

in which $\chi(t)$ is defined as in Eq. (D.6) and the solitary wave has been limited to its significant part 2λ defined as in Eq. (D.3).

In dimensionless form, Eq. (F.1) reads,

$$u'_w(x'_s, z', t') = \frac{H'}{c'^2}\mathrm{sech}^2\chi(t')\left[1 + (\kappa^2(z' + 1)^2(2\tanh^2\chi(t') + \sinh^2\chi(t'))\right] \quad (\text{F.2})$$

In the implementation of this procedure in the model, source strengths defined based on Eq. (F.1) correspond to Poisson equation's (10) for ϕ. For $\frac{\partial\phi}{\partial t}$, $\frac{\partial u_w}{\partial t}$ is used instead of u_w.

Appendix F.2. Second-order periodic waves

For a wave of period T and height H, the horizontal velocity calculated from Stokes theory in water of depth h_o, up to *second-order* in H/L, reads (Dean and Dalrymple [17]),

$$
\begin{aligned}
u_w(x_s, z, t) &= \frac{Hg}{2c}\frac{\cosh k(h_o + z)}{\cosh kh_o}\cos(kx_s - \omega t) \\
&\quad - \frac{3H^2\omega k}{16}\frac{\cosh 2k(h_o + z)}{\sinh^4 kh_o}\cos 2(kx_s - \omega t)
\end{aligned} \quad (\text{F.3})
$$

in which, $\omega = 2\pi/T$, is the wave circular frequency, $c = \omega/k$, is the wave celerity and the wavenumber k is given by the linear dispersion relation (28).

To avoid initial singularity during a "cold start", the velocity (F.2) is multiplied by a damping function of a form similar to Eq. (E.1).

Again, source strengths defined by (F.3) correspond to the Poisson equation for ϕ. For $\frac{\partial\phi}{\partial t}$, $\frac{\partial u_w}{\partial t}$ is used instead of u_w.

6. References

1. Abramowitz, M. and Stegun, I.A. *Handbook of Mathematical Functions.* Dover Pub. Inc. New York, 1965.
2. Brebbia, C.A. *The Boundary Element Method for Engineers,* John Wiley and Sons, U.K., 1978.
3. Broeze, J., *Numerical Modelling of Nonlinear Free Surface Waves With a 3D Panel Method.* Ph.D. Dissertation, Enschede, The Netherland, 1993.

4. Brorsen, M. and Larsen, J. Source Generation of Nonlinear Gravity Waves with the Boundary Integral Method. *Coastal Engineering* **11**, 93-113, 1987.
5. Baker, G.R., Meiron, D.I. and Orszag, S.A. Generalized Vortex Method for Free-Surface Flow Problems. *J. Fluid Mech.* **123**, 477-501, 1982.
6. Camfield, F.E. and Street, R.L. Shoaling of Solitary Waves on Small Slopes. *ASCE,* **WW95**, 1-22, 1969.
7. Carrier, G.F. Gravity Waves on Water of Variable Depth. *J. Fluid Mech.* **24** (4), 641-659, 1966.
8. Carrier, G.F. and Greenspan, H.P. Water Waves of Finite Amplitude on a Sloping Beach. *J. Fluid Mech.* **4** (1), 97-110, 1958.
9. Cao, Y., Beck, R.F. and Schultz, W.W. An Absorbing Beach for Numerical Simulations of Nonlinear Waves in a Wave Tank. *Proc. 8th Intl. Workshop Water Waves and Floating Bodies* (St. John's, Newfoundland May 23-26) pps. 17-20, 1993.
10. Cointe, R. Numerical Simulation of a wave Channel. *Engng. Analysis with Boundary Elements* **7** (4), 167-177, 1990.
11. Cointe, R. Quelques aspects de la simulation numérique d'un canal à houle. *Thèse de Docteur de l'Ecole Nationale des Ponts et Chaussées,* 284 pps., 1989.
12. Cooker, M. The Interaction Between Steep Water Waves and Coastal Structures. *Ph.D. Dissertation.* School of Mathematics, University of Bristol, England, 1990.
13. Cooker, M. A Boundary-integral Method for Water Wave Motion Over Irregular Beds. *Engng. Analysis with Boundary Elements* **7** (4), 205-213, 1990.
14. Cooker, M. and Peregrine, D.H. Violent Water Motion at Breaking-Wave Impact. In *Proc. 22nd Intl. Conf. on Coastal Engineering* (ICCE22, Delft, The Netherland, July 90). Vol **1** , pps. 164-176. ASCE edition, 1991.
15. Cooker, M.J., Peregrine, D.H., Vidal, C. and Dold, J.W. The Interaction Between a Solitary Wave and a Submerged Semicircular Cylinder. *J. Fluid Mech.* **215**, 1-22, 1990.
16. Dean, R.G. Stream Function Representation of Nonlinear Ocean Wave. *J. Geophys. Res.* **70**, 4561-4572, 1965.
17. Dean, R.G. and Dalrymple R.A. *Water Wave Mechanics for Engineers and Scientists* Prentice-Hall, 1984.
18. Dold, J.W. and Peregrine, D.H. An Efficient Boundary Integral Method for Steep Unsteady water Waves. *Numerical methods for Fluid Dynamics II* (ed. K.W. Morton and M.J. Baines), pp. 671-679. Clarendon Press, Oxford, 1986.
19. Dommermuth, D.G. and Yue, D.K.P. Numerical Simulation of Nonlinear Axisymmetric Flows with a Free Surface. *J. Fluid Mech.* **178**, 195-219, 1987.
20. Dommermuth, D.G., Yue, D.K.P., Lin, W.M., Rapp, R.J., Chan, E.S. and Melville, W.K. Deep-Water Plunging Breakers : a Comparison between Potential Theory and Experiments. *J. Fluid Mech.* **189**, 423-442, 1988.
21. Driscoll, A.M., Dalrymple, R.A. and Grilli Harmonic Generation and Transmission Past a Submerged Rectangular Obstacle. In *Proc. 23rd Intl. Conf. on Coastal Engineering* (ICCE23, Venice, Italy, Oct. 92) Vol. **1**, pps. 1142-1152. ASCE edition, 1993.
22. Engquist, B. and Majda, A. Absorbing Boundary Conditions for the Numerical

Simulation of Waves. *Math. Comp.* **31,** 629-651, 1977.

23. Fenton, J.D. and Rienecker, M.M. A Fourier Method for Solving Nonlinear Water-Wave Problems : Application to Solitary-Wave Interactions. *J. Fluid Mech.* **118,** 411-443, 1982.

24. Freilich, M.H. and Guza, R.T. Nonlinear Effects on Shoaling Surface Gravity Waves *Phil. Trans. R. Soc. Lond.* **A311,** 1-41, 1984.

25. Goring D.G. Tsunamis - The Propagation of Long Waves onto a Shelf. *W.M. Keck Laboratory of Hydraulics and Water Resources, California Institute of Technology, Report No.* **KH-R-38,** 1978.

26. Gravert, P. Numerische Simulation Extremer Schwerewellen im Zeitbereich mit Direkter Randelementmethode und Zeitschrittverfahren (Ph.D. Dissertation). *Fortschritt-Berichte VDI Verlag Düsseldorf, Reihe 7 No.* **132.** 1987.

27. Greenhow, M. Wedge Entry into Initially Calm Water. *Applied Ocean Res.* **9,** 214-223, 1987.

28. Griffiths, M.W., Easson, W.J. and Greated, C.A. Measured Internal Kinematics for Shoaling Waves with Theoretical Comparisons *J. Waterways, Port Coastal and Ocean Engng.* **118** (3), 280-299, 1992.

29. Grilli, S. Wave Overturning Induced by Moving Bodies. Application to Slender Ship Wave Resistance. *Invited paper* in *Proc. 1st Intl. Conf. on Computational Modelling of Free and Moving Boundary Problems* (Southampton, England, July 91)(ed. L.C. Wrobel & C.A. Brebbia) Vol. **1,** pp. 75-90. Comp. Mech. Pub., de Gruyter, Southampton, 1991.

30. Grilli, S. Modeling of Nonlinear Wave Motion in Shallow Water. Chapter 3 in *Computational Methods for Free and Moving Boundary Problems in Heat and Fluid Flow* (eds. L.C. Wrobel and C.A. Brebbia), pps. 37-65, Comp. Mech. Pub., Elsevier Applied Sciences, London, UK, 1993.

31. Grilli, S. and Horrillo, J. Generation and Absorption of Fully Nonlinear Periodic Waves in a numerical Wave Tank. *J. Engng. Mech.,* (submitted).

32. Grilli, S., Losada, M.A. and Martin, F. Kinematics of Solitary Wave Breaking over Submerged Structures : Comparison between Nonlinear Computations and Experimental Results. In *Proc. 4th Intl. Conf. on Hydraulic Engineering Software* (HYDROSOFT92, Valencia, Spain, July 92) (eds. W.R. Blain and E. Cabrera), Fluid Flow Modelling, pp. 575-586. Comp. Mech. Pub., Elsevier Applied Science, 1992.

33. Grilli, S., Losada, M.A. and Martin, F. Wave Impact Forces on Mixed Breakwaters. In *Proc. 23rd Intl. Conf. on Coastal Engineering* (ICCE23, Venice, Italy, October 92) Vol. **1,** pps. 1161-1174. ASCE edition, 1993.

34. Grilli, S., Losada, M.A. and Martin, F. Characteristics of Solitary Wave Breaking Induced by Breakwaters. in *J. Waterway, Port, Coastal, and Ocean Engng.,* **120** (1), 74-92, 1994.

35. Grilli, S., Losada, M.A., Martin, F. and Svendsen, I.A. Nonlinear Shoaling and Impact of Waves on Coastal Structures. In *Proc. 9th Engng. Mech. Conf.* (College Station, Texas, May 92) (eds. L.D. Lutes and J.M. Niedzwecki), pp. 79-82. ASCE edition, 1992.

36. Grilli, S. Skourup, J. and Svendsen, I.A. An Efficient Boundary Element Method for Nonlinear Water Waves. *Engng. Analysis with Boundary Elements* **6** (2), 97-107, 1989.

37. Grilli, S. and Subramanya, R. Quasi-singular Integrations in the Modelling of Nonlinear Water Waves. *Engng. Analysis with Boundary Elements,* **13** (2), 181-191, 1994.

38. Grilli, S. and Subramanya, R. Recent Advances in the BEM Modelling of Nonlinear Water Waves. Chapter 4 in *Boundary Element Applications in Fluid Mechanics* (ed. H. Power), pps. 91-122. Advances in Fluid Mechanics Series. Comp. Mech. Pub., Southampton, UK, 1995.

39. Grilli, S. and Subramanya, R. Numerical Modeling of Wave Breaking Induced by Fixed or Moving Boundaries. *Computational Mechanics* (in press).

40. Grilli, S., Subramanya, R., Kirby, J.T. and Wei, J. Comparison of Modified Boussinesq and Fully Nonlinear Potential Models for Shoaling Solitary Waves. In *Proc. Intl. Symposium on Waves - Physical and Num. Modelling* (Vancouver BC, Canada, Aug. 1994) (eds. M. Isaacson and M. Quick), Vol. **1**, pps. 524-533, IAHR, 1994.

41. Grilli, S., Subramanya, R., Svendsen, I.A. and Veeramony, J. Shoaling of Solitary Waves on Plane Beaches. *J. Waterway Port Coastal and Ocean Engng.,* **120** (6), 609-628, 1994.

42. Grilli, S. and Svendsen, I.A. The Modelling of Nonlinear Water Wave Interaction with Maritime Structures. *Advances in Boundary Elements* (Proc. 11th Intl. Conf. on Boundary Elements, Cambridge, Massachusetts, USA), Vol. **2** (ed. C.A. Brebbia, J.J. Connor), pp. 253-268. Comp. Mech. Publ. Springer Verlag, Berlin, 1989.

43. Grilli, S. and Svendsen, I.A. The Modelling of Highly Nonlinear Waves : Some Improvements to the Numerical Wave Tank. *Advances in Boundary Elements* (Proc. 11th Intl. Conf. on Boundary Elements, Cambridge, Massachusetts, USA), Vol. **2** (ed. C.A. Brebbia, J.J. Connor), pp. 269-281. Comp. Mech. Pub.. Springer Verlag, Berlin, 1989.

44. Grilli, S. and Svendsen, I.A. Computation of Nonlinear Wave Kinematics during Propagation and Runup on a Slope. *Water Wave Kinematics,* (Proc. NATO-ARW, Molde, Norway, May 89) (ed. A. Torum and O.T. Gudmestad) NATO ASI Series E: Applied Sciences Vol. **178,** 387-412. Klüwer Academic Publishers, 1990.

45. Grilli, S. and Svendsen, I.A. Long Wave Interaction with Steeply Sloping Structures. In *Proc. 22nd Intl. Conf. on Coastal Engineering* (ICCE22, Delft, The Netherland, July 90) Vol. **2**, pps. 1200-1213. ASCE edition, 1991.

46. Grilli, S. and Svendsen, I.A. Corner Problems and Global Accuracy in the Boundary Element Solution of Nonlinear Wave Flows. *Engng. Analysis with Boundary Elements,* **7** (4), 178-195, 1990.

47. Grilli, S. and Svendsen, I.A. The Propagation and Runup of Solitary Waves on Steep Slopes. *Center for Appl. Coastal Res., University of Delaware, Res. Rep. No.* **91-4,** 1991.

48. Grilli, S., Svendsen, I.A. and Subramanya, R., Breaking Criterion and Characteristics for Solitary Waves on Plane Beaches. *J. Waterway Port Coastal and Ocean Engng.* (submitted).

49. Hall, J.V. and Watts, J.W. Laboratory Investigation of the Vertical Rise of Solitary Waves on Impermeable Slopes. *Beach Erosion Board, US Army Corps of Engineer, Tech. Memo. No.* **33,** 14 pp, 1953.

50. Hibberd, S. and Peregrine, D.H. Surf and Run-up on a Beach : a Uniform Bore, *J. Fluid Mech.* **95** (2), 323-345, 1979.

51. Isaacson, M. de St. Q. Nonlinear Effects on Fixed and Floating Bodies. *J. Fluid Mech.* **120,** 267-281, 1982.

52. Israeli, M. and Orszag, S.A. Approximation of Radiation Boundary Conditions. *J. Comp. Phys.* **41,** 115-135, 1981.

53. Jansen, P.C.M. A Boundary Element Model for Nonlinear Free Surface Phenomena. *Delft University of Technology, Department of Civil Engineering Report No.* **86-2,** 1986.

54. Kim, S.K., Liu, P.L.-F. and Liggett, J.A. Boundary integral Equation Solutions for Solitary Wave Generation Propagation and Run-up. *Coastal Engineering* **7,** 299-317, 1983.

55. Kirby, J.T. Intercomparison of Truncated Series Solutions for Shallow Water Waves, *J. Waterways, Port, Coastal and Ocean Engng.* **117** (2), 143-155, 1991.

56. Klopman, G. Numerical Simulation of Gravity Wave Motion on Steep slopes. *Delft Hydraulics Report No.* **H195,** 1988.

57. Kobayashi, N. DeSilva, G.S. and Watson, K.D. Wave Transformation and Swash Oscillation on Gentle and Steep Slope. *J. Geoph. Res.* **94** (C1), 951-966, 1989.

58. Larsen, J. and Dancy, H. Open Boundaries in Short Waves Simulations — A New Approach. *Coastal Engng.* **7,** 285-297, 1983.

59. Le Mehauté, B. Progressive Wave Absorber. *J. Hyd. Res.* **10(2),** 153-169, 1972.

60. Lin, W.M., Newman, J.N. and Yue, D.K. Nonlinear Forced Motion of Floating Bodies. In *Proc. 15th Intl. Symp. on Naval Hydrody., Hamburg, Germany,* 1984.

61. Liu, P.L.-F., Cho, Y.S. and Kim, S.K. A Computer Program for Transient Wave Run-up *Research Report* School of Civil and Environmental Engng., Cornell University, 1992.

62. Longuet-Higgins, M.S. and Cokelet, E.D. The Deformation of Steep Surface Waves on Water - I. A Numerical Method of Computation. *Proc. R. Soc. Lond.* **A350,** 1-26, 1976.

63. Mei, C.C *The Applied Dynamics of Ocean Surface Waves (2nd ed.).*   World Scientific, New Jersey, 1989.

64. Nakayama, T. Boundary Element Analysis of Nonlinear Water Wave Problems. *Intl. J. Numer. Meth. Engng.* **19,** 953-970, 1983.

65. New, A.L., McIver, P. and Peregrine, D.H. Computation of Overturning Waves. *J. Fluid Mech.* **150,** 233-251, 1985.

66. Ohyama, T. and Nadaoka, K. Transformation of a Nonlinear Wave Train Passing Over a Submerged Shelf Without Breaking. *Coastal Engng.* **24,** 1-22, 1994.

67. Ohyama, T. and Nadaoka, K. Development of a Numerical Wave Tank for Analysis of Nonlinear and Irregular Wave Fields. *Fluid Dyn. Res.* **8,** 231-251, 1991.

68. Orlanski, I. A Simple Boundary Condition for Unbounded Hyperbolic Flows. *J. Comp. Phys.* **21,** 251-269, 1976.

69. Otta, A.K., Svendsen, I.A. and Grilli, S.T. The Breaking and Runup of Solitary Waves on Beaches. In *Proc. 23rd Intl. Conf. on Coastal Engineering* (ICCE23, Venice, Italy, October 92) Vol. **2**, pps. 1461-1474. ASCE edition, 1993.

70. Otta, A.K., Svendsen, I.A. and Grilli, S.T. Unsteady Free Surface Waves in Region of Arbitrary Shape. *CACR, University of Delaware, Res. Rep. No.* **92-10,** 153pp, 1992.

71. Pedersen, G. and Gjevik, B. Run-up of Solitary Waves *J. Fluid Mech.* **135,** 283-299, 1983.

72. Peregrine, D.H. Long Waves on Beaches, *J. Fluid Mech.* **27** (4), 815-827, 1967.

73. Peregrine, D.H. Breaking Waves on Beaches, *Ann. Rev. Fluid Mech.* **15,** 149-178, 1983.

74. Peregrine, D.H. Computation of Breaking Waves. Chapter in *Water Wave Kinematics* (ed. A. Torum and O.T. Gudmestad), NATO ASI Series E: Applied Sciences Vol. **178,** 475-490. Klüwer Academic Publishers, 1990.

75. Romate, J.E., The Numerical Simulation of Nonlinear Gravity Waves in Three Dimensions using a Higher Order Panel Method. *Ph.D. Dissertation. Department of Applied Math., University of Twente, The Netherland,* 1989.

76. Romate, J.E. The Numerical Simulation of Nonlinear Gravity Waves. *Engng. Analysis with Boundary Elements* **7** (4), 156-166, 1990.

77. Seo, Seung Nam and Dalrymple, R.A. An Efficient Model for Periodic Overturning Waves. *Engng. Analysis with Boundary Elements* **7** (4), 196-204, 1990.

78. Skjelbreia, J.E. Observations of Breaking Waves on Sloping Bottoms by Use of Laser Doppler Velocimetry, *W.M. Keck Laboratory of Hydraulics and Water Ressources, California Institute of Tech., Report No.* **KH-R-48.** 1987.

79. Skourup, J. Analytical Second-order Wavemaker Theory Verified in a Numerical Wave Flume. In *Proc. Coastal 95 Conf.* (Cancun, Mexico, Sept. 95) (eds. Brebbia, Traversoni & Wrobel), pps. 167-174. Comp. Mech. Pub., Southampton, 1995.

80. Skourup, J., Grilli, S. and Svendsen, I.A. Modelling of Steep and Breaking Waves by the Boundary Element Method. *Inst. Hydrodyn. and Hydraulic Engng., Tech. University of Denmark, Progress Rept. No.* **68,** 59-71, 1988.

81. Skyner, D.J., Gray, C. and Greated, C.A. A Comparison of Time-stepping Numerical Prediction with Whole-field Flow Measurements in Breaking Waves, Chapter in *Water Wave Kinematics* (ed. A. Torum and O.T. Gudmestad), NATO ASI Series E: Applied Sciences Vol. **178,** 491-508, 1990.

82. Sobey, R.J. and Bando, K. Variations on Higher-Order Shoaling, *J. Waterways, Port, Coastal and Ocean Engng.* **117** (4), 348-368, 1991.

83. Sommerfeld, A. *Partial Differential Equations in Physics.* Academic Press, New York, 1949.

84. Stiassnie, M. and Peregrine , D.H. Shoaling of Finite Amplitude Surface Waves on Water of Slowly-varying Depth, *J. Fluid Mech.* **97,** 783-805, 1980.

85. Subramanya, R. and Grilli, S.T. Kinematics and Properties of Fully Nonlinear Waves Shoaling and Breaking over a Gentle Slope. In *Proc. Intl. Symposium on Waves - Physical and Num. Modelling* (Vancouver BC, Canada, Aug. 1994) (eds. M. Isaacson

and M. Quick), Vol. **2**, pps. 1106-1115, IAHR, 1994.

86. Subramanya, R. and Grilli, S.T. Domain Regridding in the Computation of Non-linear Waves. In *Proc. 2nd Intl. Workshop on Bound. Elements in Fluid Mech.* (Southampton, UK, July 1994) (eds. H. Power, C.A. Brebbia and D.B. Ingham), pps. 139-150, Comp. Mech. Pub., Southampton, 1994.

87. Svendsen, I.A. and Grilli, S. Nonlinear Waves on Steep Slopes. *J. Coastal Research* **SI 7**, 185-202, 1990.

88. Svendsen, I.A., Otta, A.K. and Grilli, S. Unsteady Free Surface Waves. In *Proc. I.U.T.A.M. Symp. on Breaking Waves* (Sidney, Australia, July 91) (eds. M.L. Banner and R.H.J. Grimshaw) pp. 229-236, Springer-Verlag, Berlin, 1992.

89. Synolakis, C.E. The Runup of Solitary Waves. *J. Fluid Mech.* **185**, 523-545, 1987.

90. Tanaka, M. The Stability of Solitary Waves. *Phys. Fluids* **29** (3), 650-655, 1986.

91. Tanaka, M., Dold, J.W., Lewy, M. and Peregrine, D.H. Instability and Breaking of a Solitary Wave. *J. Fluid Mech.* **185**, 235-248, 1987.

92. Vinje, T. and Brevig, P. Numerical Simulation of Breaking Waves. *Adv. Water Ressources* **4**, 77-82, 1981.

93. Wei, J., Kirby, J.T, Grilli, S.T. and R., Subramanya, A Fully Nonlinear Boussinesq Model for Surface Waves. I. Highly Nonlinear Unsteady Waves. *J. Fluid Mech.*, **294**, 71-92, 1995.

94. Xü, H. and D.K.P., Yue, Numerical Study of Three Dimensional Overturning Waves. In *Proc. 7th Intl. Workshop on Water Waves and Floating Bodies.* (Val de Reuil, France, May 1992)(ed. R. Cointe), Fluid Flow and Computational Aspects, pp. 303-307, 1992.

95. Yeung, R.W. Numerical Methods in Free Surface Flows. *Ann. Rev. Fluid Mech.* **14**, 395-442, 1982.

96. Zaroodny, S. J. and Greenberg, M. D. On a vortex sheet approach to the numerical calculation of water waves, *J. Comp. Phys.* **11**, 440-446, 1973.

97. Zelt, J.A. and Raichlen, F. A Lagrangian Model for Wave Induced Harbour Oscillations, *J. Fluid Mech.* **213**, 203-228, 1990.

Modeling Tsunamis with Marker and Cell Methods

Peter E. Raad

FUNDAMENTALS

The numerical solution of differential equations from an Eulerian perspective requires the approximation of derivatives of a given function by the values of that function at discrete points of interest within a chosen domain. This domain can extend in one or more spatial dimensions, and in transient problems, involves time as well. Hence, one always begins by dividing up the computational domain into discrete elements which are joined by grid (or collocation) points. Then, the differential equation which describes the continuous behavior of some dependent variable f as a function of at least one independent variable x is transformed into an algebraic statement (i.e., a system of one or more algebraic equations) relating the values that f takes at chosen points of x. All numerical methods then become means of representing complex variations in the independent variable in terms of a limited sample of function evaluations at a finite number of points. And so methods essentially differ in the manner in which they represent the variations of a function, and are normally judged by their fidelity in reproducing the underlying physics which one set out to investigate to begin with.

Spatial Discretization

In some methods, "points" are intersections of grid lines, and in others, they are just small regions or cells. In some methods, the connection between adjacent divisions is structured, and a simple Cartesian indexing scheme is sufficient to identify the spatial position of a grid point. In other methods, the connection is unstructured, and the problem of identifying the grid connectivity is left to the analyst.

Numerical Methods

The majority of numerical investigatoins in fluid dynamics have been done with finite difference and finite volume methods, but lately, finite element and spectral collocation methods have seen an increase in popularity. While most phenomena in fluid dynamics are more amenable to an Eulerian approach, a moslty Lagrangian perspective can be more beneficial in some cases, and this idea has lead to the development of specialized methods such as the vortex method.

What is central to our concern at this point is the realization that all methods are inherently good, with each possessing certain intrinsic strengths and weaknesses. While at the onset several approaches may appear equally suitable, decisions made up front normally have serious implications on the cost and difficulty of obtaining a desired solution. Ultimately then, the issue is not which method gives the right answer, because they all *eventually do*[†], but rather, which method is the most effective for a particular problem of interest. While this should be an obvious point to everyone, familiarity often grows into blind allegiance, and soon people are caught tightning screws with hammers! Consequently, and before proceeding further, it would be worth our while to examine the essentials of the numerical simulation of water waves with the objective of identifying the building blocks necessary in designing a successful approach. Our aim should be to identify the flow characteristics and to use our experience with various numerical methods to

---

† Thanks to Joel Ferziger for illuminating this truth for me.

identify a "best" candidate to model each of those characteristics. And in doing so, we shall then answer the following questions: What are the characteristics of a flow with surface waves? And, why is the flow such a challenge to model?

Guiding Flow Characteristics

The flows of interest here involve an incompressible fluid which undergoes large, transient, nonhomogeneous, nonisotropic motions within a three-dimensional spatial domain. In general, the motion may not possess or exhibit any obvious symmetries.

- Naturally, we can choose to use a Lagrangian or an Eulerian approach. But, since we seek to determine how the flow evolves over time, we need to compute the velocity and pressure fields over an entire spatial domain. Hence, it appears that an Eulerian approach is mandatory to keep the problem tractable.

- We could choose to solve directly for the "primitive" variables or indirectly by first solving for the vorticity and the stream function. But, since we ultimately seek a three-dimensional solution, we must turn down the vortivity--stream function approach in favor of a primitive variable approach.

- We are dealing with the flow of an incompressible fluid whose governing Partial Differential Equations (PDEs) lack a temporal derivative of the pressure. In order to advance the pressure field in time, we could choose to introduce an artificial compressibility or to use a two-step (or projection) scheme. Since the use of artificial compressibility involves the choice of an arbitrary constant, and the value of that constant is problem dependent, we choose to use a two-step scheme‡ .

- The temporal dynamics of the flow does not favor the use of the more expensive implicit time discretization since one must use small time steps anyway to track the tumultuous motion of the free surface with fidelity. Hence, we choose the simple Euler explicit scheme, and accept its first-order accuracy. Of course, with this choice one inherits a stability condition which must be obeyed, but it turns out, as we will see later, that limitations placed on the maximum time step by other factors in the simulation are more severe.

Now, we must decide on the spatial discretization, namely, should we use finite differences, finite volume, finite elements, or spectral methods? As we agreed previously, choices determine overall cost and it behooves us to make the *right* choice, but it is better to defer answering this question until a choice is made as to how to handle the free surface itself. Through that discussion and the process of elimination, an optimum choice will become more apparent.

Defining and Tracking the Free Surface

Computing in fluid dynamics is tedious and expensive enough even when the domain in which the computations are to be carried out is fixed. Having the domain change throughtout a solution complicates matters very significantly. Several types of problems that involve a changing domain exist. On one end of the spectrum, there are problems with a moving boundary whose motion is either prescribed or constrained in some fashion. For example, a piston moving back and forth in a cylinder represents a complication, but the path of motion is known *a priori* and its extent is either

---

‡ This particular preference in terminology may be directly attributable to the author's home state!

prescribed or can be calculated from the laws of rigid body dynamics. Other examples of moving boundary problems include a paddle (or wave maker) generating waves in a channel according to a prescribed motion, and a read/write head flying over a spinning magnetic disk and reacting to the fluid dynamics of the air bearing that separates the two surfaces of the head and disk.

On the other end of the spectrum is the flow of a fluid whose boundary is undergoing significant deformations that in general are unsteady and nonuniform. These types of problems include the case where only one fluid is of interest and the presence of the other fluid is neglected, as well as the case where an interface exists between two (or more) immiscible fluids. In either case, solving the problem must begin with determining *where* to solve it, i.e., where to apply the boundary conditions. At this point, the road bifurcates and a decision must be made between essentially two approaches. One approach is to force a part of the domain boundary to follow the contour of the free surface (or interface), and the other approach is to introduce a mechanism that permits the tracking of the free surface within a larger domain whose boundaries remain fixed.

The first approach involves a mapping procedure; which can take on the form of a separate first step, as in FD/FV, or as an integral part of the methodology, as in FE. Naturally, mapping the physical domain into a computational domain also requires mapping the governing equations and boundary conditions, as well as calculating several "metrics" (including a Jacobian) which essentially carry the information that links the two domains together. As the physical domain deforms, changes may need to be made frequently to ensure that an accurate relationship is maintained between the physical and computational domains. If grid mapping is used, the metrics need to be updated. If unstructured gridding is used, a new grid is required to avoid problems with grid skewness. In either case, the need to maintain accuracy requires a frequent reassessment of how well the chosen portion of the domain boundary describes the contour of the free surface as the latter changes. The extent of the added complexity is a function of the severity of the surface deformations. It is worth noting at this point that the mapping approach and its associated difficulties are not unique to incompressible flow problems, but are faced as well in gas dynamics problems that involve shock capturing. There, however, the interface is a region of high gradients (or shock) which is located *inside* the domain and toward which grid points must be attracted in order to maintain an appropriate grid resolution.

An important property of the mapping approach is that it maintains a fully Eulerian representation of the flow problem at hand. The biggest advantage of this approach is that it allows the implementation of the free surface or interface boundary conditions *right on* the interface without ambiguity or compromise. By the same token, however, handling multiphase flows with this approach involves solving distinct problems in each single-phase domain and interfacing shared information between the different subdomains. Accurate communication of information between adjacent subdomains across their common interface(s) is not a trivial matter and can present serious accuracy and bookkeeping difficulties.

The second approach involves the introduction of a complementary implement to track the free surface as it deforms within a larger computational domain whose grid is fixed. The implement may be Eulerian or Lagrangian. An Eulerian implement would employ a special function which is advected with the flow and whose purpose is to allow the determination of where the interface lies at each point in time. Furthermore, the value of the function at each spatial position of interest determines the appropriate fluid properties and equations to use. Since cells through which the interface passes contain by definition more than one fluid (even though the effects of one of which may be neglected), fluid properties must be adjusted to reflect the "contents" of each of those cells. If mere averaging is used, the interface diffuses ouwardly and is soon lost, violating the immiscible nature of the fluids present. Thus, special care must be taken to ensure that the "sharpness" of the interface is preserved allthewhile minimizing the numerical instabilities normally associated with computing across discontinuities. The main advantage of the use of an advected function is the ease with which it can be calculated just as any other conservable property one wishes to follow through the flow. The main disadvantages are the need to essentially destroy the interface and reconstruct it at each time step, and the serious difficulty in calculating information that derives from the shape of the interface (e.g., radius of curvature). The first disadvantage often leads in practice to unexpected changes in the shape of the interface, or the violation of the conservation of mass and/or conservation of the advected function itself. The second disadvantage can severely limit the applicability and usefulness of a method when stresses are produced at the interface.

A Lagrangian implement to track the free surface would employ massless markers that advect with the flow and reflect the presence or absence of the fluid with which they move. Instead of placing the markers throughout the fluid of interest, one may choose to place them only on the free surface (or interface), and in that case, they might earn the more specialized name of surface markers. In either case, the Lagrangian markers interrogate the underlying Eulerian grid for their velocities and move over it accordingly. The markers allow us to infer the character of the underlying Eulerian cells and determine which fluid properties and equations should be used. The character (or type) of a cell indicates whether a cell is empty or contains one or more fluids. Since the correct determination of the cell type depends on the markers, these must be managed to ensure that they preserve the information they carry. When markers are "sprinkled" throughout one fluid, their motion may force them to become too sparse in some regions, potentially resulting in an inadvertent and erroneous change in the character (i.e., flag) of an underlying cell. Ensuring that markers do not become too sparse would require that markers be initially distributed in sufficient (and perhaps nonuniform) densities across the computational domain. However, it is unrealistic to expect such *a priori* knowledge.

Boundary Conditions

The Navier-Stokes and Continuity equations are uniformly applicable to a wide range of physical problems, and so it is not surprising that boundary conditions are thought of as the distinguishing element between different problems and their application is given great importance. In the solution of Elliptic or Parabolic equations, the values of a field variable within the computational domain depend heavily on the values of that variable on the entire boundary of the domain. For Hyperbolic equations, only upstream boundary conditions determine the variable field within the domain, but inaccuracies in the numerical application of a downstream boundary condition can generate numerical errors that reflect back into the interior of the domain and corrupt the solution everywhere. Thus, great attention must be paid to the selection of boundary conditions and their accurate application. Indeed, for flow problems with a free surface, the point of application of free surface boundary conditions in addition to the above factors essentially determine the capability, usefulness, and range of applicability of a given method.

The governing differential equations involve spatial derivatives of the fluid velocity and pressure. Hence, in each spatial direction, boundary conditions are required for the pressure and each of the velocity components. Velocity boundary conditions are relatively straight forward to specify from the physics. Along solid boundaries, slip or no-slip velocity boundary conditions may be used for the tangential component while an impermeability boundary condition may be used for the normal component. A velocity profile may be specified at an inlet, while at an outlet, a (nonreflective) continuity condition may be applied. At the free surface, the velocity boundary conditions may be specified by various combinations of the continuity equation and the normal and tangential stress conditions. Pressure conditions are slightly more challenging. Normally, the fluid pressure on solid boundaries is not ascertainable from the physics, and so it is usually preferable to avoid having to prescribe a pressure boundary condition on a solid boundary. As a result, the use of a "staggered" grid has been the preferred. In a staggered grid, as shown in Fig. 1, the pressure is located at the center of a cell, while the velocity components are defined on cell faces. This configuration is especially suitable for a finite volume discretization since it allows for a natural numerical representation of the continuity equation on a given cell. In addition to avoiding the need to define a pressure on a solid boundary, the use of a staggered grid approach eliminates the potential of oscillatory solutions by allowing the coupling of the velocity and pressure solutions in adjacent cells. On a non-staggered grid, by discretizing the pressure Poisson equation with central difference approximations, it is possible to obtain two uncoupled networks of pressure points whose solutions differ by a non-zero constant (Peyret and Taylor, 1983). A surmountable disadvantage of the use of a staggered grid is that only one velocity component is defined on each boundary. But, if a locally one-dimensional (or factored) approach is used for the temporal discretization, special care must be taken to ensure the accuracy of the boundary conditions in a staggered grid approach (Kim and Moin, 1985).

We now see that the simulation of incompessible flow problems with a free surface is especially challenging. The most significant of the reasons behind the difficulties can be summarized as follows:

1. The boundary conditions on the free surface can be nonlinear and involve the pressure which must then be calculated accurately to avoid spurious oscillations.

2. The Navier-Stokes equations are nonlinear and must be solved along with the continuity equation without the benefit of a temporal derivative for the pressure.

3. The domain changes shape and size continually, may develop complicated surface connections, and even break up into disconnected bodies of fluid or form crossing interfaces.

4. The interface must be tracked and described accurately to enable the calculation of derived quantities such as the surface tension force which depends on the radius of curvature.

As stated by Hirt and Nichols (1981), "three types of problems arise in the numerical treatment of free boundaries: (1) their discrete representation, (2) their evolution in time, and (3) the manner in which boundary conditions are imposed on them." Thus far, we have examined the fundamentals behind these problems. Next, we will proceed to review the origins of the marker and cell method and its developments to the present day. In the process, we will make reference to the fundamental issues we have covered above in order to classify and assess the merits of the various approaches. To conclude, we will introduce our own method which uses markers and cells, present its methodology, and demonstrate its capabilities.

HISTORICAL PERSPECTIVE

The concept of using markers and cells to address the challenges presented by incompressible fluid flow problems with a free surface was first introduced by Harlow and Welch (1965) and Welch et al. (1966). They introduced massless markers that move with the fluid and a novel finite difference solution algorithm for the velocity field. The massless markers are initially distributed throughout the fluid with a prescribed cell density. As markers move in and out of cells, the flags of those cells change accordingly. Cells which are void of markers are considered *empty* while cells containing markers are designated as either *full* or *surface*, depending on the status of the neighboring cells. If a cell containing markers is adjacent to at least one empty cell, then it must be a surface cell. If, on the other hand, it is surrounded by surface or full cells, then it is flagged as full. Implicit in this logic is the requirement that for a full cell to become empty, it must first become a surface cell for at least one time step. Conversely, empty cells must pass through the surface stage on the road to being full. We can now explain a point which we previously identified as a disadvantage of distributing markers over the entire fluid. The sparsity of markers can induce an erroneous and inadvertent change in the flag of a cell from full to empty in one time step, which results in an irreversible logical error. This issue points to the difficulty in using a single implement both to define the location of the free surface and to track its movement. The free surface pressure boundary conditions are applied at the centers of the surface cells. For any surface cell, the cell center may be very close to the free surface represented by the outermost markers, or it may be as much as one-half of the cell width on either side of the free surface. Consequently, some of the free surface pressure boundary conditions are applied at points very near the free surface, and some are not.

Since 1965, a number of authors have presented extensions and modifications of the original marker and cell method. Some of the investigations addressed the numerical methodology used to solve the governing equations, others the scheme by which the free surface is located, and still others the velocity or pressure boundary conditions.

In 1967, Chorin presented a method for calculating steady-state solutions of the equations of motion of an incompressible fluid. He introduced the use of an artificial density and sought a steady-state solution for which the rates of change of this artificial density and of the velocities all vanished. In doing so, Chorin thereby avoided the necessity of solving the pressure Poisson equation presented in the MAC method. Viecelli (1969) adapted Chorin's method of solving simultaneously for the pressure and velocity fields into MAC's scheme of tracking the free surface of a fluid with markers. In his paper, Viecelli also presented a technique for dealing with arbitrarily shaped external boundaries. The Simplified Marker and Cell (SMAC) method, presented in 1970 by Amsden and Harlow, was a significant revision of the MAC method and has been widely referenced. In SMAC, Amsden and Harlow present an alternate pressure potential Poisson equation and a scheme for applying the normal stress condition on the free surface of the fluid. By avoiding the need in the MAC method for second-order derivatives of the velocity on the boundary, they significantly simplified the calculation of the pressure potential field that is used to compute the velocity field. Viecelli (1971) showed the equivalence between his approach to simultaneously advance the pressure and velocity fields and the two-stage approach used in the marker and cell method. Hirt and Cook (1972) noted Viecelli's conclusion that the two-stage approach of the marker and cell method and the simultaneous approach to the solution of the pressure and velocity fields are equivalent. In their paper, Hirt and Cook opted to use the simultaneous procedure for the reason that its use simplifies the application of boundary conditions. Nakayama and Romero (1971) extended the application of the SMAC method to fluid flows that are almost three dimensional. In 1975, Hirt et al. used the simultaneous approach in an extension of the original MAC method (which they called SOLA) that made use of upwinding in order to control the temporal stability of the solution. Three decades after its inception, extensions of the original MAC method continue to appear in the literature. In 1994, Tome and McKee presented a method which they named GENSMAC that extends the applicability of SMAC to general curved domains. Foster and Metaxas (1995) presented a software package based on the MAC method which they developed for the "modelling and animation of viscous incompressible fluids" in three-dimensions.

Important extensions to the MAC method have been advanced by researchers at the University of Tokyo in the area of water wave simulation. In 1985, Miyata and Nishimura presented a MAC-based method for simulating nonlinear ship waves, followed in 1986 by the publication of the TUMMAC (Tokyo University Modified MAC) method by Miyata for the simulation of breaking waves. Additional investigations by Miyata and his co-workers addressed ship waves in the vicinity of a hull (Miyata et al., 1985, 1987, 1990). Recently, Miyata et al. (1992) used a curvilinear coordinate system instead of the regular fixed grid system to improve accuracy near the hull and to enable them to reach higher Reynolds number solutions. However as previously discussed, the curvilinear mapping approach is limiting when one is interested in simulating a wave past the breaking stage.

The clear benefit of the marker and cell method rested from the onset in its elegant ability to track the advance of a free surface even when the latter undergoes large contortions. Unfortunately however, the Eulerian marker implement does not in itself provide for a precise implementation of free surface boundary conditions, and this was recognized early on. Hirt and Shannon (1968) and Nichols and Hirt (1971), for example, addressed the free surface boundary conditions and presented improvements. However, at the heart of the matter was the seemingly unavoidable fact at the time that the free surface pressure boundary condition was applied at the center of a cell rather than right on the free surface itself. Similar discrepancies existed in locating the point of application of the boundary conditions for the velocity. These shortcomings became especially troublesome when the method was applied to the simulation of surface waves in shallow water, where it lead to surface oscillations and less than desirable temporal and spatial accuracy. Several authors introduced modifications which improved the treatment of the free surface, but at the expense of limiting the applicability of the resulting methods to single-valued free surfaces, which in the case of waves would prohibit breaking. Chan and Street (1970) and Chan et al. (1970) published the SUMMAC method for plane wave problems, which improved the application of the pressure boundary condition and introduced the extrapolation of pressure and velocity components from the fluid interior to obtain improved pressure and velocity boundary conditions. Hirt et al. (1970) explored the use of a Lagrangian technique for flows that do not undergo large distortions, while Amsden (1973) modified the SMAC method to incorporate the use of surface markers and line segments for single valued free surfaces. Another approach to the simulation of single valued free surface fluid flow problems, known as the height function method, was used by many researchers, including Chan et al. (1970), Hirt et al. (1975), Nichols and Hirt (1976), Bulgarelli et al. (1984), Loh and Rasmussen (1987), and Lardner and Song (1992).

Like the marker and cell method, the height function is one of several procedures developed for tracking the free surface and enabling the application of free surface boundary conditions. The line segment (Chan and Street, 1970; Nichols and Hirt, 1971; Amsden, 1973), volume of fluid (Nichols and Hirt, 1975; Hirt and Nichols, 1981; Nichols et al., 1980; Torrey et al., 1985; Ashgriz and Poo, 1991), and surface marker (Chen et al., 1991) methods are different solution procedures developed expressly for the representation of the free surface, but do not limit it to being single-valued. We shall now briefly summarize the basis of each of these procedures.

The Height Function

In the height function method, the location of one point of the free surface in each surface cell is specified by a height function. The height function specifies the height above a reference line of the point on the free surface that lies on the vertical centerline of the surface cell. For a computational domain represented by a rectangular mesh of cells of width dx and height dy, the height function can be represented as H(x,t). The time evolution of the free surface is determined by solving the kinematic equation

$$\frac{\partial H}{\partial t} + u \frac{\partial H}{\partial x} = v$$

for the height function H, where u and v represent the x and y components of velocity. The free surface pressure boundary condition in each surface cell is applied at the point specified by the height function. This approach for representing and advancing the free surface is extremely efficient. The discrete representation of the free surface requires only one value per surface cell, and the equation that governs the time evolution of the free surface is simple. The height function method has also been applied to three-dimensional free surface fluid flow problems. The major limitation of the method is that it is applicable only to single-valued free surfaces. The methods described below all can be applied to multi-valued free surfaces.

The Line Segment Method

Chan and Street (1970) developed the line segment method to improve the accuracy of the application of the free surface pressure boundary conditions in the marker and cell method. The line segment method differs from the marker and cell method in that markers distributed along the free surface are employed in addition to the markers that are distributed throughout the fluid. The surface markers are advanced at the same time and in the same manner as the other markers. The free surface is represented by line segments connecting the surface markers. Chan and Street developed a finite difference form for the pressure Poisson equation that employs an irregular star technique and allows the application of the free surface pressure boundary conditions at points nearer the free surface rather than simply at the centers of the surface cells. In order to determine the leg lengths for the irregular star, the intersections of line segments connecting markers on the free surface and lines connecting cell centers must be determined. Nichols and Hirt (1971) also adopted the use of line segments and surface markers to determine the location of a point on the free surface in each surface cell. As an alternative to the irregular star approach of Chan and Street, Nichols and Hirt developed an interpolation/extrapolation technique to determine the pressure boundary condition at the center of the surface cell based on the pressures at the center of a neighboring full cell and at the point chosen to represent the free surface in the surface cell. Extrapolation is used if the center of a surface cell lies outside the region occupied by markers; otherwise, interpolation is used. In some situations, there is no intersection inside a surface cell of a line connecting cell centers and a line segment. Hence, the free surface pressure boundary condition cannot be satisfied at a point that is on one of the line segments representing the free surface. Hirt and Nichols (1981) pointed out that the use of line segments in connection with problems in which free surfaces "fold over" or collide leads to difficult problems that apparently have not been resolved.

The Volume of Fluid Method

Rather than relying on either height functions or markers to define the fluid filled region, the volume of fluid (VOF) method utilizes a function F that represents the fractional volume of fluid in a cell. A unit value of F indicates that a cell is full of fluid, while a zero value indicates that a cell is empty. If the value of F is between zero and one, the cell contains a free surface. The shape of the free surface within a cell is approximated by a horizontal or vertical straight line. The orientation and position of the line depends on the value of F in the cell and on the local gradients of F. The free surface pressure boundary condition is applied in each surface cell at the intersection of the straight line representing the free surface and a line connecting cell centers. The time evolution of F is governed by the equation

$$\frac{\partial F}{\partial t} + u\frac{\partial F}{\partial x} + v\frac{\partial F}{\partial y} = 0$$

The VOF method requires less computer storage and computing time for advancing the free surface and reflagging cells than the MAC method, since the function F is used both to track the movement of the free surface and to flag the cells. The VOF method is powerful and computationally efficient.

Recently, Ashgriz and Poo (1991) referred to two major problems associated with the use of a volume of fluid function to represent the free surface. One is how to identify the shape and location of the free surface within the surface cells. The other is how to advance the volume of fluid field. Ashgriz and Poo developed the use of sloped line segments fitted at the boundaries between surface cells to describe the free surface and the use of the trapezoidal areas associated with these sloped line segments to advance the volume of fluid field. For the movement of a circular region of fluid across a computational domain at a constant velocity, they compare their flux line-segment model for advection and interface reconstruction (FLAIR) with the donor-acceptor technique used in the VOF method and find that FLAIR is more accurate for advancing the volume of fluid field. They note that the error in the area of the circle after twenty computational steps is less with FLAIR than with VOF. Ashgriz and Poo do not describe how or at what point the free surface pressure boundary condition is applied in a surface cell.

The Surface Marker Method

The surface marker (SM) method (Chen et al., 1991) is a modification of the original marker and cell method that employs only one string of surface markers for defining the extent of the fluid filled region and for locating and advancing the free surface. Recently, Johnson et al. (1994) used the SM method as the basis of a method to treat impact between a fluid front and a solid obstacle. In the SM method, the surface markers are consecutively numbered and initially are equally spaced along the free surface of the fluid. If a surface cell is adjacent to a boundary cell, a marker is also placed in the boundary cell. This added marker is the image of the marker that is on the free surface and closest to the obstacle. The distance between adjacent markers is managed. If, after marker movement, the distance between any two consecutively numbered markers is greater than an assigned value, a marker is added halfway between the markers in question prior to the start of the next computational cycle. Only surface cells and their neighbors are involved in the cell reflagging phase. The surface pressure boundary conditions are applied at the centers of the surface cells just as they are in the original marker and cell method. Consequently, the SM method provides no improvement of the treatment of the free surface pressure boundary conditions. Since only one string of markers is employed and no more than three rows of cells along the free surface are reflagged, the computer storage and computing time requirements of the SM method are, however, much less than those of the marker and cell method. In addition, the difficulty in using the marker and cell method recently noted by Ashgriz and Poo (1991) that is related to the possible occurrence of regions with high or low marker densities is eliminated. In the SM method, markers are not used at all in full cells; and, since a surface marker is added when the distance between neighboring markers exceeds a predefined limit, the surface markers can never become too sparse to properly describe the free surface, as they can in the marker and cell method. During a solution, markers that were on the surface can become engulfed in the fluid. As was shown in Chen et al. (1991), engulfed markers can be useful, for example, in delineating recirculation regions and visualizing the history of mixing.

THE SURFACE MARKER AND MICRO CELL METHOD

In all of the finite difference methods mentioned above, the same cells are employed to define *both* the locations of the discrete field variables and the location of the free surface. The idea that lead to the development of the surface marker and micro cell (SMMC) method is that it is not necessary that the cells employed to locate the discrete field variables be identical to the cells employed to locate the free surface. In the new method, two types of cells are used. In addition to the standard finite difference/volume cells, hereafter referred to as macro cells, new cells termed micro cells also are introduced. In the SMMC method, surface markers are used to advance the free surface; and micro cells, which are smaller than the macro cells, are used to locate the free surface, making it possible to apply the free surface pressure boundary conditions more efficiently and accurately.

The Computational Cycle

The field variables are advanced from an arbitrary time t_0 to the next time $t_0 + \delta t$ by the following steps.

1.   Reflag the micro and macro cells.

2.   Assign tentative velocity boundary conditions.

3.   Compute a tentative velocity field, $\tilde{u}$ and $\tilde{v}$, at $t_0 + \delta t$ from the velocity field at t_0 by use of

$$\frac{\partial \tilde{u}}{\partial t} = -\frac{\partial u^2}{\partial x} - \frac{\partial uv}{\partial y} + v\left(\frac{\partial^2 u}{\partial x^2} + \frac{\partial^2 u}{\partial y^2}\right) + g_x \tag{1}$$

$$\frac{\partial \tilde{v}}{\partial t} = -\frac{\partial v^2}{\partial y} - \frac{\partial uv}{\partial x} + v\left(\frac{\partial^2 v}{\partial x^2} + \frac{\partial^2 v}{\partial y^2}\right) + g_y, \tag{2}$$

where (u,v) are the (x,y) velocity components; (g_x, g_y) are the (x,y) gravity components, respectively; and v is the kinematic viscosity of the fluid.

4.   Calculate the incompressibility deviation function D defined by

$$D = \partial \tilde{u}/\partial x + \partial \tilde{v}/\partial y \ . \tag{3}$$

5.   Solve for the pressure potential field $\psi = \delta t \, p / \rho$, where p is pressure and ρ is the fluid density, by use of a new equation developed below.

6.   Compute the final velocity components at $t_0 + \delta t$ by use of new equations introducedbelow.

7.   Assign the final velocity boundary conditions.

8.   Advect the free surface and manage surface markers.

9.   Assign new fluid cell velocities.

The general outline of this computational cycle is similar to those presented in Amsden and Harlow (1970), Chen et al. (1991, 1995), and Johnson et al. (1994). However, Steps 1 and 5-9 are either new or significantly altered.

As far as vorticity is concerned, the tentative velocity field computed by use of Eqs. (1) and (2) is correct. In addition, since vorticity is not a function of pressure, the vorticity associated with the final velocity field at $t_0 + \delta t$ will be the same as the vorticity associated with the tentative velocity field. The tentative velocity field does not, however, necessarily satisfy the continuity equation. A measure of the failure of the tentative velocity field to satisfy the continuity equation is provided by the incompressibility deviation function D defined by Eq. (3). In order to obtain a final velocity field that does satisfy the continuity equation, a pressure potential field, ψ, must be solved for and used to adjust the tentative velocity field.

The New Mesh of Macro and Micro Cells

The computational domain is comprised of macro computational and boundary cells. A boundary cell defines the location of a rigid, stationary obstacle, while a computational cell that contains any amount of fluid is called a fluid cell. A fluid cell may be either a surface or full cell. If it contains surface markers and has at least one empty macro cell neighbor, it is flagged as surface, while if it has no empty macro cell neighbors, it is flagged as full. Near the free surface, it is possible for a macro cell that contains surface markers to be flagged as full. Otherwise, full macro cells do not contain markers. The interfaces of the macro cells are called macro grid lines. In general, the fluid free surface passes through the macro surface cells. The staggered grid concept is used to locate the discrete field variables and functions of those variables. The pressure potential $\psi_{i,j}$ and the incompressibility deviation function $D_{i,j}$ are located at the center of a computational macro cell; the discrete velocities $u_{i,j}$ and $u_{i-1,j}$ are located at the middle points of the right and left faces of the cell, respectively; and the discrete velocities $v_{i,j}$ and $v_{i,j-1}$ are located at the middle points of the top and bottom faces, respectively.

In the new method, the micro cells are created by subdividing each macro cell into N by N micro cells, where N is an odd integer chosen by the analyst. The value of N is chosen as an odd integer so that the center of the macro cell, where $\psi_{i,j}$ and $D_{i,j}$ are defined, coincides with the center of a micro cell. However, at any given time during a simulation, only micro cells near the free surface are involved in the computations. Micro cells, like macro cells, require flags. The definitions of micro boundary, empty, surface, and full cells are similar to those of corresponding macro cells. For example, a micro fluid cell is flagged as surface or full depending on its condition: a surface micro cell is a micro fluid cell that contains surface markers and has at least one empty micro cell neighbor, while a full micro cell is defined as a micro fluid cell that has no empty micro cell neighbors.

As a typical example, nine macro cells near the free surface are shown in Fig. 2, four of which have been subdivided into micro cells. The macro and micro grid lines are represented by thick and thin lines, respectively. Of the four macro cells that contain micro cells, the one in the center of Fig. 2 is a full macro cell, and the other three are surface macro cells. In these four macro cells, the surface and full micro cells are designated as s and f, respectively, while the empty micro cells are blank. As shown in Fig. 2, one macro cell can contain three different kinds of micro cells at the same instant. It is not necessary, however, for micro cells to be used in each macro cell. For example, no micro cells are used with five of the macro cells shown in Fig. 2. Each of these five macro cells is simply designated with either an E for empty or an F for full. By selectively superimposing micro cells on the macro cells, a better definition of the location of the free surface is possible than can be achieved by the use of macro cells alone.

The Reflagging Algorithm

Since surface markers are used to track the movement of the free surface, the reflagging algorithm introduced by Chen et al. (1991) is extended to accommodate the reflagging of micro cells. At a given instant, each macro and micro cell possesses a *full*, *surface*, or *empty* flag. During each computational cycle, the reflagging of micro cells is done prior to the reflagging of macro cells and is similar to it. Before moving the surface markers, all micro cells that contain markers are assigned a temporary *check* flag. After all the markers have been moved and markers have been added as required, the micro cells that now contain markers are flagged as surface. Next, the flag of any remaining check micro cell is changed to empty if the micro cell in question has an empty micro cell neighbor; otherwise, it is changed to full. Finally, the flag of any surface micro cell that now does not have at least one empty micro cell neighbor is changed to full. It must be noted that this micro cell reflagging procedure requires that no marker be allowed to travel in any direction in one time step a distance larger than the dimension of a micro cell in that direction. If nine micro cells per macro cell are used, this consideration imposes no additional restriction on the size of the computational time step, because the use of small time steps is necessary to insure the accuracy of the explicit schemes that are employed to solve the governing fluid flow equations. Hirt and Nichols (1981) state that: "Typically, δt is chosen equal to one-fourth to one-third of the minimum cell transit time." If twenty-five micro cells per macro cell are used, δt must be less than one-fifth of the minimum cell transit time, which is slightly less than the smaller value they suggested.

Free Surface Pressure Boundary Conditions

Boundary conditions are required for the solution of the pressure potential field, ψ. Ideally, the boundary conditions along the free surface would be applied at points that are directly on the free surface. In the SMMC method, the free surface pressure boundary conditions are applied at the centers of surface micro cells. The result is that the free surface pressure boundary conditions are applied very accurately, with the important added advantage that the locations of the points of application are determined very efficiently. For example, for the situation shown in Fig. 3, with nine micro cells per macro cell, point m is the center of macro cell (i,j); and points e, n, w, and s are neighboring points to the east, north, west, and south of point m, respectively. The locations of points e, n, w, and s and their respective distances δe, δn, δw, and δs from point m all are required for the computation of the pressure potential at point m. For the example of Fig. 3, points e and s are simply the centers of the full macro cells to the east and south of cell (i,j). Consequently, $\delta e = \delta x$ and $\delta s = \delta y$. On the other hand, points n and w represent points on the free surface and are the centers of the first surface micro cells to the north and west of the center of cell (i,j), respectively. For this particular example, the distances δn and δw are equal to $2\delta y/3$ and $2\delta x/3$, respectively. Points n and w are very near the free surface, and the determination of the lengths δn and δw is simple and efficient. If more than nine micro cells per macro cell were used, the corresponding points of application of the free surface pressure boundary conditions would, in general, be even closer to the free surface, and the determination of the required distances between the free surface and the center of cell (i,j) would still be just as efficient. The determination of "unequal leg lengths" such as δn and δw is required only for some of the macro fluid cells that are near the free surface.

New Pressure Potential Equation

The SMMC method introduces a new equation governing the pressure potential field. Since the objective of the use of the pressure potential field is to provide a final velocity field that satisfies continuity, the new equation is derived by substituting the final velocity equations directly into the continuity equation.

The general expression for the unequal-legs pressure equation used in the SMMC method is derived by starting with the usual final velocity equations that make use of the pressure potential field ψ:

$$u = \bar{u} - \partial\psi/\partial x, \qquad v = \tilde{v} - \partial\psi/\partial y. \qquad (4),(5)$$

Writing Eqs. (4-5) in difference form for cell (i,j) yields:

$$u_{i,j} = \tilde{u}_{i,j} - (\psi_e - \psi_{i,j})/\delta e_{i,j}, \qquad (6)$$

$$v_{i,j} = \tilde{v}_{i,j} - (\psi_n - \psi_{i,j})/\delta n_{i,j}, \qquad (7)$$

$$u_{i-1,j} = \tilde{u}_{i-1,j} - (\psi_{i,j} - \psi_w)/\delta w_{i,j}, \qquad (8)$$

and
$$v_{i,j-1} = \tilde{v}_{i,j-1} - (\psi_{i,j} - \psi_s) / \delta s_{i,j}, \tag{9}$$

where ψ_e, ψ_n, ψ_w, and ψ_s represent the pressure potentials at points e, n, w, and s, respectively. Next, these four equations (6)-(9) are substituted into the continuity equation

$$(u_{i,j} - u_{i-1,j})/\delta x + (v_{i,j} - v_{i,j-1})/\delta y = 0 \tag{10}$$

for the control volume that coincides with cell (i,j). The resulting equation can be written as

$$C\psi - \frac{\psi_e}{\delta x \, \delta e} - \frac{\psi_n}{\delta y \, \delta n} - \frac{\psi_w}{\delta x \, \delta w} - \frac{\psi_s}{\delta y \, \delta s} = -D \tag{11}$$

where $C = (\delta e^{-1} + \delta w^{-1}) / \delta x + (\delta n^{-1} + \delta s^{-1}) / \delta y$.

In the SMMC method, Eq. (11) is used for cells near the free surface for the determination of the pressure potential field in Step 5 of the computational cycle. It should be noted that Eq. (11) reduces to the standard finite difference form of the pressure potential Poisson equation for a full macro cell that is not near the free surface.

Calculation of the Pressure Potential Field

Consider the sketch shown in Fig. 4 of fluid sloshing in a container. The computational region consists of twenty five macro cells. The container consists of the square region of nine macro cells in the center of Fig. 4. A single layer of boundary cells completely surrounds the container. The free surface at the instant under consideration is indicated by a dashed line. Micro cells, nine per cell, are shown in five of the macro cells that the free surface passes through.

In general, the pressure potential must be calculated in all full macro cells as well as in any other macro cell whose center micro cell is full. Therefore, in the example of Fig. 4, $\psi_{i,j}$ must be calculated in all six of the macro cells in the bottom two rows of the cavity. The application of the new pressure potential equation, Eq. (11), requires the determination of the leg lengths. A leg length in the direction of a free surface is determined by counting the number of full micro cells between the center of the cell under consideration and the free surface. For the example under consideration, the unequal leg lengths are

$$\delta e_{2,3} = \delta x/3, \delta n_{2,3} = \delta y/3, \delta w_{3,3} = 2\delta x/3,$$

$$\delta n_{3,3} = 2\delta y/3, \delta e_{4,3} = \delta x/3, \delta n_{4,3} = \delta y/3.$$

All other required leg lengths are equal to either δx or δy. Using the preceding leg lengths in Eq. (11) and forcing the gradient of the pressure normal to a solid boundary to vanish leads to the six required pressure potential equations which can be written in matrix form as $\mathbf{M} \, \Psi = \mathbf{R}$. The resulting square matrix $\mathbf{M}$ is positive definite and symmetric. Hence, the PCG method, which is adaptable to parallel and vector processing, can be efficiently applied to obtain an accurate pressure solution. After the pressure potential field has been calculated, the final velocities $u_{i,j}$ and $v_{i,j}$ must be computed. Micro cells also contribute to the implementation of the new set of final velocity equations (Eqs. (6)-(9)).

Discussion

The SMMC method represents an entirely new approach to the application of free surface pressure boundary conditions. In addition to the other significant differences between the SMMC method and the line segment method, the new unequal-legs pressure equation is different than either the "irregular star" approximation that was presented by Chan and Street (1970) or the interpolation/extrapolation technique presented by Nichols and Hirt (1971). Chan and Street used Taylor series expansions to derive their irregular star approximation. Consequently, the continuity equation is not satisfied in full irregular star cells. If the technique of Nichols and Hirt is used, there can be only one free surface location in a surface cell. Consequently, it is not possible to satisfy the free surface pressure boundary conditions at two points in the same surface cell, such as points a and b in cell (2,3) in Fig. 4. In contrast, the continuity equation will be satisfied in every cell in which the pressure potential is computed in the SMMC method, and the free surface pressure boundary conditions can easily be satisfied at points such as a and b in surface cell (2,3). In addition, in spite of the fact that cell (4,3) in Fig. 4 has no empty neighbors and, thus, is flagged as a full cell, the new method also allows the imposition of free surface pressure boundary conditions at points d and e, the centers of the two surface micro cells that are found in cell (4,3). Whereas the irregular star approximation leads to an unsymmetrical system of equations, the new pressure equation leads to a symmetric set of equations. Finally, the use of micro cells allows the efficient determination of the points of application of the pressure boundary conditions and of the associated leg lengths that are required in the new pressure equation.

ADDITIONAL READING

The review articles by Floryan and Rasmussen (1989), Laskey et al. (1987), Hyman (1984), and Yeung (1982) are recommended for additional information on the treatment of moving interfaces and free surface flows. The book by Lemos (1992) is recommended for readers interested specifically in the application of marker and cell methods to the simulation of wave breaking.

REFERENCES

Amsden, A. A. and F. H. Harlow, 1970, "The SMAC Method: A Numerical Technique for Calculating Incompressible Fluid Flows," Los Alamos Scientific Laboratory Report LA-4370.

Amsden, A. A. and Harlow, F. H., 1970, "A Simplified MAC Technique for Incompressible Fluid Flow Calculations," *Journal of Computational Physics*, 6, pp. 322 - 325.

Amsden, A. A., and Hirt, C. W., 1973, "A Simple Scheme for Generating General Curvilinear Grids, *Journal of Computational Physics*, 11, pp. 348-359.

Amsden, A.A., 1973, "Numerical Calculation of Surface Waves: A Modified ZUNI Code with Particles and Partial Cells," Los Alamos Scientific Laboratory Report LA-5146.

Anilkumar A. V., Lee, C. P., and Wang, T. G., 1991, "Surface-Tension-Induced Mixing Following Coalescence of Initially Stationary Drops," *Physics of Fluids A*, 3, pp. 2587-2591.

Ashgriz, N. and Poo, J. Y., 1991, "FLAIR: Flux Line-Segment Model for Advection and Interface Reconstruction," *Journal of Computational Physics*, 93, pp. 449-468.

Benson, D. J., 1992, "Momentum Advection on a Staggered Mesh," *Journal of Computational Physics*, 100, pp. 143-162.

Bernard, R. S. and Kapitza, H., 1992, "How to Discretize the Pressure Gradient for Curvilinear MAC Grids," *Journal of Computational Physics*, 99, pp. 288-298.

Besson, O., Bourgeois, J., Chevalier, P. A., and Touzani, R., 1991, "Numerical Modelling of Electromagnetic Casting Processes," *Journal of Computational Physics*, 92, pp. 482-507.

Boris, J. P., and Book, D. L., 1973, "Flux-Corrected Transport. I. SHASTA, A Fluid Transport Algorithm That Works," *Journal of Computational Physics*, 11, pp. 38-69.

Brackbill, J. U., and W. E. Pracht, 1973, "An Implicit, Almost-Lagrangian Algorithm for Magnetohydrodynamics, *Journal of Computational Physics*, 13, pp. 455-482.

Brackbill, J. U., Kothe, D. B., and Zemach, C., 1992, "A Continuum Method for Modelling Surface Tension," *Journal of Computational Physics*, 100, pp. 335-354.

Brackbill, J. U., Kothe, D. B., Ruppel, H. M., 1988, "FLIP: A Low-Disspation, Particle-in-Cell Method for Fluid Flow," *Computer Physics Communications*, 48, n1, pp. 25-38.

Broeze, J. and Romate, J. E., 1992, "Absorbing Boundary Conditions for Free Surface Wave Simulations with a Panel Method," *Journal of Computational Physics*, 99, pp. 146-158.

Brunet, J. P., Greenberg, A., and Mesirov, J. P., 1992, "Two-Dimensional, Viscous, Incompressible Flow in Complex Geometries on a Massively Parallel Processor," *Journal of Computational Physics*, 101, pp. 185-206.

Bulgarelli, U., Casulli, V., and Greenspan, D., 1984, "Pressure Methods for the Numerical Solutions of Free Surface Fluid Flows," Pineridge Press, Swansea, UK.

Bush, M. B., and N. Phan-Thien, 1985, "Three Dimensional Viscous Flows with a Free Surface: Flow Out of a Square Die," *Journal of Non-Newtonian Fluid Mechanics*, 18, pp. 211-218.

Chan, R. K.-C. and Street, R. L., 1970, "A Computer Study of Finite-Amplitude Water Waves," *Journal of Computational Physics*, 6, pp. 68-94.

Chan, R. K.-C., R. L. Street, and J. E. Fromm, 1970, "The Digital Simulation of Water Waves -- an Evaluation of SUMMAC," Proc. Second Internat. Conf. Num. Meth. Fluid Dynamics, Springer-Verlag, Berkeley, pp. 429-434

Chen, S., 1991, "The SMU Method: a Numerical Scheme for Calculating Incompressible Free Surface Fluid Flows by the Surface Marker Utility," Ph.D. dissertation, Southern Methodist University, Dallas, Texas.

Chen, S., Johnson, D. B., and Raad, P. E., 1991, "The Surface Marker Method," *Computational Modeling of Free and Moving Boundary Problems, Vol. 1, Fluid Flow*, pp. 223-234, W. de Gruyter, New York.

Chen, S., Johnson, D. B., and Raad, P. E., 1995, "Velocity Boundary Conditions for the Simulation of Free Surface Fluid Flow," *Journal Computational Physics*, 116, pp. 262-276.

Chesshire, G. and Henshaw, W. D., 1990, "Composite Overlapping Meshes for The Solution of Partial Differential Equations," *Journal of Computational Physics*, 90, pp. 1-64.

Chorin, A. J., 1967, "A Numerical Method for Solving Incompressible Viscous Flow Problems," *Journal of Computational Physics*, 2, pp. 12-26.

Christodoulou, K.N. and Scriven, L E., 1992, "Discretization of Free Surface Flows and Other Moving Boundary Problems," *Journal of Computational Physics*, 99, pp. 39-55.

Cole, S. L. and Strayer, T. D., 1992, "Low-Speed Two-Dimensional Free-Surface Flow Past a Body," *Journal of Fluid Mechanics*, 245, pp. 437-447.

Daly, B. J. and W. E. Pracht, 1968, "Numerical Study of Density-Current Surges," *Physics of Fluids*, 11, pp. 15-30.

Daly, B. J., 1969, "A Technique for Including Surface Tension Effects in Hydrodynamic Calculations," *Journal of Computational Physics*, 4, pp. 97-117.

Dear, J. P. and Field, J. E., 1988, "High-Speed Photography of Surface Geometry Effects in Liquid/Solid Impact," *Journal of Applied Physics*, 63.

Dias, F. and Christodoulides, P., 1991, "Ideal Jets Falling Under Gravity," *Physics of Fluids A*, 3, pp. 1711-1717.

Dias, F. and Tuck, O., 1991, "Weir Flows and Waterfalls," *Journal of Fluid Mechanics*, 230, pp. 525-539.

Dias, F., Keller, J., and Vanden-Broeck, J. M., 1988, "Flows over Rectangular Weirs," *Physics of Fluids*, 31, pp. 2071-2076.

Duraiswami, R. and Prosperetti, A., 1992, "Orthogonal Mapping in Two Dimensions," *Journal of Computational Physics*, 98, pp. 254-268.

Floryan, J. M. and Rasmussen, H., 1989, "Numerical Methods for Viscous Flows with Moving Boundaries," *ASME Applied Mechanics Reviews*, 42.

Foster, N., and Metaxas, D., 1995, "Visualization of Dynamic Fluid Simulations: Waves, Splashing, Vorticity, Boundaries, Buoyancy," *Engineering Computations*, 12, pp. 109-124.

Ghia, U., Ramamurti, R., and Ghia, K.N., 1988, "Solution of The Neumann Pressure Problem in General Orthogonal Coordinates Using The Multigrid Technique," *AIAA Journal*, 26, pp. 538-547.

Greenspan, H. P., and R. E. Young, 1978, "Flow over a Containment Dyke," *Journal of Fluid Mechanics*, 87, pp. 179-192.

Gresho, P. M. and R. L. Sani, 1987, "On the Pressure Boundary Conditions for the Incompressible Navier-Stoke Equations," *International Journal for Numerical Methods in Fluids*, 7, pp. 1111-1145.

Gupta, M. M., 1991, "High Accuracy Solutions of Incompressible Navier-Stokes Equations," *Journal of Computational Physics*, 94, pp. 343-359.

Harlow, F. H. and Amsden, A. A., 1968, "Numerical Calculation of Almost Incompressible Flow," *Journal of Computational Physics*, 3, pp. 80-93.

Harlow, F. H. and Amsden, A. A., 1971, "A Numerical Fluid Dynamics Calculation Method for All Flow Speeds," *Journal of Computational Physics*, 8, pp. 197-213.

Harlow, F. H. and Amsden, A. A., 1974, "Multifluid Flow Calculations at All Mach Number," *Journal of Computational Physics*, 16, pp. 1-19.

Harlow, F. H. and Welch, J. E., 1965, "Numerical Calculation of Time-Dependent Viscous Incompressible Flow of Fluid with Free Surface," *Physics of Fluids*, 8, pp. 2182-2189.

Harlow, F. H., 1988, "PIC and Its Progeny," *Computer Physics Communications*, 48, pp. 1-10.

Hirt, C. W. and J. L. Cook, 1972, "Calculating of Three-Dimensional Flows Around Structures and Over Rough Terrain," *Journal of Computational Physics*, 10, pp. 324-340.

Hirt, C. W. and Nichols, B. D., 1981, "Volume of Fluid (VOF) Method for the Dynamics of Free Boundaries," *Journal of Computational Physics*, 39, pp. 201 - 225.

Hirt, C. W., 1968, "Heuristic Stability Theory for Finite-Difference Equations," *Journal of Computational Physics*, 12, pp. 339-355.

Hirt, C. W., Amsden, A.A., and J.L. Cook, 1974, "An Arbitrary Lagrangian-Eulerian Computing Method for All Flow Speed," *Journal of Computational Physics*, 14, pp. 227-253.

Hirt, C. W., and B. D. Nichols, 1980, "Adding Limited Compressibility to Incompressible Hydrocodes," *Journal of Computational Physics*, 34, pp. 390-400.

Hirt, C. W., and F. H. Harlow, 1967, "A general Corrective Procedure for the Numerical Solution of Initial-Value Problems," *Journal of Computational Physics*, 2, pp. 114-119.

Hirt, C. W., and J. L. Cook, 1972, "Calculating of Three-Dimensional Flows Around Structures and Over Rough Terrain," *Journal of Computational Physics*, 10, pp. 324-340.

Hirt, C. W., and J. P. Shannon; 1968, "Free Surface Stress Conditions for Incompressible Flow Calculations," *Journal of Computational Physics*, 2, pp. 403-411.

Hirt, C. W., B. D. Nichols, and N. C. Romero, 1975, "SOLA - A Numerical Solution Algorithm for Transient Fluid Flows," Los Alamos Scientific Laboratory Report LA-5852.

Hirt, C. W., J. L. Cook, and T. D. Bulter, 1970, "A Lagrangian Method for Calculating the Dynamics of an Incompressible Fluid with Free Surface," *Journal of Computational Physics*, 5, pp. 103-124.

Howison, S. D., Ockendon, J. R., and Wilson, S. K., 1991, "Incompressible Water-Entry Problems at Small Deadrise Angles," *Journal of Fluid Mechanics*, 222, pp. 215-230.

Hwang, W. S., and R. A. Stoehr, 1986, "Computer Aided Fluid Flow Analysis of The Filling of Casting Systems," Proceedings of the NUMIFORM Conference, Gothenburg, pp. 361-366.

Hyman, J. M., 1984, "Numerical Methods for Tracking Interfaces," *Physica D*, 12, pp. 396-407.

Ibrahim, E. A. and Przekwas, A. J., 1991, "Impinging Jets Atomization," *Physics of Fluids A*, 3, pp. 2981-2987.

Jagannathan, S., 1988, "Non-Linear Free Surface Flows and an Application of the Orlanski Boundary Condition," *International Journal for Numerical Methods in Fluids*, 8, pp. 1051-1070.

Jeng, Y. N. and Chen, J. L., "Truncation Error Analysis of the Finite Volume Method for a Model Steady Convective Equation," *Journal of Computational Physics*, 100, pp. 64-76.

Johnson, D. B., Chen, S., and Raad, P. E., 1994, "Simulation of Impacts of Fluid Free Surfaces with Solid Boundaries," *International Journal for Numerical Methods in Fluid*, 19, pp. 153-176.

Joly, P. and Eymard, R., 1990, "Preconditioned Biconjugate Gradient Methods for Numerical Reservoir Simulation," *Journal of Computational Physics*, 91, pp. 298-309.

Kallinderis, Y., 1992, "Numerical Treatment of Grid Interfaces for Viscous Flows," *Journal of Computational Physics*, 98, pp. 129-144.

Karageorghis, A., 1992, "The Method of Fundamental Solutions for the Solution of Steady-State Free Boundary Problems," *Journal of Computational Physics*, 98, pp. 119-128.

Kaushik, S. and Hagelstein, P. L., 1992, "The Application of the Preconditioned Biconjugate Gradient Algorithm to NLTE Rate Matrix Equatiopns," *Journal of Computational Physics*, 101, pp. 360-367.

Kelly, D. W., Mills, R.J., and Reizesm, J.A., 1988, "A Posteriori Error Estimates in Finite Difference Techniques," *Journal of Computational Physics*, 74, pp. 1214-232.

Kim, J., and Moin P., 1985, "Application of a Fractional-Step Method to Incompressible Navier-Stokes Equations," *Journal of Computational Physics*, 59, pp. 308-323.

Korobkin, A., 1992, "Blunt-Body Impact on a Compressible Liquid Surface," *Journal of Fluid Mechanics*, 244, pp. 437-453.

Ku, H.C., Taylor, T. D., and Hirsh, R. S., 1987, "Pseudospectral Methods for Solution of The Incompressible Navier-Stokes Equations," *Computational Fluids*, 15, pp. 195-214.

Kwak, D., 1986, "A Three-Dimensional Incompressible Navier-Stokes Flow Solver Using Primitive Variables," *AIAA Journal*, 24, pp. 390-396.

Lardner, R. W. and Song, Y., 1992, "A Comparison of Spatial Grids for Numerical Modelling of Flows in Near-Coastal Seas," *International Journal for Numerical Methods in Fluids*, 14, pp. 109-124.

Laskey, K. J., Oran, E. S., and Boris, J. P., 1987, "Approaches to Resolving and Tracking Interfaces and Discontinuities," Naval Research Laboratory Report 5999, Arlington, VA.

Le, H. and Moin, P., 1991, "An Improvement of Fractional Step Methods for the Incompressible Navier-Stokes Equations," *Journal of Computational Physics*, 92, pp. 369-379.

Lemos, C. M., 1992, *Wave Breaking - A Numerical Study*, Springer-Verlag, Berlin.

Loh, C. Y. and Rasmussen, H., 1987, "A Numerical Procedure for Viscous Free Surface Flows," *Applied Numerical Mathematics*, 3, pp. 479-495.

Lu, W. Q. and Chang, H. C., 1988, "An Extension of the Biharmonic Boundary Integral Method to Free Surface Flow in Channels," *Journal of Computational Physics*, 77, pp. 340-360.

Mansutti, D., Graziani, G., and Piva, R., 1991, "A Discrete Vector Potential Model for Unsteady Incompressible Viscous Flows," *Journal of Computational Physics*, 92, pp. 161-184.

Mavriplis, D. J., 1990, "Adaptive Mesh Generation for Viscous Flows Using Delaunay Triangulation," *Journal of Computational Physics*, 90, pp. 271-291.

Mekias, H. and Vanden-Broeck, J. M., 1991, "Subcritical Flow with a Stagnation Point due to a Source beneath a Free Surface," *Physics of Fluids A*, 3, pp. 2652-2658.

Mekias, H., and Vanden-Broeck, J. M., 1989, "Supercritical Free-Surface Flow with a Stagnation Point due to a Submerged Source," Physics of Fluids A, 1, pp. 1694-1697.

Miyata, H., 1986, "Finite-Difference Simulation of Breaking Waves," *Journal of Computational Physics*, 65, pp. 179-214.

Miyata, H., and Lee, Y.-G., 1990, "Vortex Motions about a Horizontal Cylinder in Waves," *Ocean Engineering*, 17, pp. 279-305.

Miyata, H., and Nishimura, S., 1985, "Finite-Difference Simulation of Nonlinear Ship Waves," *Journal of Fluid Mechanics*, 157, pp. 327-357.

Miyata, H., and Nishimura, S., and Masuko, A., 1985, "Finite-Difference Simulation of Nonlinear Waves Generated by Ships of Arbitrary Three-Dimensinoal Configuration, *Journal of Computational Physics*, 60, pp. 391-436.

Miyata, H., Sato, T., and Baba, N., 1987, "Difference Solution of a Viscous Flow with Free Surface Wave About an Advancing Ship," *Journal of Computational Physics*, 72, pp. 393-421.

Miyata, H., Zhu, M., and Watanabe, O., 1992, "Numerical Study on a Viscous Flow with Free-Surface Waves About a Ship in Steady Straight Course by a Finite-Volume Method," *Journal of Ship Research*, 36, pp. 332-345.

Montigny-Rannou, F., and Y. Morchoisne, 1987, "A Spectral Method with Staggered Grid for Incompressible Navier-Stokes Equations," *International Journal for Numerical Methods in Fluids*, 7, pp. 175-189.

Moriarty, J. A. and Schwartz, L. W., 1991, "Unsteady Spreading of Thin Liquid Films with Small Surface Tension," *Physics of Fluids A*, 3, pp. 733-742.

Nakayama, P. I., and N. C. Romero, 1971, "Numerical Method for Almost Three-Dimensional Incompressible Fluid Flow and a Simple Internal Obstacle Treatment, *Journal of Computational Physics*, 8, pp. 230-240.

Nichols, B. D. and Hirt, C. W., 1971, "Improved Free Surface Boundary Conditions for Numerical Incompressible-Flow Calculations," *Journal of Computational Physics*, 8, pp. 434-448.

Nichols, B. D. and Hirt, C. W., 1975, "Methods for Calculating Multi-Dimensional, Transient, Free Surface Flows Past Bodies," *Proceedings of the First International Conference on Numerical Ship Hydrodynamics*, Gaithersburg, MD, 253-277.

Nichols, B. D. and Hirt, C. W., 1976, "Calculating Three-Dimensional Free Surface Flows in the Vicinity of Submerged and Exposed Structures," *Journal of Computational Physics*, 12, pp. 234 - 246.

Nichols, B. D., 1970, "Recent Extensions to the Marker-and-Cell Method for Incompressible Fluid Flows," Proc. Second Internat. Conf. Num. Meth. Fluid Dynamics, Springer Verlag, Berkeley, pp. 371-376.

Nichols, B. D., and C. W. Hirt, 1973, "Calculating Three-Dimensional Free Surface Flows in the Vicinity of Submerged and Exposed Structures," *Journal of Computational Physics*, 12, pp. 234-246.

Nichols, B. D., C. W. Hirt, and R. S. Hotchkiss, 1980, "SOLA-VOF: A Solution Algorithm for Transient Fluid Flow with Multiple Free Boundaries," Los Alamos Scientific Laboratory Report LA-8355.

Nicolaides, R.A., 1991, "Analysis and Convergence of The MAC Scheme: 1.The Linear Problem," NASA Contractor Report 187536, ICASE Report No. 91-28.

Osher, S. and Sethian, J. A., 1988, "Front Propagating with Curvature-Dependent Speed: Algorithms Based on Hamilton Jacobi Formulations," *Journal of Computational Physics*, 79, pp. 12-49.

Patankar, S. V., 1980, Numerical Heat Transfer and Fluid Flow, Hemisphere, Washington, D.C.

Pelekasis, N. A., Tsamopoulos, J. A., and Manolis, G. D., 1992, "A Hybrid Finite-Boundary Element Method for Inviscid Flows with Free Surface," *Journal of Computational Physics*, 101, pp. 231-251.

Poo, J.Y. and Ashgriz, N., 1989, "A Computational Method for Determining Curvatures," *Journal of Computational Physics*, 84, pp. 483-491.

Raad, P. E., S. Chen, and D. B. Johnson, 1995, "The Introduction of Micro Cells to Treat Pressure in Free Surface Fluid Flow Problems," *ASME Journal of Fluids Engineering*, in press.

Ramaswamy, B., 1990, "Numerical Simulation of Unsteady Viscous Free Surface Flow," *Journal of Computational Physics*, 90, pp. 396-430.

Ramaswamy, B., and M. Kawahara, 1987, "Arbitrary Lagrangian-Eulerian Finite Element Method for Unsteady, Convective, Incompressible Viscous Free Surface Fluid Flow," *International Journal for Numerical Methods in Fluids*, pp. 1053-1075.

Rausch, R. E., 1990, "The Use of Parallel Processing in Cavity Filling Simulation," MS thesis, Southern Methodist University, Dallas, Texas.

Reid, D., Chan, A., and Al-Mudares, M., 1992, "A Direct Solution of Poisson's Equation in a Three-Dimensional Field-Effect Transistor Sturcture," *Journal of Computational Physics*, 99, pp. 79-83.

Renardy, M. and Renardy, Y., 1991, "On the Nature of Boundary Conditions for Flows with Moving Free Surfaces," *Journal of Computational Physics*, 93, pp. 325-335.

Romate, J. E., 1992, "Absorbing Boundary Conditions for Free Surface Waves," *Journal of Computational Physics*, 99, pp. 135-145.

Rosenfeld, M., Kwak, D., and Vinokur, M., 1991, "A Fractional Step Solution Method for the Unsteady Incompressible Navier-Stokes Equations in Generalized Coordinate Systems," *Journal of Computational Physics*, 94, pp. 102-137.

Rothberg, R. H., Walker, W. F., and Chapman, A. J., 1990, "Enchanced Boundary Pressure Update for Incompressible Flow Simulation," *Journal of Computational Physics*, 88, pp. 495-498.

Schmidt, G.H. and Jacobs, F.J., 1988, "Adaptive Local Grid Refinement and Multi-grid in Numerical Reservoir Simulation," *Journal of Computational Physics*, 77, pp. 140-165.

Seldner, D. and Westermann, T., 1988, "Algorithms for Interpolation and Localization in Irregular 2D Meshes," *Journal of Computational Physics*, 79, pp. 1-11.

Sharif, M. A. R., and Busnaina, A. A., 1988, "Assessment of Finite Difference Approximations for the Advection Terms in the Simulation of Practical Flow Problems," *Journal of Computational Physics*, 74, pp. 143-176.

Shyy, W. and Vu, T. C., 1991, "On the Adoption of Velocity Variable and Grid System for Fluid Flow Computation in Curvilinear Coordinates," *Journal of Computational Physics*, 92, pp. 82-105.

Smith, B. L., 1988, "A General Numerical Procedure for the Treament of Moving Interfaces in Implicit Continuous Eulerian (ICE) Hydrodynamics," *Journal of Computational Physics*, 77, pp. 384-424 .

Soh, W. Y. and Goodrich, J. W., 1988, "Unsteady Solution of Incompressible Navier-Stokes Equations," *Journal of Computational Physics*, 79, pp. 113-134.

Sotiropoulos, F. and Abdallah, S., 1991, "The Discrete Continuity Equation in Primitive Variable Solutions of Incompressible Flow," *Journal of Computational Physics*, 95, pp. 212-227.

Steinberg, S. and Roache, P., 1992, "Variational Curve and Surface Grid Generation," *Journal of Computational Physics*, 100, pp. 163-178.

Tadjbakhsh, I. and Keller, J. B., 1960, "Standing Surface Waves of Finite Amplitude," *Journal of Fluid Mechanics*, 8, pp. 442-451.

Theoclossiou, V.M., 1986, "An Efficient Algorithm for Solving The Incompressible Fluid Flow Equations," *International Journal for Numerical Methods in Fluids*, 6, pp. 557-572.

Thompson, J. F., 1988, "Composite Grid Generation Code for General 3-D Regions -- The Eagle Code," *AIAA Journal*, 26, n3, pp. 271-272.

Tidd, D. M., Thatcher, R. W., and Kaye, A., 1988, "The Free Surface Flow of Non-Newtonian Fluids Trapped By Surface Tension," *International Journal for Numerical Methods in Fluids*, 8, pp. 1011-1027.

Tome, M. F., and McKee, S., 1994, "GENSMAC: A Computational Marker and Cell Method fot Free Surface Flows in General Domains," *Journal of Computational Physics*, 110, pp. 171-186.

Torrey, M. D., Cloutman, L. D., Mjolsness, R. C., and Hirt, C. W., 1985, "NASA-VOF2D: A Computer Program for Incompressible Flows with Free Surfaces," Los Alamos National Laboratory Report LA-10612.

Tsai, W. and Yue, D. K. P., 1991, "Features of Nonlinear Interactions between a Free Surface and a Shed Vortex Shear Layer," *Physics of Fluids A*, 3, pp. 2485-2488.

Tyvand, P. A., 1992, "Unsteady Free-Surface Flow due to a Line Source," *Physics of Fluids A*, 4, pp. 671-676.

Unverdi , S. O. and Tryggvason, G., 1992, "A Front-Tracking Method for Viscous, Incompressible, Multi-Fluid Flows," *Journal of Computational Physics*, 100, pp. 25-37.

Vanden-Broeck, J. M. and Dias, F., 1991, "Nonlinear Free-Surface Flows Past a Submerged Inclined Flat Plate," *Physics of Fluids A*, 3, pp. 2995-3000.

Vanden-Broeck, J. M. and Keller, J. B., 1989, "Pouring Flows with Separation," *Physics of Fluids A*, 1, pp. 156-158.

Vanden-Broeck, J. M., 1991, "Cavitating Flow of a Fluid with Surface Tension past a Circular Cylinder," *Physics of Fluids A*, 3, pp. 263-266.

Vanden-Broeck, J. M., 1991, "Elevation Solitary Waves with Surface Tension," *Physics of Fluids A*, 3, pp. 2659-2663.

Vanden-Broeck, J.-M., 1987, "Free-Surface Flow over an Obstruction in a Channel," *Physics of Fluids*, 30, pp. 2315-2317.

Viecelli, J. A., 1969, "A Method for Including Arbitrary External Boundaries in the MAC Incompressible Fluid Computing Technique," *Journal of Computational Physics*, 4, pp. 543-551.

Viecelli, J. A., 1971, "A Computing Method for Incompressible Flows Bounded by Moving Walls," *Journal of Computational Physics*, 8, pp. 119-143.

Vinokur, M., 1989, "An Analysis of Finite-Difference and Finite-Volume Formulations of Conservation Laws," *Journal of Computational Physics*, 81, pp. 1-52.

Vradis, G., Zalak, V., and Bentson, J., 1992, "Simultaneous Variable Solutions of the Incompressible Steady Navier-Stokes Equations in General Curvilinear Coordinate System," *ASME Journal of Fluids Engineering*, 114, pp. 299-305.

Wang, G.X. and Matthys, E. F., 1992, "Numerical Modelling of Phase Change and Heat Transfer During Rapid Solidification Processes: Use of Control Volume Intergral with Element Subdivision," *International Journal of Heat and Mass Transfer*, 35, pp. 141-153.

Warming, R. F., and B.J. Hyett, 1974, "The Modified Equation Approach to the Stability and Accuracy Analysis of Finite-Difference Method," *Journal of Computational Physics*, 14, pp. 159-179.

Welch, J. E., F. H. Harlow, J. P. Shannon, and B. J. Daly, 1965, "The MAC Method: A Computing Technique for Solving Viscous, Incompressible, Transient Fluid-Flow Problems Involving Free Surfaces," Los Alamos Scientific Laboratory Report LA-3425.

Welch, J. E., Harlow, F. H., Shannon, J. P., and B. J. Daly, 1965, "The MAC method," Los Alamos Scientific Laboratory Report, LA-3425.

Wesseling, P., 1988, "Cell -Centered Multigrid for Interface Problems," *Journal of Computational Physics*, 79, pp. 85-91.

Yeung, R. W., 1982, "Numerical Methods in Free Surface Flows," *Annual Review of Fluid Mechanics*, 14, pp. 395-442.

Young, D. P., Melvin, R. G., Bieterman, M. B., Johnson, F. T., Samant, S. S., Bussoletti, J. E., "A Locally Refined Rectangular Grid Finite Element Method: Application to Computational Fluid Dynamics and Computational Physics," *Journal of Computational Physics*, 92, pp. 1-66.

Young, J. A., and C. W. Hirt, 1972, "Numerical Calculation of Internal Wave Motions," *Journal of Fluid Mechanics*, 56, pp. 265-276.

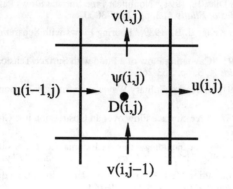

Fig. 1 Staggered Grid.

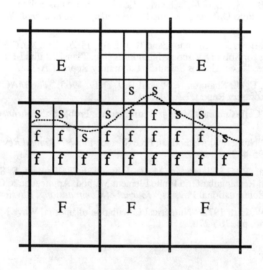

Fig. 2 Macro and Micro Cells.

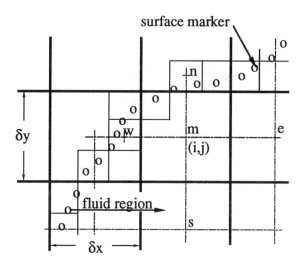

Fig. 3 Points used in pressure potential approximation for cell (i,j).

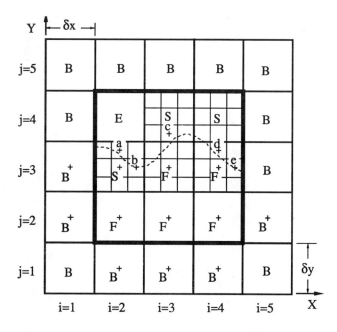

Fig. 4 Micro cells used for calculation of leg lengths in pressure potential equation

Contributed
Discussion Papers

PHYSICAL MODEL TEST OF HOKKAIDO NANSEI-OKI EARTHQUAKE TSUNAMI AROUND CAPE AONAE

Kenji Noguchi[a], Shinji Sato[b], Shigenobu Tanaka[c]

1. Introduction

On July 12, 1993, a tsunami generated by the Hokkaido Nansei-Oki earthquake caused devastation on Okushiri Island . The inundation heights exceeded over 10 m at various sites of the island, which were Inaho located on the west side of north end, along the southwest coast and Hamatsumae district located on the southeastern part. One of these area with high tsunami runup, Hamatsumae district was of particular interest since it is located in an area sheltered from tsunami source area by Cape Aonae. Moreover, the shape of the coastline around Hamatsumae was convex to the sea, which was not believed to produce large inundation. Numerical simulations were performed to reproduce the propagation and the inundation of the tsunami (for example, Takahashi et al.; 1994). They simulated the distribution of inundation heights to some extent, but failed to reproduce the anomalous inundation height at Hamatsumae. Since the tsunami attacked in midnight immediately after the earthquake occurred at 10:17pm, observations were scarce. Physical model tests were therefore performed in the present study to investigate the mechanism of tsunami transformation in this area in detail.

2. Tsunami inundation height in Okushiri Island

Several groups measured tsunami runup height immediately after the tsunami. These data were summarized by Bernard and González (1993, fig.1). According to this figure, the runup heights reached 20 m along the southwest coast of the island and Hamatsumae district located on the southeast part. Maximum inundation height recorded at Monai on the southwest coast reached 31.7 m. The occurrence of large inundation height at Monai can be explained that it is located on the west side of the island facing to the tsunami source area and that it is located in a V-shape bay. However, Hamatsumae district is located in an area sheltered by Cape Aonae and is convex to the sea. It is therefore to be studied why the inundation height was large. Photo1 shows a collapse of retaining wall for falling stone along a road at Hamatsumae. Seaward collapse of the wall suggests this was caused by

[a]Researcher, Coastal Engineering Division, Public Works Research Institute, Ministry of Construction, 1 Asahi, Tsukuba, 305, Japan
[b]D. Eng.,Senior Researcher, ditto
[c]M. Eng.,Head, ditto

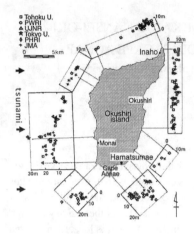

fig.1 Tsunami inundation height
(after Bernard and Gonzáles, 1993)

photo 1 Seaward overturn of wall at Hamatsumae

receding waves.

3. Implementation of physical model test

There exists a broad shallow sea, called Okushiri Spur, in front of the coast between Cape Aonae and Hamatsumae with a gentle slope for southeast. Murotsu rock island is located in this area. It is considered that the tsunami propagation was influenced by the complex bottom topography on the Okushiri Spur, especially in the shallow sea area surrounded by Cape Aonae, Murotsu rock islands and Hamatsumae. The area which was reproduced in the model was selected to be a rectangular area shown in fig.2. The area was selected on the basis of numerical simulation of tsunami refraction so that it contained the whole source area which might affect tsunami transformation around the southern part of the Island. The experimental basin has 35m length, 30m width and 0.6m depth. The physical model reproduced topographies from 500 m beneath the sea level to 20 m above the sea level with non-distorted 1/1100 scale.

A pneumatic tsunami generator is installed in the wave basin(fig.3). It generates a single positive wave by suddenly releasing valves attached to the air chamber . The physical model tests were performed for 8 cases with incident tsunami height ranging between 2.7 and 7.8 mm. The wave heights were measured by capacitance-type wave gages. At the same time, video pictures were taken from three angles. Their snapshots are (1) the whole propagation of the tsunami from the top, (2) swash motion around Hamatsumae and (3) from northeast Hamatsumae on the coast for Murotsu island. In fig.4, the triangle symbol denotes measuring point. Measuring points are intensively situated in the shallow area around

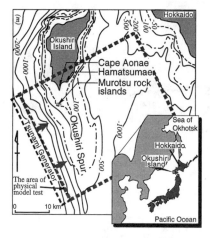

fig.2 Reproduced area in physical model test

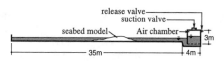

fig.3 Pneumatic tsunami generator

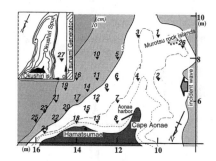

fig.4 Measuring points

Hamatsumae.

4. Results of experiments

4.1 Tsunami wave profiles

The propagation of tsunami with incident wave height being 3.9 mm will be described in detail since it agreed the initial water elevation in the tsunami source area estimated by a fault model (Takahashi et al., 1994). The wave 3.9 mm height is scaled to 4.3 m in prototype. The incident wave profile measured in front of the tsunami generator is shown in fig.5. The time 't' is the elapsed time from the initiation of data sampling.

Tsunamis may break and be dispersed into several short waves while traveling over shallow area. In the Nihonkai Chubu earthquake tsunami in 1983, the dispersed tsunami hit a coast in Akita prefecture and caused armor blocks piled near the shoreline scattered in a wide area (Shuto, 1985). In the present experiments, the tsunami was observed to break on the Okushiri Spur. The broken tsunami was then dispersed into several short waves after it passed over the Okushiri Spur. Fig.6 shows wave profiles observed at offshore measuring points. The dispersion of the tsunami was observed at all the points except for Point 4 which was located on the top of the Okushiri Spur. Point 6 was located at the point where two waves refracted from both sides of Murotsu Island cross each other. Nonlinear interaction occurred between the two waves might affect a peculiar tsunami profile at Point 6. In

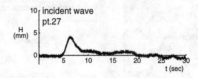

fig.5 Incident wave profile

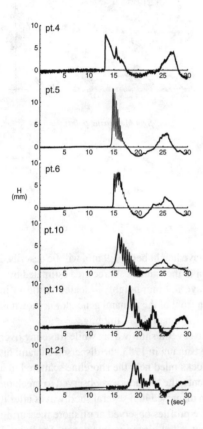

fig .6 Wave profiles in the offshore area

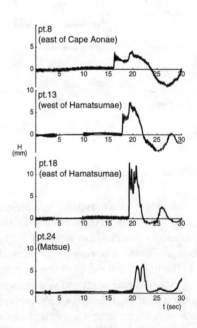

fig.7 Wave profiles along shoreline

this paper, the each wave of dispersed short waves is called individual wave. Fig.6 illustrates that the distance of individual waves will gradually increase while traveling over shallow area since each individual wave travels with different celerity.

Fig.7 shows profiles along shoreline. Point 8 is located east of Cape Aonae and sheltered by the cape. Point 13 and Point 18 are located respectively to the west and to the east of Hamatsumae. These four profiles have remarkable two peaks. According to video frame analysis, the second peak is due to reflected wave from land. Point 18, which was located east of Hamatsumae, recorded the highest wave height.

4.2 Amplified wave height at a concave part in the crest line

Four snapshots of video frames are shown in photo 2 (1) to (4). The model of Okushiri island is located on the bottom and tsunami is incident from the right. These photos were taken respectively at t=14.20s, 14.26s, 16.20s, and 18.60s. In the prototype scale, these times corresponded to t=7.7min, 7.8min, 8.9min and 10.23min respectively. Photo 2 (1) (t=14.20s) shows a scene immediately after breaking over the Okushiri Spur. The tsunami was dispersed into several short waves after breaking. In photo 2 (2), the crest line of the first wave were bent owing to refraction and created a concave part. Wave height of this concave part became higher than those on both sides due to nonlinear interaction occurred between two wave components crossing with a thin angle. This part gradually caught up the surrounding parts, since waves with large wave height propagate with large celerity. In photo 2 (3), the white area in the crestline indicates the area with large wave heights, which was gradually shifted towards the island and approached the shoreline. At last, in photo 2

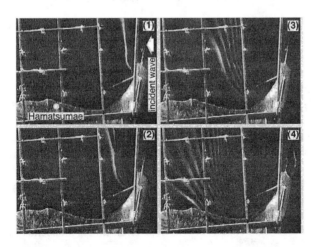

photo.2 Snapshots of tsunami propagation (1)t=14.20s, (2)t=14.26, (3) t=16.20s, (4) t=18.60s

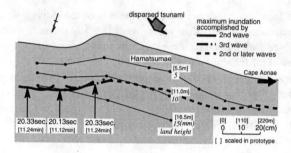

fig.8 Distribution of maximum inundation heights

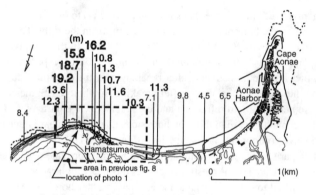

fig.9 Inundation heights around Hamatsumae by Shuto(1994)

(4), the part with large wave height reached the east of Hamatsumae and hit the shoreline.

4.3 Observation of tsunami run-up

Tsunami run-up along the island was complex since individual waves dispersed on the Okushiri Spur were observed to hit the island successively. Tsunami run-up around Hamatsumae was analyzed in detail from video frames. The maximum inundation lines are illustrated in fig.8 with various line types depending on which wave produced the maximum inundation height. The thin line with dots shows the contour line of land height. The maximum inundation height in this area was 15mm, which is scaled to 16.5m in prototype. The lines depicted by arrows indicate the maximum inundation line produced at t=20.13s and 20.33s. The solid lines show the maximum inundation lines produced by the second wave at t=20.13s and at t=20.33s. The chain line shows that produced by the third wave at t=20.33s. The dashed line indicates the envelope of maximum inundation lines produced by the second or the following waves. Shuto (1994) summarized the inundation heights surveyed around Hamatsumae in detail. Their data are shown in fig.9. The area surrounded by broken line corresponds to the area shown in the fig.8 Inundation heights eastward of convex part to sea exceeded 16m. In the west of Hamatsumae village, inundation heights were about 10m. It is noticed that the physical model generally reproduced the large inundation heights observed around Hamatsumae.

5. Conclusions

The major conclusions of the present study can be summarized as follows:
1) It is considered that the tsunami generated by the Hokkaido Nansei-Oki earthquake experienced breaking and dispersion over shallow sea around Cape Aonae.
2) The crest line of the first wave formed a concave part with large wave height. The part was gradually shifted toward the shoreline along the crestline.
3) The maximum inundation height is produced by the second or the following dispersed wave of tsunami, not by the first wave.
4) Results of physical model tests can reasonably explain the distribution of inundation heights obtained by the field survey.

Acknowledgments
The authors would like to express their sincere thanks to Professor Shuto, Tohoku University, for his valuable suggestions on the design of the physical model.

6. Reference

1. N. Shuto, *Surveyed data, Tsunami Engineering Technical Report*, **No.11**, (Disaster Control Research Center, Tohoku University,1994), 120p.
2. T. Takahashi, N. Shuto, F. Imamura and H. Matsutomi, *The measured and computed Hokkaido Nansei-Oki Earthquake tsunami of 1993*, (Coastal Eng. 1994, ASCE), 886-900.
3. E. Bernard and F. I. González, *Tsunami Devastates Japanese Coastal Region*, (EOS, Transactions, American Geophysical Union, 1993), **Vol.74**, No.37, Sep., 14, pp.417,432.
4. Shuto, N (1985), *The Nihonkai-Chubu Earthquake Tsunami on the North Akita Coast*, (Coastal Eng. in Japan, JSCE), **Vol.28**, 255-264.

ANALYTIC SOLUTIONS OF SOLITARY WAVE RUNUP ON THE CONICAL ISLAND AND ON THE REVERE BEACH

U. KÂNOĞLU, C. E. SYNOLAKIS

School of Engineering, University of Southern California
Los Angeles, California 90089-2531, USA

We apply our analytic solutions for piecewise linear two and three–dimensional bathymetries to study the evolution of solitary waves on the conical island and on the Revere Beach, and we compare our results with the laboratory data. In the three–dimensional problem, we find adequate quantitative agreement for both the time histories of surface elevation and runup; in the two–dimensional problem, we find good agreement for the time histories of surface elevation and excellent agreement for the maximum runup, improving for smaller waveheights.

1 Introduction

We present an analytic method of solution for long wave evolution on piecewise linear two and three–dimensional bathymetries. We use the general solution method of Kânoğlu (1996) to determine the relationship between the amplification factor and the incident wave amplitude which allows for the determination of close–form asymptotic results, also known as runup laws. We compare our results with the laboratory experiments. Details of the laboratory results and the analytical solutions for benchmark problem 2 and 3 can be found in Kânoğlu (1996), and the laboratory results also in Briggs et al. (1995) and Liu et al. (1995). Here we will give only brief description of our method and how we apply it in two workshop problems.

2 General Method

We use the linear shallow water wave equation that describes a propagation problem in water of variable depth as our field equation. Using the undisturbed water depth as our characteristic lenght scale, we introduce dimensionless variables. Our method of solution consists of representing a given topography by a series of constant or linearly varying depth segments and writing the solution for η in terms of two unknowns in each segment. Continuity of the surface elevation and the surface slope provide two boundary conditions at each interface, so that for an m–segment topography there are $(2m - 1)$ equations. Instead of solving the entire system, our methodology, Kânoğlu

(1996), allows us to characterize each segment by a (2×2) matrix which incorporates its topographic character. Our solution produces a product of (2×2) matrices allowing direct and explicit inversion and determination of quantities of interest. Each additional segment adds one matrix and one inverse matrix to the formulation. Our method also allows special boundary conditions such as perfect reflection condition at the shoreline, as in the Revere Beach case.

3   Application of Analytic Theory

3.1   Benchmark Problem 2, Conical Island

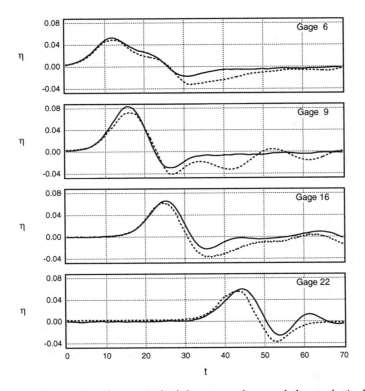

Figure 1. Comparison between the laboratory data and the analytical solution for the time histories of the surface elevation for a $H = 0.05$ target solitary wave, at four different locations. Dotted lines represent the laboratory results. The offset of the laboratory data was removed for gage 16.

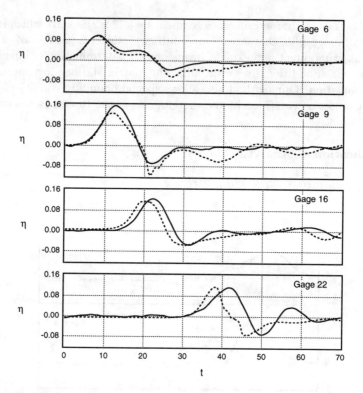

Figure 2. Comparison between the laboratory data and the analytical solution for the time histories of the surface elevation for a $H = 0.10$ target solitary wave, at four different locations. Dotted lines represent the laboratory results.

We use the linearized equations of motion for long wave propagation in polar coordinates as a field equation and divide the slope of the island into m constant–depth segments. The solution in each individual constant–depth segment is known. The field equation has eigenfunctions in terms of Bessel functions. Outside of the toe of the conical island, the solution is the sum of the incoming wave represented in the form of a Fourier–Bessel series, while the scattered waves are represented by Hankel functions. At the edge of the each segment, we consider continuity of the surface elevations and the normal component of the mass flux, and the perfect reflection at the shoreline. Applying the boundary conditions, we set a system of $(2m - 1)$ equations. Instead of solving this system of equations, we use products of (2×2) matrices to get the solution as described two–dimensional case. We specify an initial condition using the Fourier transform of a solitary wave, as in Synolakis (1987).

For the cases, $H = 0.05$, 0.10 and 0.20 target solitary waves, the time histories of surface elevation and the maximum runup values are calculated using IMSL routines. The comparisons between the laboratory experiments and the analytical solutions are shown in figure 1–4. The overall agreement is quite good. For larger H, nonlinear effects become important and the wave may break nearshore. As discussed in Liu et al. (1995) and Yeh et al. (1994), the runup on the back of the island is enhanced, because of the collision of the two trapped waves. For $H = 0.10$, the maximum runup heights at the lee side of the island are larger than of the front, because of the constructive effects of refraction and diffraction. Our analytical model did not reproduce this feature of the physical manifestation of the problem.

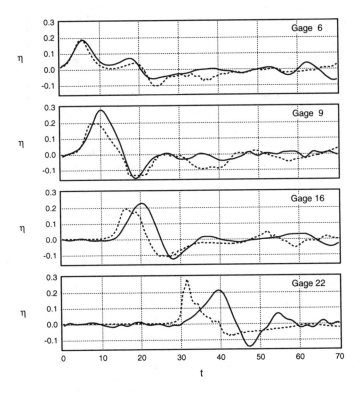

Figure 3. Comparison between the laboratory data and the analytical solution for the time histories of the surface elevation for a $H = 0.20$ target solitary wave, at four different locations. Dotted lines represent the laboratory results.

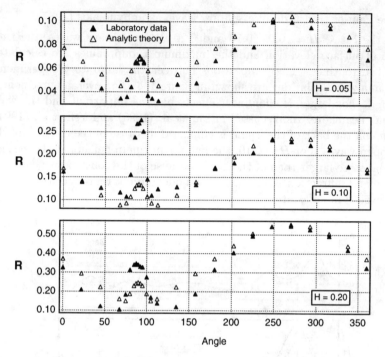

Figure 4. Comparison for the maximum runup between the laboratory data and the analytical results for three different incident wave heights.

3.2   Benchmark Problem 3, Revere Beach

As described in the previous section, the new formalism is applied to find the time histories of the surface elevation and an analytical expression for the maximum runup. The laboratory time histories of the free–surface displacements and our analytical predictions are shown in figure 5. We observe that, the linear theory predicts well the shape of the free surface displacements. Applying the general method with $\partial \eta / \partial x = 0$ boundary condition at the shoreline, and following the solution method outlined at Synolakis (1987) and extended by Kânoğlu (1996), we found that the maximum runup is given by

$$\mathbf{R} = 2\, h_w^{-1/4}\, H, \qquad (1)$$

where h_w is the dimensionless water depth at the wall. Figure 6 compares predictions using this result with the laboratory data. The agreement is gratifying.

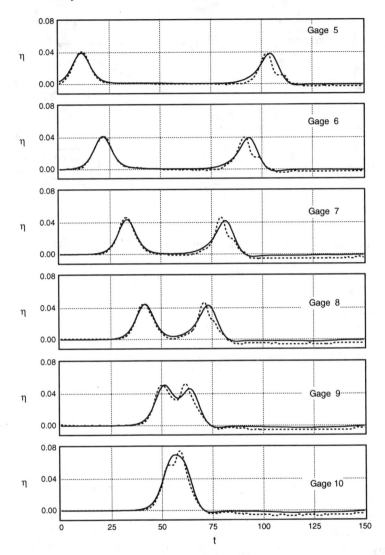

Figure 5. Comparison between the laboratory data and the analytical solution for the time histories of the surface elevation for a $H = 0.05$ target solitary wave, at six different locations. Dotted lines represent the laboratory results. The offset of the laboratory data was removed for gage 7. To faciliate the referencing of results between the theory and the experiments, the gage 4 data was used.

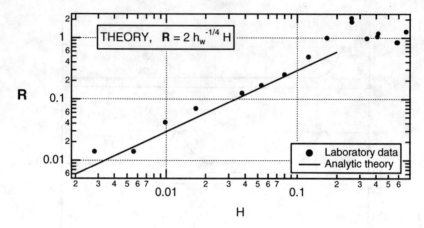

Figure 6. Comparison for the maximum runup between the laboratory data and the analytical expression (1). Additional laboratory data for the benchmark problem are provided in Kânoğlu (1996).

Acknowledgments

We are grateful to the National Science Foundation of the United States for its support through grant BCS–9201326 and a PYI grant.

References

1. Briggs, M. J., Synolakis, C. E., Harkins, G. S., Green D. R. 1994 Laboratory experiments of tsunami runup on a circular island. *PAGEOPH* **144**, 569-593.

2. Kânoğlu, U. 1996 *The runup of long waves around piecewise linear bathymetries.* Ph.D. Thesis, University of Southern California, Los Angeles, California, 90089.

3. Liu, P. L.-F., Cho, Y.-S., Briggs, M., Kanoglu, U., Synolakis, C. E. 1995 Wave runup on a conical island. *J. Fluid Mech.* **302**, 259-285.

4. Synolakis, C. E. 1987 The runup of solitary waves. *J. Fluid Mech.* **185**, 523-545.

5. Yeh, H. H., Liu, P. L.-F., Briggs, M. J., Synolakis, C. E. 1994 Propagation and amplification of tsunamis at coastal boundaries. *Nature* **372**, 353-355.

Application of Linear Theory to the Computation of Runup of Solitary Wave on a Conical Island

Koji Fujima
Department of Civil Engineering
National Defense Academy of Japan

1. Outline

Because the runup process of solitary wave is strongly nonlinear, it might be difficult to reprodece the runup of solitary wave by linear theory. However, if the results of linear analysis are compared with the results of nonlinear analysis (which may be done numerically), effect of nonlinearity will be made clear. Further, in the case of wave having small wave height, linear theory may reproduce the behavior of wave. As long as the linear theory is satisfactorily enough, linear theory seems a strong tool to analize the behavior of wave.

In the present discussion, approximate linear solution for transient incident wave is estimated by summing up the solution for stationary waves corresponding to the Fourier components of transient wave.

2. Approximate solution for transient incident wave

The linear solution for the interaction of stationary incident wave train with conical island was developed by Fujima et al. (1995). Yuliadi (1995) showed that the linear theory could sucecessfully predict the experimental data obtained by loading the incident wave whose ratio of wave height to water depth are less than 0.03.

In order to obtain the approximate solution for the transient incident wave, the most simple method is used. Namely, the incident wave profile is resolved to the summation of Fourier components by FFT, then corresponding solution to the wave of each component are summed up. This procedure can be applied to the problems of wave having an arbitrary two-dimensional initial wave profile by using two-dimensional FFT.

For the case of solitary wave, incident wave profile can be assumed to be;

$$\eta = H \mathrm{sech}^2 \left(\sqrt{\frac{3H}{4h_1^3}} (x - c(t - t_0)) \right) \tag{1}$$

where H = solitary wave height, h_1 = water depth, and t_0 = time when the wave crest was located at $x = 0$.

Table 1: Wave conditions in benchmark experiments

	h_1	H	t_0	measured H
Case A	32cm	1.5cm	25.5s	1.4cm
Case B	32cm	3.1cm	24.5s	2.9cm
Case C	32cm	6.5cm	23.7s	5.8cm

3. Computations for benchmark problems

At the Second International Workshop on Long Wave Runup Models, trajectory curves of wave-paddle used in the ecperiments were given to participants. By fitting the given curves by curves of *tanh*-function, H and t_0 in Eq. (1) can be estimated. The estimated wave heights of cases A, B and C are 1.5cm, 3.1cm and 6.5cm, respectively. The estimated time of t_0 are 25.5s for case A, 24.5s for case B and 23.7s for case C, respectively. However, Briggs et al. (1995) reported that the measured incident wave heights were 1.4cm, 2.9cm and 5.8cm respectively. The measured values are slightly smaller than the estimated values. In the present paper, computations were conducted by using both the measured and the estimated values of H. Here, it should be noted that the wave breaking was observed before the wave reached the coastline of the island in the experiment of case C.

Initial wave profile at $t = t_0$ is given at 16,382 points with the interval of $\Delta x = 10$cm. However, Fourier components whose amplitude is less than 1 percent of maximum amplitude are ignored. Numbers of mode considered in cases A, B and C are about 650, 900 and 1200, respectively.

Figures 1, 3 and 5 show the time histories of water elevation at several points in cases A, B and C, respectively. In the figures, the circles represent the experimental data, the dotted lines the theoretical values predicted by using the estimated values of H, and the solid lines the theoretical values predicted by using the measured values of H. Because the present solution is obtained by the summation of stationary solutions, small oscillation of water elevation exists when the incident wave does not reach the island yet. However, the magnitude of these unreasonable oscillation is small enough compared with the physical wave height.

Figures 2, 4 and 6 show the runup height distribution in cases A, B and C, respectively. Here, the theoretical runup height is obtained as the maximum water level at the coastline. Since the present theory adopts the Eulerian coordinates, the boundary condition at the wave tip is not satisfied perfectly. However, the solutions of the linear theory in Eulerian coordinates agree with those in Lagrangian coordinates for the first approximation. Therefore, the maximum water level at the coastline can be regarded as the runup height.

Figure 7 shows the variation of spatial distribution of water elevation in case A computed by using measured H.

In case A, the linear theory reproduces the measured time histories of water elevation and runup height distribution satisfactorily.

However, if the theoretical results for cases B and C are compared with the experimental results, following characteristics of linear theory are found.

1. Because wave celerity is underestimated, arrival time of incident wave to the island is delayed.

2. Runup height tends to be overestimated. When the measured values of H are used for computations, overestimation become small. However, the linear theory still tends to overestimate the runup height.

   In the real phenomena, the water particle runs up on the fore slope of island but may rundown along the other route on the island. On the other hand, in the linear theory, the water particle cannot run up on a dry land. Thus, the behavior of reflected wave predicted by the linear theory differs from the actual phenomena. This might cause the overestimation of runup height distribution by linear theory to some extent.

3. In the experiments of case B, the runup height at the lee of the island is higher than that at the front of the island. The linear theory cannot reproduce this fact.

   The high runup height at the lee of the island in case B might be caused by the collision of the waves whose front surface become steep. The effects of nonlinearity and dispersion could not be ignored to reproduce the high runup height at the lee of the island in case B.

4. In case C which is the large wave case, the disagreement between the prediction and the experimental results is observed most remarkably not on the lee or front of the island, but on the region $\theta = 0°$ to $45°$ and $\theta = 145°$ to $180°$. This phenomena will be discussed in the next section.

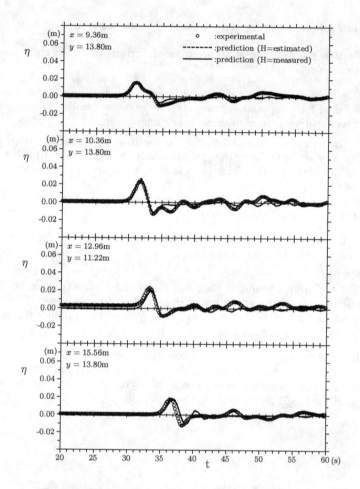

Figure 1: Time history of water elevation in case A

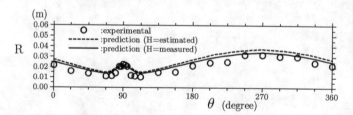

Figure 2: Runup height distribution in case A

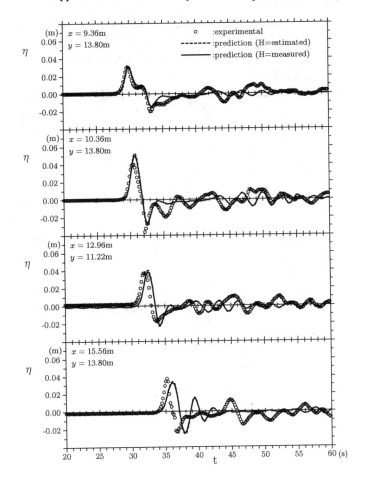

Figure 3: Time history of water elevation in case B

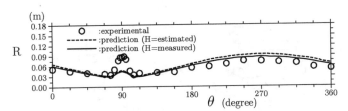

Figure 4: Runup height distribution in case B

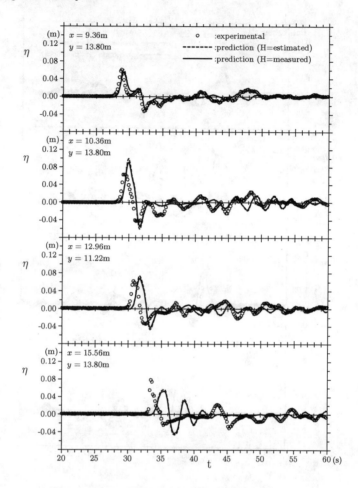

Figure 5: Time history of water elevation in case C

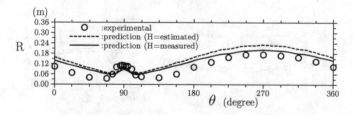

Figure 6: Runup height distribution in case C

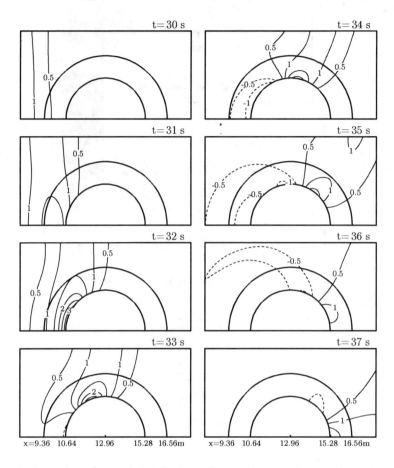

Figure 7: Variation of spatial distribution of water elevation in case A computed by using the measured H

4. Computations for NDA experiments

Model experiments of solitary wave runup on a conical island were conducted also in a wave tank of NDA (National Defense Acedemy) shown in Fig. 8 (Fujima et al., 1995; Yuliadi, 1995). In the NDA experiments, the radius of island base was 1.65m, the slope was 0.246 and the water depth h_1 was 20 and 30cm.

The positions of runup were measured on the slope of the island model by visual observation. These data were converted to runup height R. The incident wave height were determined by analyzing the wave records at the offshore point. Wave conditions

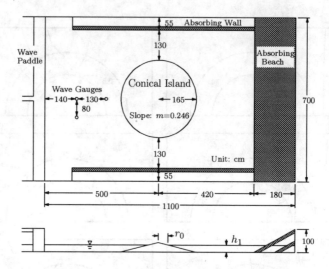

Figure 8: Experimental setup of NDA

Table 2: Wave conditions in NDA experiments

	h_1	H
Case A	29.6cm	1.1cm
Case B	29.6cm	2.2cm
Case C	29.6cm	3.5cm
Case D	19.7cm	1.1cm
Case E	19.7cm	2.1cm
Case F	19.7cm	3.7cm

of NDA experiments are listed on Table 2.

Figures 9 to 14 show the comparisons of measured runup height distribution with the predicted one. In the cases A and D, the predicted runup heights agree well with the experimental ones except for just the lee of the island. These results are similar to the result for case B in the benchmark experiments. These results show that the severe collision of the waves at the lee of the island can be occured in the case of solitary wave having small wave height.

In cases B and E, agreement is rather good. In cases C and F, the results are similar to the result of case C in the benchmark experiments. Namely, the measured runup heights are quite smaller than the predicted ones in the regions where θ is from $0°$ to $45°$ and θ is from $145°$ to $180°$. In these cases, runup water has a large inertia force, and plunges into the regions stated above. This 'plunging' may decrease the runup height in these regions. The same phenomena was possibly occured in the case C of the benchmark experiments.

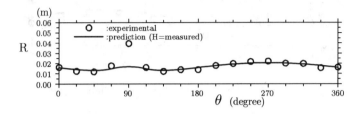

Figure 9: Runup height distribution in case A

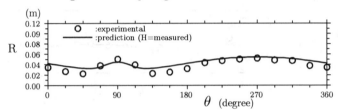

Figure 10: Runup height distribution in case B

5. Conclusions

In the case of wave having small wave height, the linear theory can reproduce both the variation of water elevation and the runup height distribuion. However, there are some cases in which the runup height at the lee of the island become quite large due to the collision of the waves. The linear theory can not reproduce such a high runup height.

In the case of wave having large wave height, the linear theory tends to underestimate the wave celerity and overestimate the runup height. When the 'plunging' is occured, disagreement between the prediction and the experimental results becomes most remarkable in the regions where θ is from $0°$ to $45°$ and θ is from $145°$ to $180°$.

1. K. Fujima, D. Yuliadi, C. Goto, K. Hayashi and T. Shigemura, *Coastal Engineering in Japan* **Vol.38, No.2** (1995), pp.111-132.
2. D. Yuliadi, *Theory and Experiments of Long Wave Run-up on a Conical Island* (Master Thesis of National Defense Academy of Japan) (1995), 124p.
3. M. J. Briggs, C. E. Synolakis, G. S. Harkins and D. R. Green, *Pure and Applied Geophisics* **Vol.144, Nos.3/4** (1995), pp.569-593.

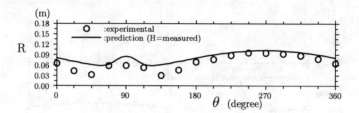

Figure 11: Runup height distribution in case C

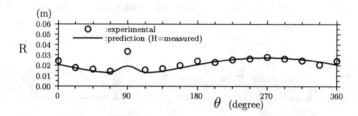

Figure 12: Runup height distribution in case D

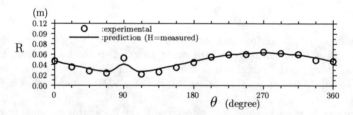

Figure 13: Runup height distribution in case E

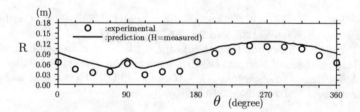

Figure 14: Runup height distribution in case F

Simulation of Wave-Packet Propagation
Along Sloping Beach by TUNAMI-Code

F. Imamura
Disaster Control Research Center
Tohoku University

1. Introduction

1.1. Governing equation

The main numerical code of the tsunami simulation, TUNAMI (Tohoku University Numerical Analysis Model for Investigation of tsunami), is introduced to analyze the problem of wave-packet propagation. Numerical model for the Near-field (TUNAMI-N) with shallow water theory for a long wave including nonlinear and friction terms is applied. The bottom friction is expressed by Manning's law with a roughness which value is selected depending on a roughness condition of a bottom.

1.2. Numerical scheme and computational condition

By using a "central" difference with the staggered numerical points for water level and discharges, the staggered leap-frog scheme, linearlized long wave theory without friction and diffusion terms is, for example, descretized as follows:

$$
\begin{cases}
\dfrac{\left[\eta_{i+\frac{1}{2}}^{n+\frac{1}{2}} - \eta_{i+\frac{1}{2}}^{n-\frac{1}{2}}\right]}{\Delta t} + \dfrac{\left[M_{i+1}^n - M_i^n\right]}{\Delta x} + O(\Delta x^2) = 0 \\[2ex]
\dfrac{\left[M_i^{n+1} - M_i^n\right]}{\Delta t} + g\dfrac{\left(h_{i+\frac{1}{2}} + h_{i-\frac{1}{2}}\right)}{2}\dfrac{\left[\eta_{i+\frac{1}{2}}^{n+\frac{1}{2}} - \eta_{i-\frac{1}{2}}^{n+\frac{1}{2}}\right]}{\Delta x} + O(\Delta x^2) = 0
\end{cases}
\tag{1}
$$

where η is the water elevation, M the discharge, g the gravity acceleration, Δx and Δt the grid sizes in the directions of x and t, and $O(\Delta x^2)$ is the truncation error of the second order approximation, which is the difference between the partial derivative and its finite differential representation. For dealing with discrete values in numerical computations, $\eta(x, y)$ and $M(x, t)$ are expressed for the case of the leap frog scheme as

$$
\begin{cases}
\eta(x, t) = \eta\left((i + \frac{1}{2})\Delta x, (n + \frac{1}{2}\Delta t)\right) = \eta_{i+\frac{1}{2}}^{n+\frac{1}{2}} \\[1.5ex]
M(x, t) = M(i\Delta x, n\Delta t) = M_i^n
\end{cases}
\tag{2}
$$

The point schematics for the numerical scheme are illustrated in Fig.1. Grid points are alternatively calculated for water level and discharge. The points for water depth, h , is the same as those for water elevation, h. Since a central scheme applied to a convection term causes an instability due to a negative numerical dissipation, a space derivative for a convection is approximated by either forward or backward difference depending on the sign of coefficient C, celerity. This is called the " upwind" scheme which is widely used because of its simplicity and applicability. The leap-frog scheme has the truncation error of $O(\Delta x^2)$, while the order of convection term, upwind scheme, is $O(\Delta x)$. For this problem, spatial grid size, Dx , and time step, Δt, are 0.05 m and 0.01 second which is satisfied with CFL condition. Reproduction time is 20 seconds.

1.3. Runup condition

One of difficult problems in tsunami simulation is runup condition in order to estimate a wave front. A wave front is located between the dry and submerged cells. And whether a cell is dry or submerged is judged as follows.

$$D = \begin{cases} h + \eta > 0, & \text{then the cell is submerged} \\ h + \eta \le 0, & \text{then the cell is dry} \end{cases} \tag{3}$$

A discharge across the boundary between the two cells is calculated if a ground level in a dry cell is lower than a water level in a submerged cell. In the contrary case, a discharge is considered to be zero. The problem in runup condition is how to calculate values of water elevation and/or discharge at a wave front. There are several approximations proposed for moving boundary condition in the Eulerian description. In the leap-frog scheme, grid points are alternatively located for velocity and water level. Assuming that a water level is already computed as a computational cell. If a water level is higher than ground level in a next cell, the water may flow into the landward cell. TUNAMI model evaluates the discharge by applying directly the momentum equation keeping a total depth as zero on a first dry cell. The total depth at a point of discharge is given by the difference between the ground level on a dry cell and the water level on a wet one.

2. Problem identification

2.1. Experimental condition

The wave basin used for the experiments in shown in Fig.2. The dimension of the basin are 13.4 m long 5.5 m wide , and 0.9 m deep, and the slope of the beach , b, is 15. The central section of the beach is constructed of a 2.4 m long, 3 m wide, and 1.27 cm thick tempered glass plate. As shown in Fig.2, the glass plate has a water-tight seal to provide a dry observation section under the basin. Temporal water-surface variation, at ST.1 (x=5.5 m, y= 0.4 m) and temporal velocity variation,u (long-shore) and v (cross-shore) , at ST.2 (x=6.89 m, y=0.1 m) were measured.

2.2. Boundary condition at wave-maker

Using a wave paddle, a packet of waves was directly generated at one end of the beach. The wave paddle is wedge-shaped, hinged offshore and swings in the long-shore direction about the axis perpendicular to the beach surface. The paddle motion creates the long shore velocity that varies linearly from zero value at the hinge. Wave paddle motion was controlled to push the water mass with a single positive stroke from the initial position, which was perpendicular to the undisturbed shoreline. The avoid excessive high-frequency noise, the paddle trajectory was set to that of a hyperbolic tangent function of time, which achieves a gradual start and cessation of paddle motion, viz.

$$\frac{\theta}{\theta_{max}} = \frac{\tanh\left(6.6\frac{t}{t_{max}} - 3.3\right) + \tanh 3.3}{2\tanh 3.3} \tag{4}$$

where q is the angle of the wave paddle from the initial position of 0 (i.e. the position perpendicular to the undisturbed shoreline), θ_{max} is the stroke angle, t_{max} is the stroke time, and t is time. The paddle hinge is located at 82.0 cm along the beach surface from the acquiescent shoreline. The water depth for all cases is 68.7 cm in the offshore region where the basin bottom is horizontal as shown in Fig.2. Defining the coordinates that the x-axis points in the long-shore direction from the initial wave-paddle position, the y-axis points horizontally offshore, and the z-axis points upward, the following cases are considered :

$$\text{CaseA} ; \quad \theta_{max} = 6.5 \quad , t_{max} = 1.5 \text{ (sec)}$$
$$\text{CaseB} ; \quad \theta_{max} = 12.87 \quad , t_{max} = 1.5 \text{ (sec)}$$

The problem in this case is how to install a rotational motion of the paddle into the Cartesian coordinate. When we use a boundary condition at the paddle as a vertical wall, we have difficulty that the direction of velocity component perpendicular to the wall being zero is varied in time. And a control volume near the paddle for a mass conservation is also changed according to the motion of the paddle. Moreover, the calculation of the convection term with up-wind scheme become complicated and difficult to solve. Therefore, the simple and practical method with a step approximation of paddle motion as shown in Fig.3 is introduced here. The problem of direction of velocity as well as calculation of convection term by using variable grid shown in a shadow area is to be solved. The length of variable grid should be longer than the distance between initial and final paddles, otherwise an area of variable grid reaches to zero or negative volume.

3. Numerical result and discussion

3.1. Time histories

Time histories of water level at ST.1 and velocities at ST.2 in case A and B are shown in Figs.4 and 5. Solid line and dashed one in these figures are computed and experimental results respectively. Regarding water levels, the behavior of edge wave like a wave trains with the second or third wave as a largest one is obviously noticed in both results. However, higher frequency components at beginning of the history of computed result is more significant than that in experimental one, suggesting that frequency dispersion effect due to the topography in the computation is smaller. And it is observed that the third wave is largest in the experiments, while the second is in the computational results. This is also caused by the dispersion relationship which would be related with an accuracy of wave refraction with a discretized bottom slope in a computation discussed by Sayama et al.(1988) and Lee & Shuto (1995). Resolution of a topography slope affects route of wave propagation on a slope. Although the magnitude of u-component in the computation is larger than that in the experiment, the agreement in the velocity is fairly good compared with results in the water level. This is contradictory to our experience of the wave computation in which a water surface is well simulated but a velocity field is not. It should be reminded that the stations to compare a water level and velocities between experiments and computations are different. In the comparison of velocities, the higher frequency component are also observed at the beginning of the first wave in the computation, which is same as the case of the water level. The good correlation of phase of the water level and u-component could be found, meaning the progressive waves in the long-shore direction. However, the v-component is not correlated with those specially at the beginning of propagation. After propagation of two main wave lengths, the v-component is shifted 90 phase with the u-component, which indicates a circular shaped velocity field in the horizontal space like a current ellipse due to a reflection on a shoreline with a inclined wave angle. This phenomenon is discussed more detail in the next section by introducing the trajectory of velocity.

3.2. *Trajectory of velocity at station*

The trajectory diagram with u- and v-component is introduced to understand the velocity field in the horizontal space. Figures 6 and 7 show the diagrams at ST. 1 and 2 during the two different time periods; from 0 to the 8 seconds and 8 to 13 minutes after the wave generation. In the beginning of propagation, the trajectory in Fig.6 shows the complicated phenomena due to the combination of two wave components with different wave periods, while during the period of 8 to 13 seconds, diagram seems to be a ellipse shaped, meaning a simple incident and reflected wave components. However, the v-component at ST. 2 shown in Fig.7 involves the significant higher frequency component shown in Fig.4, complexity at ST. 2 is more remarkable than those at ST.1.

4. Conclusion

The TUNAMI code was applied to simulate the edge wave along sloping beach and provided fairly good agreement with the measured one. However, the higher frequency wave component in the computation is not found in the experiment, which would be caused by the dispersion relationship. Although the comparisons of velocity fields between computed and measured one is good, those of water level is not. The further study including an accuracy of refraction and wave-maker condition is required.

Reference

Lee Ho Jun and N.Shuto (1995), Grid size accuracy of refraction around a conical island, Int. Long Wave Runup Workshop, Friday Harbor, Washington.

Sayama,J, N.Shuto and C.Goto (1988), Errors induced by refraction in tsunami numerical simulation, 6th APD-IAHR, Kyoto, pp.257-264.

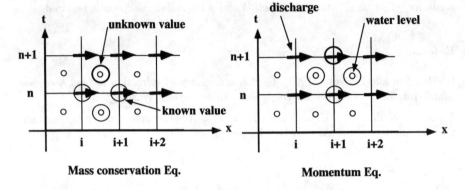

Figure 1. Arrangement of points for the computations with the staggered leap-frog scheme.

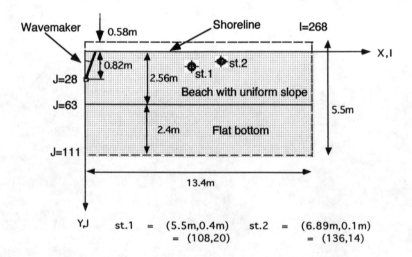

Figure 2. Coordinate system of 2D basin.

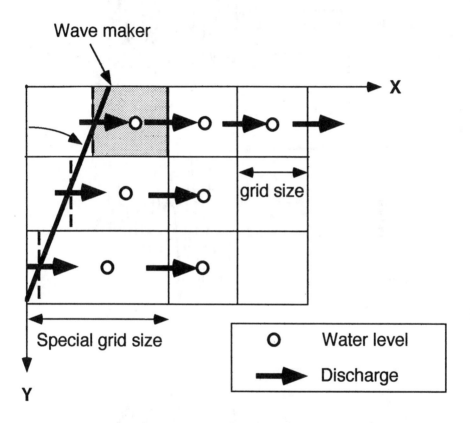

Figure 3. Approximated method of wave-maker.

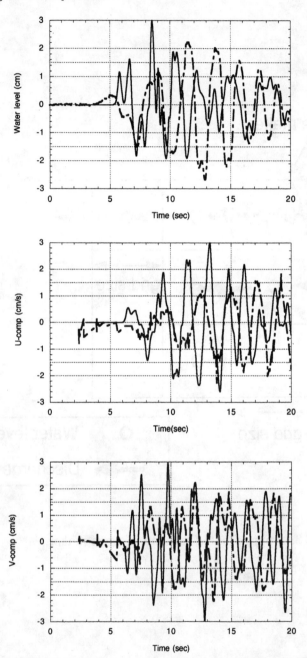

Figure 4. Comparison of water level at ST.1 and velocities at ST.2 for Case A.
Solid and dashed line show the computed and measured results respectively.

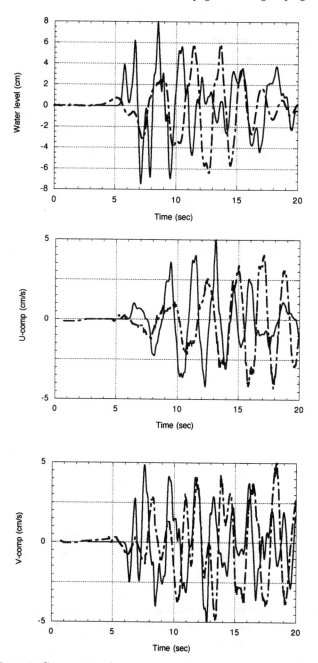

Figure 5. Comparison of water level at ST.1 and velocities at ST.2 for Case B. Solid and dashed line show the computed and measured results respectively.

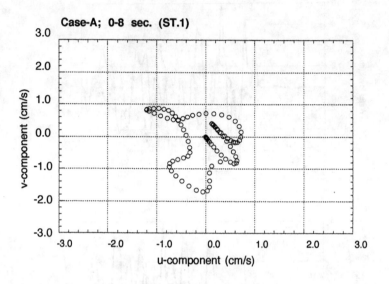

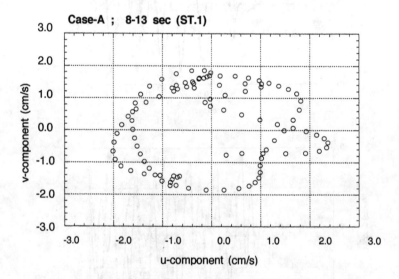

Figure 6. Trajectories of velocity components at ST.1 during the period of 0-8 seconds (upper) and 8-13 seconds (lower) for Case A.

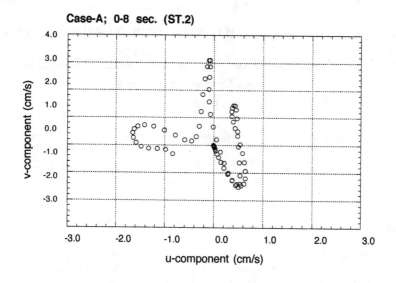

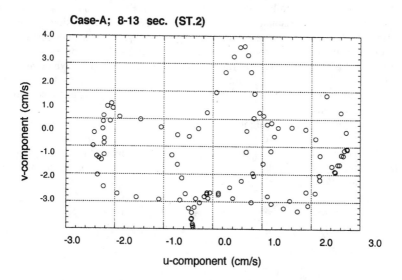

Figure 7. Trajectories of velocity components at ST.2 during the period of 0-8 seconds (upper) and 8-13 seconds (lower) for Case B.

NUMERICAL MODELING OF 3-D LONG WAVE RUNUP USING VTCS-3

Vasily Vladimirovitch Titov and Costas Emmanuel Synolakis

School of Engineering, University of Southern California

Los Angeles, California 90089

We will present solutions benchmark problems two and four, i.e., wave runup on a conical island and simulation of Hokkaido-Nansei-Oki tsunami. Our field model are the shallow-water wave equations, without friction factors or artificial viscosity. Here, we will show that—despite the severe limitation of the shallow-water wave equations—our results agree well with laboratory data and with field observations.

1. Numerical model

We use the two-dimensional shallow-water-wave (SW) equations without viscosity or bottom friction. Despite certain limitations (1), these equations have proven capable to model many important physical characteristics of tsunami propagation including wave breaking and bore runup on mild and steep beaches (2, 3). Recent studies (4) show that this approximation works reasonably well even in the case of relatively short (length to depth ratio less then 10) breaking waves. Although the equations cannot resolve the specific details of the breaking front, they adequately model —to first order—the wave behavior and give accurate estimates of runup values in a wide range of wave parameters. Details of the numerical model and results of its testing can be found in (4). Here we will only briefly describe the method focusing on the results of its application to the solution of the benchmark problems.

We will use finite-differences algorithm based on the method of fractional steps (5) to solve the SW system. This method reduces the numerical solution of the two-dimensional problem into consecutive solution of two locally one-dimensional problems. We therefore split the governing system of equations into a pair of systems, each containing only one space variable, as following:

$$h_t + (uh)_x = 0 \qquad\qquad h_t + (vh)_y = 0$$
$$u_t + uu_x + gh_x = gd_x \qquad\qquad v_t + vv_y + gh_y = gd_y \qquad (1)$$
$$v_t + uv_x = 0 \qquad\qquad u_t + vu_y = 0$$

These systems of equations are to be solved sequentially at each time step using an appropriate finite-difference method. Each of the systems in Eq. (1) can be written in characteristic form as follows

$$p_t + \lambda_1 p_x = g d_x$$
$$q_t + \lambda_2 q_x = g d_x \tag{2}$$
$$v' + \lambda_3 v'_x = 0,$$

where

$$p = u + 2\sqrt{gh},$$
$$q = u - 2\sqrt{gh}, \tag{3}$$
$$v' = v$$

are the Riemann invariants of this system and

$$\lambda_1 = u + \sqrt{gh},$$
$$\lambda_2 = u - \sqrt{gh}, \tag{4}$$
$$\lambda_3 = u$$

are the eigenvalues.

We use explicit finite-difference method to solve Eq. (2) along x and y coordinate, sequentially every time step. The first two equations in Eq. (2) constitute a one-dimensional shallow-water-wave problem, implying that every time step we will be solving a one-dimensional long wave propagation problem along each coordinate plus one more equation describing a nonlinear momentum flux in the direction normal to the coordinate along which we are solving. Therefore, we will apply the finite-difference scheme developed for one-dimensional long wave propagation and runup:

$$
\frac{\Delta_t p_i^n}{\Delta t} + \frac{1}{\Delta x_{i-1} + \Delta x_i} \left[\lambda_i^n (\Delta_{-x} + \Delta_x) p_i^n - 2\Delta t \lambda_i^n \Delta_x \left(\frac{\Delta_{-x}}{\Delta x_i} \right) \lambda_i^n p_i^n \right]
$$
$$
= \frac{g}{\Delta x_{i-1} + \Delta x_i} \left[(\Delta_{-x} + \Delta_x) d_i^n - 2\Delta t \lambda_i^n \Delta_x \left(\frac{\Delta_{-x}}{\Delta x_i} \right) d_i^n \right], \tag{5}
$$

where $f_i^n = f(x_i, t_n)$, $\Delta_t f_i^n = f(x_i, t_n + \Delta t) - f(x_i, t_n)$, $\Delta_x f_i^n = f(x_i + \Delta x_i, t_n) - f(x_i, t_n)$, $\Delta_{-x} f_i^n = f(x_i, t_n) - f(x_i - \Delta x_{i-1}, t_n)$.

The finite-difference scheme above was developed and tested for the one-dimensional shallow-water-wave equations and described in (4). The scheme allows for the spatial grid with a variable space step Δx_i. To calculate the evolution of the wave on a dry bed, we use moving boundary conditions.

2. Application of the VTCS-3 model to the benchmark problems.

2.1. Conical Island experiment (problem 2)

We performed a numerical simulation of the experiments using our numerical model. The discretized basin has a mesh size $\Delta x = \Delta y = 0.28$ m over the flat-bottom part and $\Delta x = \Delta y = 0.09$ m over sloping beach of the conical island, $\Delta t = 0.02$ sec. The boundary $y = 0$ of the computational domain is a wave generator of the numerical model.

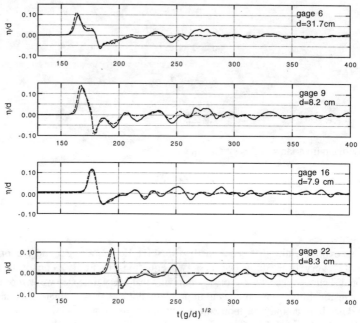

Figure 1. Time histories of surface elevation for a 0.1 target solitary wave (benchmark problem 2b). Comparison among runup computation (dashed line) and laboratory data (dots).

The movement of the generator is modeled by inputting the values of the velocity of the generator's paddle every time step. On the boundaries $x = 0$, $x = 30$ m, $y = 25$ m, the absorbing boundary conditions were specified. The moving boundary computations are performed around the conical island shoreline.

Figures 1, 2 show the comparison among the computed time series and the tide gage records of the experiment. Figures 3a,b show the comparison among the computed and measured runup values around the island for the incident waves with initial amplitudes $\eta/d = 0.1$; 0.2. Even though the experiment manifestation of the waves with the initial amplitudes 0.1 and 0.2 break on the back of the island, the results suggest that the numerical model was able to compute the runup process adequately for these breaking waves.

2.2 Hokkaido-Nansei-Oki tsunami (problem 4)

The bathymetry provided for the numerical experiment consists of several nested bathymetric arrays with three level of grid resolution. The largest array has grid size of 450 meters and contains a large area of the Sea of Japan including the source area, the Okushiri Island and West coast of the Hokkaido Island. The area of Okushiri island is covered by several nested arrays with 150 m and 50 m grid resolution. We performed computation for

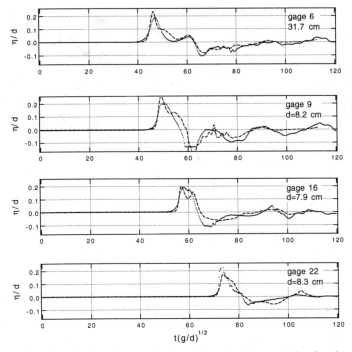

Figure 2. Time histories of surface elevation for a 0.2 target solitary wave (benchmark problem 2b). Comparison among runup computation (dashed line) and laboratory data (dots).

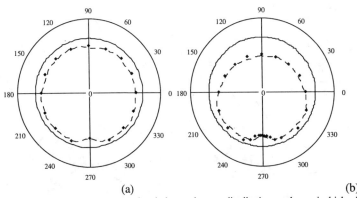

Figure 3. Comparison among computed and observed runup distribution on the conical island for (a) 0.1 initial wave amplitude and (b) 0.2 initial amplitude. Solid line indicates initial shoreline; dashed line is computed maximum runup contour, solid diamonds are observations. Incident wave came from the direction of 90 degrees.

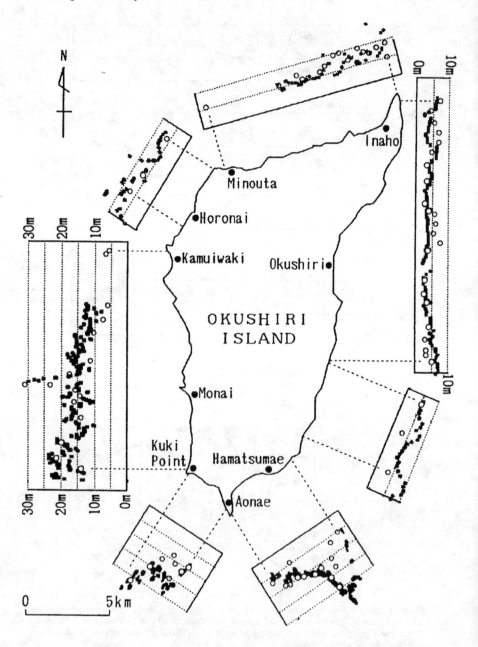

Figure 4. Runup Distribution in Okushiri Island (DCRC).

all area included in the bathymetric data, but—due to limited space of the publication—present here only the most controversial part of the computational results—the area around the Okushiri Island. Flow parameters were computed for all nested grids included in the area during the same computation cycle to provide an interchange of computed data between the grids. The runup boundary conditions were used on all coastal boundaries.

Figure 4 shows the distribution of runup heights around Okushiri island and it compares the model predictions (solid circles) with the field measurements (empty circles). The computed distribution of the maximum runup heights reproduces most features of the field measurements well. Even the extreme 31.7m measurement near Monai was computed as 29.7m. The measurement was made at the tip of a small canyon and it is undoubtedly a local effect; yet all measurements around it were consistently high with a global maximum at 31.7m, similar to the distribution of computed values shown in the figure. Most values of the computed runup are a slightly higher than measured , but then this model does not include any dissipation. It is clear, that even for this extreme tsunami event, the SW model appears to predict runup heights approximately correctly. A challenging computation for the SW theory is overland flow. We illustrate the overland flow computation near Aonae by the velocity distribution snapshot in figure 5. The snapshot corresponds to the moment of the first highest wave overtaking the Aonae cape. The flow-velocity distribution in figure 5

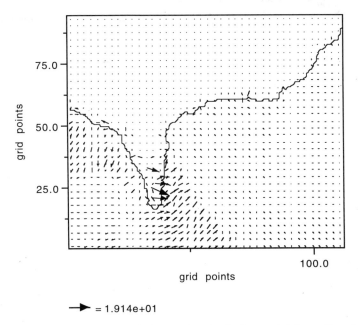

→ = 1.914e+01

Figure 5. Snapshot of the computed flow-velocity distribution around Aonae Cape for the moment 375 sec after the tsunami generation.

shows values in the range of 10 to 19 m/s over Aonae, consistent with Shimamato et al (6) estimates of flow velocities of 10–18 m/sec in Aonae. The directions of the flow over the dry bed are also consistent with the current directions estimated from observations (6,7).

We conclude that the discussed SW approximation is not only proficient in modeling laboratory data, but is also capable of quantitatively correct tsunami inundation modeling, including extreme runup heights and inundation velocities. Our results suggest that at least for numerical schemes which are stable without additional dissipation, introduction of friction factors is unnecessary.

Acknowledgments

We wish to thank the National Science Foundation of the United States for its support of this study through grant BCS–9201326 and a PYI grant.

References

1. P. L.-F. Liu, C.E. Synolakis, H.H. Yeh, *J. Fluid Mech.*, **229**, (1991), 675-688.
2. S. Hibbert and D. H. Peregrine, *J. Fluid Mech.*, **95**, (1979), 323-345.
3. N. Kobayashi, A.K. Otta and I. Roy, *J. Waterways, Port, Coastal and Ocean Eng.*, **113**(3), (1987), 282-298.
4. V. V. Titov and C. E. Synolakis, *J. Waterways, Port, Coastal and Ocean Eng.*, **121** (6), (1995), 308.
5. N.N. Yanenko, *The Method of Fractional Steps*, trans. by M. Holt, (Springer, New York, Berlin, Heidelberg, 1971).
6. T. Shimamato et al., *PAGEOPH*, **144** (3/4), (1995), 665.
7. N. Shuto and H. Matsutomi, *PAGEOPH*, **144** (3/4), (1995), 649.

Tsunami Runup on Okushiri Island

Yuichiro Tanioka, University of Michigan
Kenji Satake, Geological Survey of Japan

Abstract

Numerical computations of tsunami runup on Okushiri Island, Hokkaido, Japan are carried out using the finite-difference method. We found that the effects of grid size and initial condition on computed runup heights are significant. Use of a small grid size and a correct initial condition is essential to accurately simulate tsunami runup. Our estimation of the initial condition by inverse method from seismological and geodetic data as well as tsunami waveform data is also described briefly.

1. Computational Method

1.1. Governing Equation

The nonlinear shallow water equations with the bottom friction considered are solved by finite-difference method. We use a staggered leap-frog method. For the advection term, an upwind difference scheme is used (e.g. [1]).

1.2. Coastal Boundary Condition

We slightly modified the moving boundary condition used by Aida [2]. In the leap-flog scheme, water height and velocity are computed alternatively. After the water height is computed, the water level and the ground height are compared for each grid to determine if the grid is wet or dry. The velocity is then computed using different method depending on the type of grids. There are four types of boundaries as shown in Figure 1. The velocity at each boundary is computed as follows:

- *Type 1*: Higher grid is dry, lower grid is wet, and the water thickness of the wet grid is less than the elevation difference between the two grids. In this case, water does not flow into the higher grid, so the velocity $V = 0$ (Figure 1a).
- *Type 2*: Higher grid is dry, lower grid is wet, and the water thickness of the wet grid is larger than the elevation difference between the two grids. In this case, water flows into the higher grid. The velocity V is computed as $V = F\sqrt{gH}$ where g is the gravitational acceleration, F is the coefficient described below,

and H is the water thickness at the wet grid measured from the higher and dry grid (Figure 1b).

- *Type 3*: Higher grid is wet, lower grid is dry. In this case, water flows into the lower grid. The velocity V is computed as $V = F\sqrt{gH}$ where H is the water thickness above the wet grid (Figure 1c).
- *Type 4*: Both grids are wet but the water height at the lower grid is less than the elevation difference between the two grids. In this case, the velocity is computed in the same way as type 3 (Figure 1d).

The coefficient F corresponds to the Froude number. Yeh et al. [3] used $F = 1.43$ for the initial bore strength of their experiments. Hibberd and Peregrine [4] and Svendsen and Madsen [5] used $F = 1.45$ and 1.37, respectively, as the initial bore strength for the numerical studies. Although F for the tsunami runup might be different from these cases, we used a fixed value of $F = 1.4$.

1.3. Bathymetry Data

The computational area extends from 138°30'E to 140°33'E and from 43°18'N to 41°22'N. We used two different grid intervals , 450 m and 150 m, to examine the effect of grid size on runup computation. For the 450 m grids, we used the bathymetry data provided by Tohoku University for this workshop. For the 150 m grids, we merged the provided 150 m grid data around Okushiri Island with the interpolated data from the 450 m grids.

1.4. Initial Condition

We used two different initial conditions to examine the effect on runup. One (A in Figure 2) is provided by Tohoku University for the workshop and was estimated from the observed runup heights around Okushiri Island [6]. The other (B in Figure 2) is calculated from the fault parameters and slip distribution (Figure 3) estimated by Tanioka et al. [7] and also described in Satake and Tanioka [8]. The complex fault motion of the earthquake was estimated from the aftershock distribution, teleseismic waveforms, P wave first motions, tsunami waveforms at tide gauge stations around the Japan Sea, and geodetic data on Okushiri Island. The slip distribution on the complex fault model was determined from the joint inversion of the tsunami waveforms and geodetic data. Tanioka et al. [7] conclude that initial condition B is the best model to satisfy the seismic waves, tsunami waveforms and geodetic data.

1.5. Time Step

The time step of computation is determined to satisfy the stability condition (CFL condition) of the finite-difference computation. It is set to 1 s for the 450 m grid system and 0.5 s for the 150 m grid system. Numerical computation of tsunami are made for 20 minutes after the earthquake.

2. Effect of Grid Size on Runup

We repeated runup computation around Okushiri Island for the two different grid sizes (150 m and 450 m). We used the initial condition provided by Tohoku University for the workshop (A in Figure 2). We divided the Okushiri coast into small segments, each about 2 km long, as shown in Figure 4 and compared the observed and computed runup heights in each segment (Figure 4). The observed runup heights are taken from the Hokkaido Tsunami Survey Group [9].

The computed runup heights on the 150 m grid system match the observed heights better than those on the 450 m grid system. In the region where the large runup was observed such as Monae (segment 14 in Figure 4) and Hamatsumae (segment 20 in Figure 4), the difference between the 150m and 450m grid systems is very large. This indicates that the larger runup heights are mostly responsible to local topography. In other words, if we do not use a small grid size, large runup cannot be reproduced by numerical computation.

3. Effect of Initial Condition on Runup

We also compared runup heights from the two initial conditions, A and B (Figure 2). The grid size was fixed at 150 m. The computed runup heights are plotted with the observed heights in Figure 5. In segments 1 through 12 and 22 through 31, the computed runup heights from both initial conditions match the observed heights very well. However, in segments 13 through 21, where the large runup was observed, the computed runup from initial condition A matches the observed better than the computed runup from initial condition B. The key question is: can we simply conclude that initial condition A is better than initial condition B? We believe that the answer is no.

Initial condition A is estimated from the observed runup data [6]. Therefore, the runup heights calculated from initial condition A is supposed to match the observed heights. On the other hand, initial condition B is estimated from seismic waves, tsunami waveforms at tide gauge stations (far field tsunami), and geodetic data as described before. In other words, initial condition B is independent of the observed runup data and explains other kind of observed data, which generally constrain the earthquake source process much better than the observed runup heights. Therefore, initial condition B represents the actual vertical deformation of the ocean bottom. Then, the next question is: why are the observed runup heights between segments 13 and 21 much larger than the computed heights?

There might be two possible answers for this question. One is that the grid size was not small enough to reproduce the observed runup as described in the previous section. In fact, the maximum runup height of 30 m was observed in a very small valley near Manae and was not an average value for the corresponding grid. Figure

4 indicates that the computed runup heights become larger for the smaller grids, suggesting that if we use a small enough grid system, we may be able to reproduce such a large localized runup heights. The other possibility is that the localized tsunami source (e.g. submarine slump), which does not affect the far-field tsunami waveforms nor seismic waves, may be responsible for the large runup between the segments 13 and 21. Although there has been no evidence found so far for such a local source, the final answer awaits future studies.

We conclude this section by pointing out that the effect of initial condition on runup is significant, and the use of correct initial condition is essential to accurately compute tsunami runup.

4. Conclusion

Use of a small grid size and a correct initial condition is essential to compute tsunami runup. Study of earthquake source process from observed runup heights would be possible, but it will require the constraints from other data such as seismic waves, geodetic data, or tsunami waveforms at tide gauges.

5. References

1. W. H. Press, S. A. Teukolsky, W. T. Vetterling, and B. P. Flannery, *Numerical Recipes in FORTRAN: the art of scientific computing* (Cambridge Univ. Press, 1992).
2. I. Aida, Numerical experiments for inundation of tsunamis, Susaki and Usa, in Kochi Prefecture, *Bull. Earthq. Res. Inst.*, **52** (1977) 441-460.
3. H. H. Yeh, A. Ghazali, and I. Marton, Experimental study of bore run-up, *J. Fluid Mech.*, **206** (1989) 563-578.
4. S. Hibberd, and D. H. Peregrine, Surf and run-up on a beach: a uniform bore., *J. Fluid Mech.*, **95** (1979) 323-345.
5. I. A. Svendsen, and P. A. Madsen, A turbulent bore on a beach, *J. Fluid Mech.*, **148** (1984) 73-96.
6. T. Takahashi, T. Takahashi, N. Shuto, F. Imamura, and M. Ortiz, Source models for the 1993 Hokkaido Nansei-oki earthquake tsunami, *Pure and Applied Geophysics*, **144** (1995) 747-767. also in *"Tsunamis:1992-94"*, ed. K.Satake and F.Imamura (Birkhauser Verlag, 1995).
7. Y. Tanioka, K. Satake, and L. J. Ruff, Total analysis of the Hokkaido Nansei-oki earthquake using seismological, tsunami and geodetic data, *Geophys. Res. Lett.*, **22** (1995) 9-12.
8. K. Satake anmd Y. Tanioka, Tsunami generation of the 1993 Hokkaido Nansei-oki earthquake, *Pure and Applied Geophysics*, **144** (1995) 803-821, also in *"Tsunamis:1992-94"*, ed. K.Satake and F.Imamura (Birkhauser Verlag, 1995).

9. Hokkaido Tsunami Survey Group, Tsunami Devastates Japanese Coastal Region, *EOS, Trans. Am. Geophys. Union,* **74** (1993) 417, 432

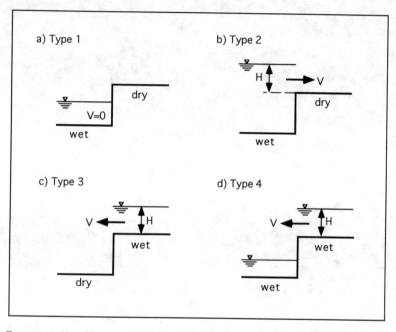

Fig. 1. Four types of boundary conditions for the coastal boundary. Thin line represents water level and thick line represents ground surface for two adjacent grids. The boundary is between the two grids. For cases 2, 3, and 4 , velocity (V) is computed as $V = F\sqrt{gH}$ where F is the coefficient (see text) and g is the gravitational acceleration.

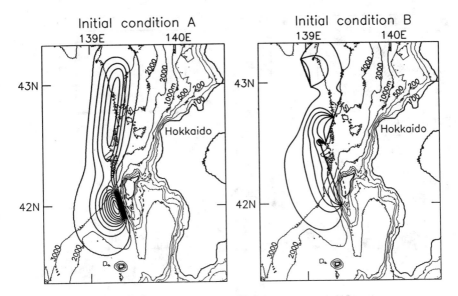

Fig. 2. Two initial conditions of the runup computation. (left) Initial condition A is provided by Tohoku University group for the workshop. (right) Initial condition B is calculated from the slip distribution (Figure 3) estimated by Tanioka et al. [7]. The solid curves indicate uplift and the dashed curves indicate subsidence. The contour interval is 50 cm.

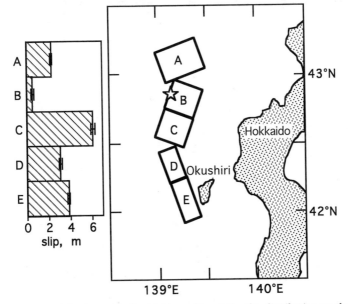

Fig. 3. The location of the five subfaults (right panel) and the slip distribution on the subfaults (right panel) estimated by Tanioka et al. [7]. A star shows the epicenter of the earthquake.

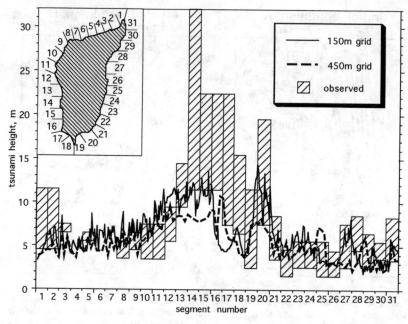

Fig. 4. Comparison of the observed tsunami runup heights around Okushiri Island (bars) and the computed tsunami heights (curves) using the two different grid sizes, 150 m (solid) and 450 m (dashed). The location of each segment is shown in inset map.

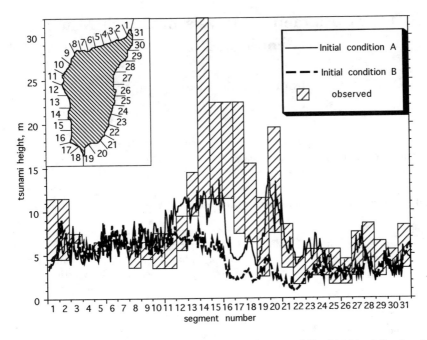

Fig. 5. Comparison of the observed tsunami runup heights around Okushiri Island (bars) and the computed tsunami heights (curves) using the two different initial conditions, A and B, shown in Figure 2. The location of each segment is shown in inset map.

Finite Element Analysis in Bench Mark Problems 2 and 3

Toshimitsu Takagi

INA Corporation, Tokyo, Japan

1. Introduction

An Analysis on propagation of long waves using the finite element method in the benchmark problem 2 and 3 have been carried out. Further a comparison is made between the computed and experiment results.

2. Benchmark Problem 2. - Runup of Solitary Waves on a Circular Island

The governing equations used in this problem is a non-linear shallow water equation, in which the bottom friction, the eddy visccsity and the Coliori force are neglected. A linear interpolation function in space, based on a triangular finite element and a three-step explicit scheme in time are employed (see the review articles).

2.1. Boundary conditions

On the boundary of the basin without the wave maker, the normal direction of velocity to the boundary set to be zero as follows:

$$u_n = 0. \tag{1}$$

On the boundary where the wave maker is placed, the water elevation η and the velocity u obtained from a solitary wave theory are specified as follows:

$$\eta(x_0, t) = H_0 sech^2 \left\{ \sqrt{\frac{3H_0}{4h^3}}(x_0 - ct) \right\}, \qquad u(x_0, t) = \frac{c\,\eta(x_0, t)}{h + \eta(x_0, t)} \tag{2}$$

where H_0 is the amplitude of incident waves, h is the water depth and c is the wave velocity defined as:

$$c = \sqrt{g(h + \eta)}. \tag{3}$$

On the shoreline around the island, a treatment of moving boundary is considered with the procedure presented by Umetsu(1995)[1]. This treatment is written in the review articles in detail.

2.2. Other conditions of the computation

The wave heights of incident solitary waves in 3 cases: case-A, B and C are 1.5, 3.0 and 6.0 cm respectively, which are determined from the experiment data. A time increment Δt is 0.002 s and a minimum depth ϵ is assumed to be 0.1 cm. Figure 1 shows the finite element mesh which composed of 76,800 elements and 38,640 nodes. The element size close to the shoreline is about 5 cm.

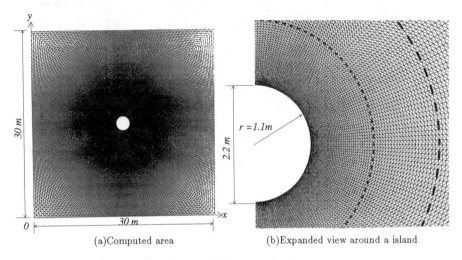

(a)Computed area   (b)Expanded view around a island

Fig. 1.   Finite element mesh

2.3. *Results and Discussion*

Figure 2 shows the comparison between the computed and experimental time histories of the water elevation at four stations: the gage A,B,C and D placed around the island and also Figure 3 shows the profiles of the water surfaces. The computed results is in good agreement with the experimental ones, especially, in the amplitudes and configuration of the first wave, recording the maximum wave height.

Table 1 and Figure 3 show the maximum vertical runup heights around the island, comparing with the experiments. It is obvious that both results are in reasonable agreement with each other, and the computed results describe the same fact as the experimental one that the runup height in the back side of the island is large partially and further it is not small comparing with that in the front.

Table 1.   Maximum wave runup heights at the front and back of island

Case No.	$H_0(cm)$	$R(cm)[R/H_0]$ in Front side		$R(cm)[R/H_0]$ in Back side	
		Computed	Experiment	Computed	Experiment
2.A	1.5	4.9 [3.3]	3.2 [2.1]	4.1 [2.7]	2.2 [1.5]
2.B	3.0	8.7 [2.9]	7.4 [2.5]	7.2 [2.4]	8.6 [2.9]
2.C	6.0	16.3 [2.7]	17.5 [2.9]	10.2 [1.7]	10.9 [1.8]

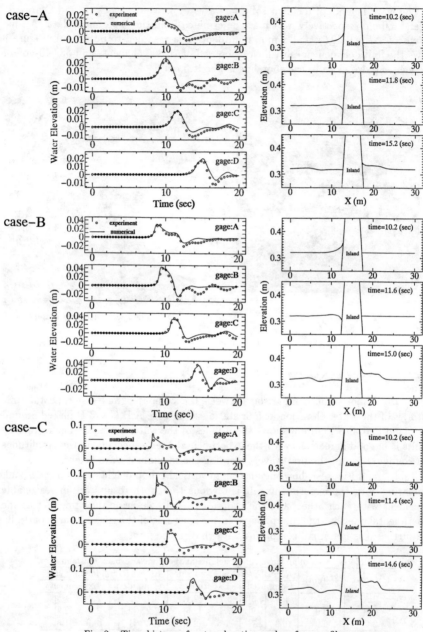

Fig. 2.   Time history of water elevation and surface profiles

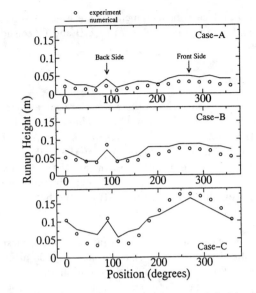

Fig. 3. Maximum wave runup heights around the island

3. Benchmark Problem 3 - Runup of Solitary Waves on a Vertical Wall

The governing equation used in this problem is the Boussinesq equations with the one dimension on the variable depth. A procedure to solve it is a semi-implicit finite element method (see the review articles).

3.1. Boundary Conditions

On the boundary in which waves are generated, the accerelation of velocity $\dot{u}$ obtained from a solitary wave theory is specified as follows:

$$\dot{u}(x_0, t) = \frac{c}{h + \eta} \left(1 - \frac{\eta}{h + \eta} \right) \dot{\eta} \tag{4}$$

where,

$$\dot{\eta} = 2c\eta \sqrt{\frac{3H_0}{4h^3}} tanh \left\{ \sqrt{\frac{3H_0}{4h^3}}(x_0 - ct) \right\}, \tag{5}$$

and η and c is obtained by equations (2) and (3).

3.2. Other conditions of the computation

The mesh size Δx and the time increment Δt are selected as respectively 5 cm and 0.01 s for case-A, and 2.5 cm and 0.005 s for case-B and C, considering the stability of computation.

3.3. *Results and Discussion*

Figure 4 shows the time histories of the water elevation at four points: gage No.5, 7, 9 and the vertical wall respectively, and also Table 2 shows the maximum vertical runup heights at the wall. The computed results in case A are in good agreement with the experiments beside the profile of wave at the wall, in which the computed one is little larger than the experimental one. In case B and C, the computed ones at gage No.5 and 7 approximately coincide with the experimental ones. However, at the gage No.9 and the wall, the disagreements between the both results are recognized. Those may be attributed to wave breaking, because that effect is not included in the present model.

Figure 5 shows the wave configurations as function of the mesh size in each case. These results describe that the Δx is 0.1 m for case A, 0.05 m for case B and 0.025 m for case C are enough to approximate the wave profiles.

Figure 6 shows two types of the process during the wave propagation; one is the result using the Boussinesq equation and the other is using the non-linear shallow water equation. It is obvious from comparison of both that results from using the shallow water equation show more tendency toward the steepness of the front of waves than one from using the Boussinesq equation. Furthermore, using the shallow water equation, the noise in the wave profile is recognized when waves approach the wall.

Table 2.   Maximum vertical runup heights at the vertical wall

Case No.	$H_0(cm)$	$R(cm)[R/H_0]$		
		Computed	Experiment	
3.A	0.85	2.2 [2.6]	1.7 [2.0]	
3.B	5.76	33.0 [5.7]	5.3 [0.92]	
3.C	15.17	–	9.6 [0.63]	

Acknowledgements

The authors would like to thank Dr.T.Umetsu for his suggestions and contributions to the bench mark problem 2.

References

1. Umetsu, T., A boundary condition technique of moving boundary simulation for broken dam problem by Three-step explicit finite element method, Advances in Hydro-Science and Engineering, Vol.II,pp.394-399,1995.

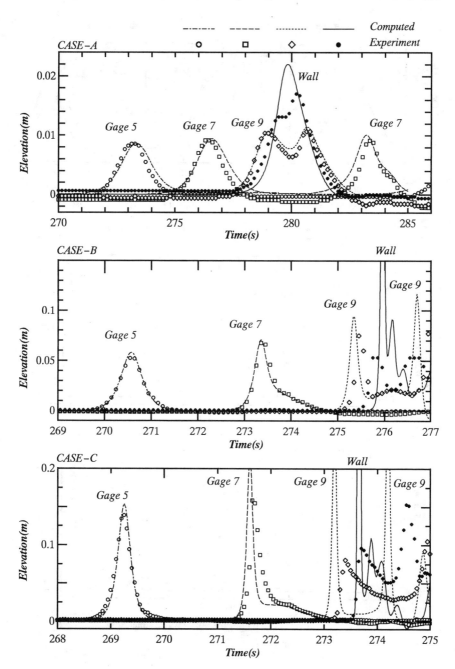

Fig. 4. Time histories of water elevations

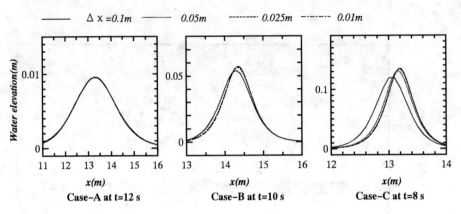

Fig. 5.   Comparison of water surface profiles as function of Δx

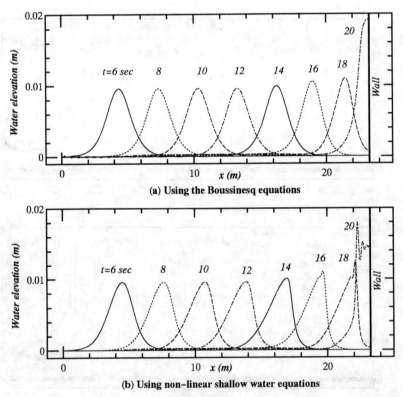

(a) Using the Boussinesq equations

(b) Using non-linear shallow water equations

Fig. 6.   Comparison of Water surface profile
between using the Boussinesq eq. and the shallow water eq.

WET2: An Eulerian–Lagrangian Shallow Water FEM Model

B. C. Beck and A. M. Baptista

Center for Coastal and Land-Margin Research
Department of Environmental Science and Engineering
Oregon Graduate Institute of Science & Technology

Abstract

Benchmark problem 3 of the International Workshop on Long-Wave Runup Models is solved using a 2D finite element model, based on the primitive form of the shallow water equations. A notable feature of the model is the Eulerian-Lagrangian treatment of advective terms, which enables the use of potentially very large Courant numbers. Results show acceptable agreement with data for case A, but increasingly poorer agreement for cases B and C, as the physics of the laboratory experiments deviates increasingly from shallow water conditions. Formal analysis helps illustrate the impact of Δx and Δt on the numerical accuracy of the model. Results shown here are a sub-set of a broader repository available through the World Wide Web.[a]

1. Introduction

Finite element models with unstructured grids have been successfully used in many coastal applications. An example is provided elsewhere in this volume [5] through an application of ADCIRC [4] to benchmark problem 4. However, ADCIRC uses a continuity-wave equation formulation that is sensitive to the choice of the wave equation parameter, and would benefit from added robustness in the presence of strong advection. While conventional primitive equation formulations are plagued by numerical oscillations, TELEMAC [2] is a model of this class that appears to both avoid oscillations and effectively handle advection, hence potentially providing an alternative to continuity-wave equation models.

Although WET2 [1] is implemented differently, it follows the same approach of TELEMAC, in that the equations are solved in primitive form, and advective terms are treated in an Eulerian-Lagrangian framework. As a part of a more systematic testing of the properties of this type of formulation, our paper describes aspects of the WET2 solution of benchmark problem 3 of the International Workshop on Long-

---

[a]http://www.ccalmr.ogi.edu/benchmarks/friday-harbor/

265

Wave Runup Models, and interprets the results based in part on formal accuracy analysis.

2. Numerical Formulation

The version of WET2 used for this benchmark is based on the shallow water equations cast in total depth form.

$$H_t + \vec{u} \cdot \nabla H + H \nabla \cdot \vec{u} = 0 \tag{1}$$

$$u_t + \vec{u} \cdot \nabla u + g H_x + \tau u = \nabla \cdot (\nu \nabla u) + g h_x \tag{2}$$

$$v_t + \vec{u} \cdot \nabla v + g H_y + \tau v = \nabla \cdot (\nu \nabla v) + g h_y \tag{3}$$

where:

H = total water depth $(\eta + h)$

η = surface elevation

h = bathymetry

u = x-velocity

v = y-velocity

τ = friction coefficient

ν = viscosity coefficient

Each solution step is split into two stages: 1) tracking through the velocity field to obtain estimations of the physical variables at the next time level $(\tilde{H}, \tilde{u}, \tilde{v})$; 2) these estimates are corrected using the remaining physical processes to obtain the final physical variables at the next time level $(H^{n+1}, u^{n+1}, v^{n+1})$.

Stage 1 uses the backwards tracking of velocity-based characteristic lines to solve Eq. (4-6). This is currently done using a 4th order Runge-Kutta integration.

$$H_t + \vec{u} \cdot \nabla H = 0 \tag{4}$$

$$u_t + \vec{u} \cdot \nabla u = 0 \tag{5}$$

$$v_t + \vec{u} \cdot \nabla v = 0 \tag{6}$$

Stage 2 uses a linear Galerkin finite element approach to solve Eq. (7-9), which are the shallow water equations written in Lagrangian form.

$$\frac{DH}{Dt} + H \nabla \cdot \vec{u} = 0 \tag{7}$$

$$\frac{Du}{Dt} + g H_x + \tau u = \nabla \cdot (\nu \nabla u) + g h_x \tag{8}$$

$$\frac{Dv}{Dt} + g H_y + \tau v = \nabla \cdot (\nu \nabla v) + g h_y \tag{9}$$

In solving the problem in stage 2, an alpha method is used for the finite difference time discretization of Eq. (7-9) resulting in the following set.

$$\frac{H^{n+1} - \tilde{H}}{\Delta t} + \alpha H^n \nabla \cdot u^{n+1} = (\alpha - 1)H^n \nabla \cdot u^n \tag{10}$$

$$\begin{aligned}\frac{u^{n+1} - \tilde{u}}{\Delta t} + \alpha g H_x^{n+1} + \alpha \tau u^{n+1} \\ -\alpha \nabla \cdot (\nu \nabla u^{n+1})\end{aligned} = \begin{aligned}(\alpha - 1)g H_x^n + (\alpha - 1)\tau u^n \\ +(1 - \alpha)\nabla \cdot (\nu \nabla u^n) + g h_x\end{aligned} \tag{11}$$

$$\begin{aligned}\frac{v^{n+1} - \tilde{v}}{\Delta t} + \alpha g H_y^{n+1} + \alpha \tau v^{n+1} \\ -\alpha \nabla \cdot (\nu \nabla v^{n+1})\end{aligned} = \begin{aligned}(\alpha - 1)g H_y^n + (\alpha - 1)\tau v^n \\ +(1 - \alpha)\nabla \cdot (\nu \nabla v^n) + g h_y\end{aligned} \tag{12}$$

Equations (10-12), are coupled such that $(H^{n+1}, u^{n+1}, v^{n+1})$ are obtained simultaneously. WET2 employs the element-by-element method of Hervouet [3], for sparse representation of the mass matrix, coupled with a conjugate gradient method for solving the system of equations resulting from the application of the finite element method in stage 2. A detailed description of the internal workings of WET2 will be published separately [1].

3. Problem 3 Simulations

3.1. Set-up

The computational domain for this problem is shortened so that the forced boundary is coincident with gage 4. The gage data at this location are fit with gaussian curves for the purpose of imposing elevation boundary conditions.

$$\eta = a_0 \exp\left(-\frac{(t - a_1)^2}{a_2}\right) + a_3 \exp\left(-\frac{(t - a_4)^2}{a_5}\right) \tag{13}$$

The parameters $(a_0, a_1, a_2, a_3, a_4, a_5)$ for each case are given in Table 1. The domain is shortened because the shallow water equations cannot propagate a solitary wave without prematurely steapening and eventually breaking. This shortened domain provides a better test of the shallow water equations and their ability to describe the physics in the shoaling region.

Table 1: Boundary condition fit parameters.

Case	a_0	a_1	a_2	a_3	a_4	a_5
A	0.00832	271.58	0.57749	n/a	n/a	n/a
B	0.05598	269.98	0.11729	n/a	n/a	n/a
C	0.10509	268.89	0.02528	0.04317	268.89	0.13927

In each of the three cases the domain is discretized in such a way that the dimensionless wavelength remains approximately constant. In all of the results presented here the dimensionless wavelength is at least 40 in the long-channel direction.

All non-linear terms, including advection, are included in the simulations. No friction or viscosity is included in any of the simulations presented.

3.2. Results

Figures 1 thru 6 show time histories of elevation at the gages requested. Each of these figures shows the data along with the results of one simulation. The simulations shown represent the best results obtained thus far. Additional results will be made available via the World Wide Web (see Sec. 4) as they are generated.

Also as requested, Table 2 provides normalized maximum runup heights at the vertical wall. The runup heights are normalized using the water depth at the location of gage 4 $(0.218m)$.

Table 2: Maximum normalized runup on the vertical wall (R/d).

	Case A	Case B	Case C
Data	0.13	2.10	1.26
Simulation	0.09	0.36	0.43

3.3. Discussion

The simulation for case A agrees reasonably well with the data, although some dissipation is present. The simulations for cases B and C show significant dissipation with respect to the data. One explanation could be that the shallow water equations cannot represent the physics of the vertical jet formed at the wall. This is because the vertical acceleration of the water is no longer much smaller than the acceleration due to gravity, hence the hydrostatic assumption is no longer valid. Another explanation for the dissipation could be the use of linear basis functions for interpolation at the feet of the characteristics during the Lagrangian tracking step. Also for the shorter period waves in cases B and C a shallow water equation model will tend to form a dissipative bore prematurely.

Formal analyis conducted on the current model [1] shows that the amplification of the solution is sensitive to the fractional step parameter α. In order to maintain stability of the solution it is often necessary to use an α greater than 0.5. This may also contribute to the dissipative nature of the simulations presented, since all of the results utilize an α of 0.6 or above.

While sensitivity is somewhat larger for cases B and C than shown in Fig. 7 for case A, the variation of Δx appears to have little effect on accuracy, in the range of adopted dimensionless wavelengths. This would exclude, for this problem, our earlier hypothesis that the observed dissipation is due to the interpolation at the feet of the characteristic lines.

The effect of time step on the solutions changes with the physical characteristics of the problem. For case A the solution steadily improves as the time step is decreased (Fig. 8). While for cases B and C the best agreement is obtained with Courant numbers greater than 1.

Figures 9 and 10 show results of the formal analysis of WET2 for case A and they tend to support the behavior shown in Figs. 7 and 8. For each set of conditions the formal analysis provides 2 roots. The arrows in Figs. 9 and 10 show the movement of these roots at Δt is decreased. Figure 9 which shows little variation in amplification factor for the range of Δx used in Fig. 7. Both Fig. 9 and Fig. 10 show both roots of the formal analysis moving toward 1.0 as Δt becomes small which supports the behavior seen in Fig. 8.

4. Utilizing The Web

Our results including those presented here are available via the World Wide Web.[b] Our presentation of the results allows the user to select results for particular gages and simulations which are returned to the user and can be viewed using most plotting packages. This type of presentation offers flexibility in viewing the results and allows updated results to be made available quickly and efficiently. Results of simulations for the other benchmarks problems are also available.[c]

5. Final Considerations

WET2 shows promising results for case A, whose conditions come closest to the shallow water conditions. The reason for the significant dissipation in the results for cases B and C still needs to be explored and linked with the results of the formal analysis. This will allow us to determine how much of the dissipation can be attributed to the numerical method and how much can be attributed to the physics.

6. Acknowledgements

This research was supported in part by the U.S. Army Research Office (Grant No. DAAL03-92-G-0065) and by the National Science Foundation (Contract No. OCE-9412028). Thanks are due to H. Yeh, P. Liu, and C. Synolakis for setting-up the benchmark problems.

7. References

1. B. C. Beck and A. M. Baptista, An Eulerian-Lagrangian Finite Element Model for Coastal Inundation by Free and Forced Waves, for *Int. J. Num.*

---

[b]http://www.ccalmr.ogi.edu/benchmarks/friday-harbor/prob3/
[c]http://www.ccalmr.ogi.edu/benchmarks/friday-harbor/

Meth. in Fluids (1996) [in preparation].

2. J.-C. Galland, N. Goutal and J.-M. Hervouet, TELEMAC: A new numerical model for solving shallow water equations. *Advances in Water Resources,* **14** (3), (1991) 138-148.

3. J.-M. Hervouet, Element by element methods for solving shallow water equations with F.E.M, *Vol. 1: Numerical Methods in Water Resources, Computational Methods in Water Resources IX, Computational Mechanics Pub.,* (1992) 761-768.

4. R. A. Luettich and J. J. Westerink, *Implementation and Testing of Elemental Flooding and Drying in the ADCIRC Hydrodynamic Model,* (Dept. of the Army, U.S. Army Corps of Engineers, Vicksburg, MS., 1995).

5. E. P. Myers and A. M. Baptista, Finite Element Solutions of the Hokkaido Nansei-Oki Benchmark. *this volume,* (1996).

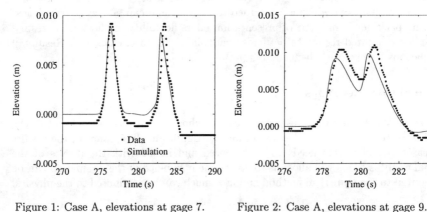

Figure 1: Case A, elevations at gage 7.   Figure 2: Case A, elevations at gage 9.

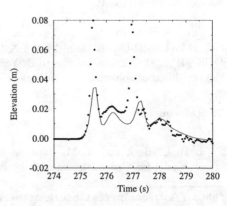

Figure 3: Case B, elevations at gage 7.   Figure 4: Case B, elevations at gage 9.

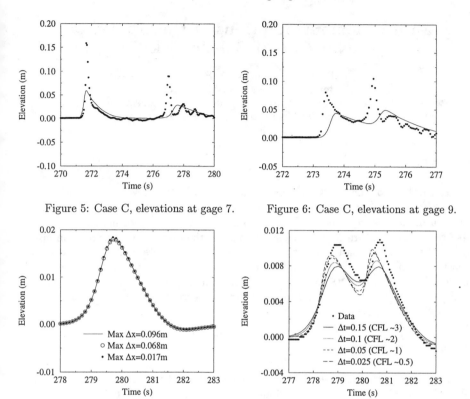

Figure 5: Case C, elevations at gage 7.

Figure 6: Case C, elevations at gage 9.

Figure 7: Case A, elev. at the vertical wall.

Figure 8: Case A, elevations at gage 9.

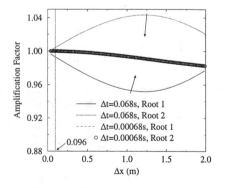

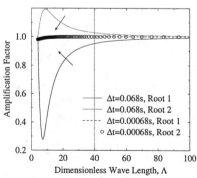

Figure 9: Case A, variation of amp. factor versus Δx with respect to Δt.

Figure 10: Case A, variation of amp. factor versus dimensionless wavelength with respect to Δt.

Finite Element Solutions of the Hokkaido Nansei-Oki Benchmark

E. P. Myers and A. M. Baptista
Center for Coastal and Land-Margin Research
Department of Environmental Science and Engineering
Oregon Graduate Institute of Science & Technology

Abstract

The Hokkaido Nansei-Oki benchmark of the International Workshop on Long-Wave Runup Models is solved for three progressively more refined unstructured grids, using a shallow water finite element model that accounts for inundation. Numerical results capture well the trends (if not all the details) of observed run-ups along the Okushiri coast. Observed records of water heights along the Hokkaido coast are reproduced well in magnitude, but periods are significantly underestimated. In general, results compare favorably with previous efforts by the authors and by others, and show the importance of unstructured grid refinement to capture localized inundation effects.

1. Introduction

This paper describes the solution of benchmark problem 4 of the International Workshop on Long-Wave Runup Models (Yeh et. al., 1995). This is the only benchmark with a field setting: the July 12, 1993 Hokkaido Nansei-Oki (henceforth HNO93) tsunami which impacted the coastlines of Okushiri, Hokkaido and other surrounding lands of the Sea of Japan.

The HNO93 tsunami has been investigated by several researchers, independently from the workshop. Of particular relevance to this paper are the work of Takahashi et. al. (1995), Satake and Tanioka (1995), Shuto and Matsutomi (1995), Shimamoto et. al. (1995), and Myers and Baptista (1995). Takahashi et. al. (1995) proposed the source mechanism adopted in the benchmark. Like Satake and Tanioka (1995), they performed a joint inversion of geodetic and tsunami data. However, their recommend source results from separate extensive forward trial-and-error simulations. Shuto and Matsutomi (1995) and Shimamoto et. al. (1995) provided two of the most detailed descriptions of the several post-tsunami surveys conducted for the event.

In their analysis, Takahashi et. al. (1995) and Satake and Tanioka (1995) used finite difference models with orthogonal grids. By contrast, Myers and Baptista

(1995) report a modeling investigation based on unstructured grids, using an earlier version (without wetting and drying) of the same finite element model adopted here (modified from Luettich et. al., 1995). In this earlier investigation we used a range of excitation mechanisms distinct from the mechanism specified in the benchmark, a coarser bathymetric data base, and a less refined grid.

2. Modeling set-up[a]

The model used in the simulations, ADCIRC, solves the depth-averaged shallow water equations on unstructured grids, using a continuity-wave equation finite element formulation. The basic algorithm is that of Westerink et. al. (1992), modified by Myers and Baptista (1995) to account for bottom motion and for transmissive boundary conditions, and extended by Luettich et. al. (1995) to include inundation.

Wetting and drying is accomplished by turning at each time step the contribution of individual elements to the global weighted residual statement on (if they are "wet") or off (if they are "dry"). Consistently, at each time step individual nodes are classified as either "dry", "interface", or "wet". Dry nodes are connected only to dry elements, and are constrained to have zero velocities and constant water levels. Interface nodes are connected to both wet and dry elements and are treated as a part of a non-normal flux land boundary. Wet nodes are connected only to wet elements and are not subject to any special constraint.

A wet element dries when the water depth at any of its nodes goes below an user-specified minimum. However, to help stability, recently wetted elements can only dry after an user-imposed lag. The wetting of elements requires a decision on whether water levels at all dry nodes will increase enough in the following time step. An element is allowed to become wet only when both the water level gradient and (when applicable) the vector sum of the water level gradient and the wind stress indicate "enough" water motion in the direction of all its dry nodes.

Numerical grids used in the simulations are unstructured, composed of linear triangles. Three grids were constructed using the semi-automatic grid generator of Turner and Baptista (1991). Transmissive boundary conditions are imposed at water boundaries, while land boundaries are allowed to inundate.

Grid refinement varies over the domain, taking advantage of the unstructured nature of the grids (Figure 1). The "coarse" grid has 34294 nodes and 66599 elements, with the smallest element being characterized by an equivalent diameter (twice the square root of the division of the elemental area by π) of 21.48 m. These numbers change to (41629, 80945, 22.40) and (60181, 117004, 10.34) nodes, elements, and

---

[a]Figures presented in this paper to describe the modeling set-up and results are a sub-set of images and animations available throught the World Wide Web (WWW):

http://www.ccalmr.ogi.edu/benchmarks/friday-harbor

The same address gives access to solutions by the Center for Coastal and Land-Margin Research of other benchmarks of the International Workshop on Long-Wave Runup Models, including Beck and Baptista (1995).

smallest equivalent diameter for the "medium" and "fine" grids, respectively. Refinement from grid to grid concentrates primarily on the near-shore Okushiri. The distribution of the equivalent diameter of the elements is shown for each grid in Figure 2.

For each grid, nodal depths are computed by linear interpolation from the bathymetric data base provided in support of the benchmark. Hence, our three grids differ from each other both in horizontal refinement and in bathymetric resolution.

Of the different options allowed by ADCIRC, we adopt a fully explicit lumped formulation. To guarantee stability and enhance accuracy, celerity-based Courant numbers are kept below 0.90 over the entire domain. This constraint forces the use of fairly small time steps: 2, 2, and 1 seconds respectively for each of the increasingly finer grids. The wave-equation parameter was selected as G=0.3, partially based on the sensitivity analysis of Myers and Baptista (1995).

3. Results

Observed and simulated run-ups along the coast of Okushiri are shown in Figure 3. The agreement is generally quite satisfactory. However, simulations do not capture the highest observed run-ups (31.7 m), first measured by Y. Tsuji in a very narrow valley north of Monai. Based on our own observations (Baptista et. al., 1993) we tend to agree with Shuto and Matsutomi (1995) that a local grid resolution of O(5 m) would probably be needed to represent this effect. While this refinement is only two-fold smaller than the one that we used, hence numerically possible, we have no supporting bathymetry to make simulations on a locally further refined grid meaningful.

Interestingly, simulations with the fine grid suggest run-ups of 30-35 m in Hamatsumae and in a narrow region south of Monai. These large run-ups are neither confirmed nor negated by observations, raising the question of whether the effect is real or purely numerical. A possibility is that, because velocities are fairly large, the effective Courant number (based on the sum of celerity and velocity) locally exceeds unity, leading temporarily to incipient instabilities.[a]

Wave arrival times for the Okushiri coast are reported by Shuto and Matsutomi (1995) and Shimamoto et. al. (1995) based on the information from watches presumably broken during the inundation process. The reliability of this information is low. Not only is it common for personal watches to be inaccurate by several minutes, but is also possible that watches keep running for several minutes after the arrival of the first wave. Predicted arrival times (Figure 4) are within two minutes of observations at South Aonae, Monai, and Hoyaishikawa, and five minutes at Hamatsumae. While certainly within the range of the observational error, these differences are substantial compared to the actual arrival times (5-10 minutes), preventing meaningful further analysis.

---

[a]http://www.ccalmr.ogi.edu/benchmarks/friday-harbor/prob4

More interestingly, predictions may explain why much longer observed arrival times were reported at North Aonae (about 21 minutes) than at South Aonae (about 5 minutes). Indeed, while the first waves in both locations appear to have arrived within a couple of minutes of each other, in North Aonae the first and second waves are small enough that inundation may have taken place only during the larger third wave. Simulations suggest that this third wave, possibly a result from reflections on the Hokkaido coast, arrived about 17 minutes after the earthquake.

Observed and simulated run-ups along the coast of Hokkaido are shown in Figure 5. Simulations appear to capture trends well, but to overpredict magnitudes. Simulated time histories of elevations at the tidal gauges of Esashi and Iwanai (Figure 6) also show overprediction of magnitude; perhaps more significantly, they also show substantial underprediction of periods.

Isolines of maximum velocity magnitudes are shown in Figure 7, for the southern part of Okushiri. Unlike the isolines of maximum elevations (not shown) and the diagrams of maximum run-up (Figure 3), the isolines of maximum velocity suggest that there is a concentration of energy in shallow water off the narrow valley where Tsuji measured 31.7m of run-up. This concentration may combine with the geometry of the valley to explain the local intensity of the run-up.

4. Discussion

The inundation algorithm of Luettich and Westerink (1995) has shown good robustness for the present application, allowing us to take advantage of unstructured local grid refinements to better characterize run-ups at both Okushiri and Hokkaido. Results are markedly improved relative to Myers and Baptista (1995) and to Satake and Tanioka (1995), both of whom strongly underpredicted the observed run-ups at Okushiri. Neither of these earlier efforts accounted for inundation or achieved the present degree of grid refinement, which arguably explains a large part of the improvement.

By contrast, there is a broad level of consistency among our results and those of Takahashi et. al. (1995). Both modeling efforts arguably capture the order of magnitude and major trends of the run-up along the coasts of Okushiri and Hokkaido, while still overpredicting impact at Hokkaido and (except for our more refined grids) underpredicting impact in southern Okushiri. Both efforts predict at Esashi and Iwanai waveforms with shorter periods than those observed. The models also appear self-consistent in their predictions of arrival times along the Okushiri coast.

While reassuring from a numerical viewpoint, this general level of self-consistency among Takahashi et. al. (1995) and our modeling efforts does not imply that the understanding of the HNO93 tsunami is satisfactory. In particular, the reliability of the source characteristics remains difficult to establish:

- observed arrival times along the Okushiri coast differ in pattern from those predicted by both models; however, the low reliability of observed arrival times precludes a firm judgement of the source on this basis alone;

- observed waveform periods at Esashi and Iwanai are about double of those predicted numerically; however, it is possible that factors other than the source characteristics are responsible; for instance, the local topography and containment conditions of the tidal gauges are not described by any of the models;

- the source characteristics proposed by Takahashi et. al. (1995) and used in the benchmark were chosen by trial and error fitting to the run-up data, and differ significantly from those derived by Satake and Tanioka (1995).

In spite of their general self-consistency, our results show some important differences relative to Takahashi et. al. (1995). In particular, our arrival times at Iwanai match well those observed, while a five minutes early arrival is predicted by Takahashi et. al. (1995). Also, our use of unstructured locally refined grids appears to capture the high run-ups of southern Okushiri better.

These differences and the degree of confidence acquired in the inundation algorithm may be substantial enough to justify performing a new tsunami inversion of the source, based jointly on tidal records at Iwanai and Esashi and on run-ups along the Okushiri and Hokkaido coasts. Although we have used this approach diagnostically before (Myers et. al., 1993), we have until now deemed the approach unfeasible for a real application because of model limitations in describing local inundation.

From a public safety viewpoint, the examination of the results is both reassuring and sobering. Indeed, models are approaching a reasonable description of the HNO93 tsunami, by properly capturing trends and orders of magnitude of impact. However, local effects and important wave properties are often still not properly captured, over two years after one of the best monitored events ever. Furthermore, post-tsunami survey data was critical to the definition of source characteristics, which shows the difficulty in forecasting impact prior to the event.

5. Acknowledgements

This research was supported in part by Oregon Sea Grant Subcontract No. R/CP-30/2-5706-01 and by the U.S. Army Research Office Grant No. DAAL03-92-G-0065. Thanks are due to R. Luettich and J. Westerink for providing the base ADCIRC algorithm, to H. Yeh, P. Liu, and C. Synolakis for setting-up the benchmark, and to B. Beck for assisting with the preparation of the TEX version of this document. Special thanks are also due to the many researchers in Japan and elsewhere who made the benchmark useful by contributing bathymetric data, raw and processed tidal gauge records, and run-up data.

7. References

1. A.M. Baptista, G.R. Priest, and Y. Tanioka, A post-tsunami survey of the 1993 Hokkaido tsunami, *EOS Trans. Amer. Geophy. Union* **74**(43) (1993) 350.

2. B.C. Beck and A.M. Baptista, WET2: An Eulerian-Lagrangian Shallow Water FEM Model, International Workshop on Long-Wave Runup Models, September 12-17 (1995), Friday Harbor, Washington.

3. R.A. Luettich and J.J. Westerink, Implementation and Testing of Elemental Flooding and Drying in the ADCIRC Hydrodynamic Model, Dept. of the Army, U.S. Army Corps of Engineers, Vicksburg, MS (1995).

4. E.P. Myers, A.M. Baptista, A.M., and Y. Wang, Non-linear Inversion of Tsunami Waveforms: Diagnostic Analysis for the 1992 Nicaragua and 1993 Hokkaido Tsunamis, *EOS Trans. Amer. Geophy. Union* **74**(43) (1993) 351.

5. E.P. Myers and A.M. Baptista, Finite Element Modeling of the July 12, 1993 Hokkaido Nansei-Oki Tsunami, *Pure and Applied Geophysics* **144**(3/4) (1995) 1070-1103.

6. K. Satake, and Y. Tanioka, Tsunami Generation of the 1993 Hokkaido Nansei-Oki Earthquake, *Pure and Applied Geophysics* **144**(3/4) (1995) 803-822.

7. T. Shimamoto, A. Tsutsumi, E. Kawamoto, M. Miyawaki, and H. Sato, Field Survey Report on Tsunami Disasters Caused by the 1993 Southwest Hokkaido Earthquake, *Pure and Applied Geophysics* **145**(3/4) 1995 665-691.

8. N. Shuto and H. Matsutomi, Field Survey of the 1993 Hokkaido-Nansei-Oki Earthquake Tsunami, *Pure and Applied Geophysics* **144**(3/4) (1995) 650-663.

9. To. Takahashi, Ta. Takahashi, and M. Ortiz, Source Models for the 1993 Hokkaido Nansei-Oki Earthquake Tsunami, *Pure and Applied Geophysics* **144**(3/4) (1995) 747-768.

10. P.J. Turner and A.M. Baptista, ACE/Gredit Users Manual: Software for Semi-automatic Generation of Two-dimensional Finite Element Grids, CCALMR Software Report SDS2(91-2), Oregon Graduate Institute of Science and Technology, Portland, Oregon, USA (1991).

11. J.J. Westerink, R.A. Luettich, A.M. Baptista, N.W. Scheffner, and P. Farrar, Tide and Storm Surge Predictions in the Gulf of Mexico Using a Wave-Continuity Equation Finite Element Model, *ASCE Journal of Hydraulic Engineering* **118**(10) (1992) 1373-1390.

12. H. Yeh, P. Liu, and C. Synolakis, International Workshop on Long-Wave Runup Models, September 12 - 17 (1995), Friday Harbor, Washington.

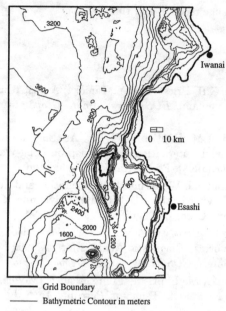

Figure 1a: The computational domain.

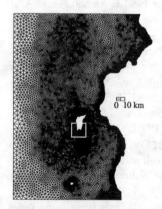

Figure 1b: Global view of the discretization with an unstructured grid.

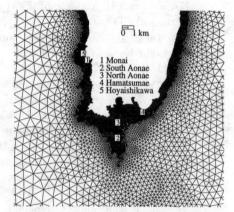

Figure 1c: Detail of grid discretization around southern Okushiri.

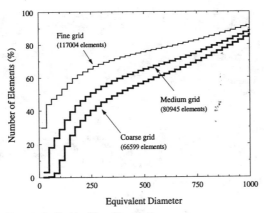

Figure 2: Equivalent diameter.

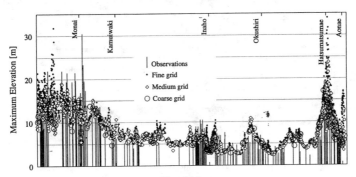

Figure 3: Maximum run-ups - Okushiri.

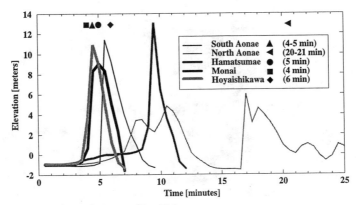

Figure 4: Arrival times - Okushiri.

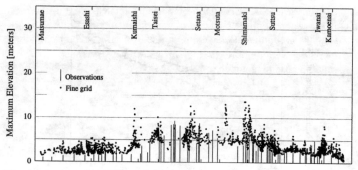

Figure 5: Maximum run-ups - Hokkaido.

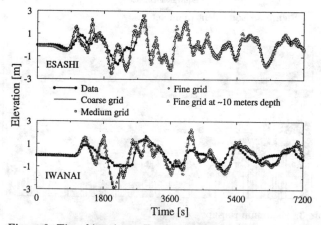

Figure 6: Time histories at Esashi and Iwanai.

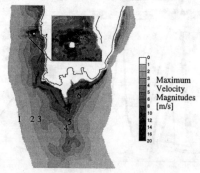

Figure 7: Maximum velocity magnitudes
in southern Okushiri.

APPLICATION OF A FINITE ELEMENT, WAVE EQUATION MODEL TO THE BENCHMARK PROBLEMS: 1 AND 3

Roy A. Walters [a]

1. Introduction

A finite element model that has been applied extensively to flow simulations in estuaries and coastal seas is applied to the wave runup problem. The finite element method offers great flexibility in the discretization of complex physical domains such as the Inland Marine Waters of Washington State which form the backdrop for this workshop. In addition, there is a wealth of theory for error estimation, h (grid size) and p (interpolation order) adaptive refinement strategies, grid generation, and other topics (see review chapter).

The purpose of this study is to explore the applicability of the finite element wave equation formulation of the shallow water equations to the wave runup problem. This formulation decouples the problem into a nonlinear wave equation for surface elevation, and a momentum equation for the horizontal velocity, a procedure that leads to high computational efficiency. In particular, there are two problems that must be dealt with: simulation of advection dominated flows, and approximation of moving boundaries. This initial discussion focuses on benchmark problem 1, an edge wave that propagates along a sloping beach, and problem 3, wave runup on a vertical wall. Sensitivity tests examine the effects of grid refinement, time-step size, bottom friction coefficient, and dissipation coefficient. The mathematical and numerical models that are used in the analysis are described in the next section. Following this, computational results for Benchmark problems 1 and 3 are presented along with some preliminary conclusions.

2. Model

The model is based on the shallow water equations which are derived by vertically integrating the Navier-Stokes or Reynolds equations[4]. They consist of the continuity equation and the horizontal components of the momentum equation (see review chapter).

The coefficient for the horizontal component of viscous stresses is $A_h(x, y, t)$. In

---

[a]U.S. Geological Survey, 1201 Pacific Ave., Suite 600, Tacoma, WA, 98402, U.S.A.

tidal models, A_h is usually assumed to vanish because bottom friction dominates. However, here it is used as a shock-capturing term to dissipate small-scale waves that sometimes appear near steep wave crests. The form for A_h adopted here is given by

$$A_h = C_l h_e{}^2 |\nabla \mathbf{u}| \tag{1}$$

where C_l is a constant coefficient with a typical range of $1 < C_l < 2$, h_e is an element length scale, $\mathbf{u}(x, y, z, t)$ is the depth-averaged horizontal velocity, and ∇ is the horizontal gradient operator $(\partial/\partial x, \partial/\partial y)$. Note that A_h becomes negligible for small gradients and with increasing grid refinement.

The continuity and momentum equations are combined to give a wave equation that replaces the continuity equation (see review chapter). The wave equation is discretized in time using a 3-level finite difference approximation with time levels $k + 1$, k, and $k - 1$ [3]. All terms are centered at time level k and the scheme can be made implicit or explicit. The explicit version uses a lumped mass matrix (no matrix solution required) and the implicit version retains a consistent mass matrix. The momentum equations are centered between levels $k+1$ and k, with the exception that the dispersion term and the advective term are evaluated at level k.

These governing equations are approximated using standard Galerkin techniques. The spatial domain is discretized by defining a set of 2-dimensional triangular elements. A standard Lagrange basis of polynomial degree p is defined on the master element [1] and this basis is used to interpolate variable quantities within each element. Expanding the dependent variables in terms of the finite element basis and numerically integrating produces an algebraic problem for the nodal unknowns. Essential conditions on η (surface elevation) or the boundary integral for discharge are specified on all external boundaries. The natural boundary condition is that $H\overline{\mathbf{u}} \cdot \mathbf{n} = 0$.

3. Results: Test Problem 1

Benchmark problem 1 is a test case for the runup of an edge wave as it propagates along a sloping beach. Four cases are considered and denoted as A, B, C, and D, where the motion of the paddle about its pivot point was specified from the corresponding experiments for each case. Numerical experiments indicated that the paddle motion could not be simulated with a fixed grid and flux boundary conditions. Hence a locally deforming grid that followed the paddle motion was used.

Several grids were used in the simulations. Results for the two most refined grids are reported here. These grids have a node spacing of 10cm in the deep water. One has a node spacing of 5 cm shoreward of the paddle pivot (10535 nodes), and the other has a transition to 2.5 cm and a node spacing of 2.5 cm shoreward of the pivot (31985 nodes). The grids are uniform in the longshore direction.

The original intention was to compare the various methods to simulate a moving boundary (see review chapter). In the end, only two methods seemed worth pursuing: a fixed grid with element switching, and a locally deforming grid. In the former, elements were activated or deactivated depending on a minimum water depth. A 30 percent hysteresis was used to prevent elements oscillating on and off in subsequent time steps. The locally deforming grid has thus far only been applied to the paddle so that all the results presented here use switched elements along the shoreline.

Results for the most refined grid are shown in Fig. 1 and the parameters used are shown in Table 1. The solution showed little sensitivity to C_l or C_D. In general, the travel time, amplitude, and wave packet size are simulated reasonably.

case	Δt	Δx_{min}	Δx_{max}	C_l	C_D
A	0.002	0.025	0.10	0.50	0.002
B	0.001	0.025	0.10	0.50	0.002
C	0.002	0.025	0.10	0.50	0.002
D	0.002	0.025	0.10	0.50	0.002

Table 1: Parameters for benchmark 1. Δt is the time step (sec), Δx_{min} is the minimum grid size (m), Δx_{max} is the maximum grid size (m), C_l is a coefficient used in the dispersion term, and C_D is the bottom friction coefficient.

4. Results: Sensitivity Tests for Problem 1

After a number of numerical experiments, it became apparent that the most important factors in this problem are the correct simulation of the paddle movement and runup at the shoreward end, and the runup of the edge wave along the beach. The use of a locally deforming grid to represent the paddle resolved that issue. Hence the sensitivity tests focused on grid refinement and the criteria for wetting/drying of elements.

In general, the wave amplitude was reduced and the wave packet length increased as the grid was refined. This brought the results in closer agreement with the observations. For the coarser grid, wave runup at the end of the paddle was impeded resulting in an initial wave amplitude about 50 percent too large. For the coarse grid or for large values of the depth criterion, wave runup at the shoreline is impeded and only the first 3 wave peaks appear in the solution. The trailing part of the wave packet was lost.

5. Results: Test Problem 3

Benchmark problem 3 is a test case for the runup of a solitary wave on a vertical wall. Three cases are considered and denoted as Case A, Case B, and Case C in order of increasing wave amplitude. The digitized paddle-motion data was differenced to determine the paddle velocity. Case A in particular suffered from digitization noise. Rather than smooth these time-series, the paddle velocity was fitted to

$$u = a_0 sech^2(t/t_0) \tag{2}$$

where the amplitude a_0 and the function width t_0 were fitted from the paddle velocity data. These values are contained in Table 2. For the model forcing conditions, two

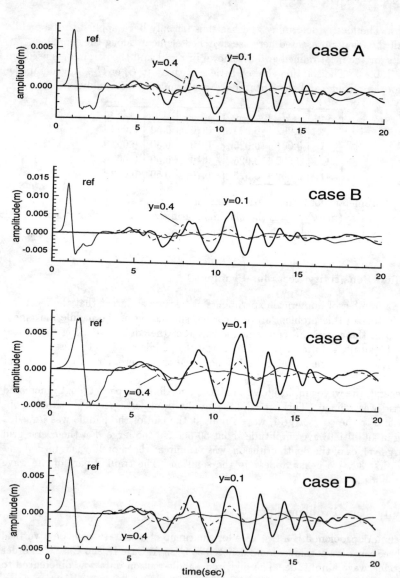

Fig. 1. Results for problem 1, cases A, B, C, and D at x=5.5 m and y=0.1 m and y=0.4 m.

methods were used: (1) the flux calculated from this data is specified in the boundary integral in the discretized wave equation, and (2) the paddle is treated as a moving boundary with locally deforming grid. For this problem, there was little difference in the results.

case	a_0	t_0
A	0.068	1.00
B	0.360	0.75
C	0.670	0.65

Table 2: Parameters for the definition of the input wave. a_0 is velocity amplitude (m/s) and $t_0(s)$ is a measure of the wave width.

case	$\triangle t$	$\triangle x$	C_D	C_l	R
A	0.005	0.028	0.002	3.00	2.79
B	0.005	0.028	0.002	3.00	1.33
C	0.005	0.028	0.002	3.00	1.10

Table 3: Parameters for benchmark 3. $\triangle t$ is the time step (sec), $\triangle x$ is the approximate grid size (m), C_D is the bottom friction coefficient, C_l is a coefficient used in the dispersion term, and R is the runup normalized to the incident wave at $x = 0$..

Initial model calculations were made with a relatively coarse grid where $\triangle x$ varied from 0.112 m at the wavemaker to 0.075 m at the wall (grid 2). The grid is a 2-dimensional grid that is 5 nodes wide and 205 nodes long (1025 nodes total) with variable spacing in the longitudinal direction. The grid was generated by subdividing each section of the flume uniformly, then applying a triangulation algorithm to the resultant nodes. For irregular domains, a 2-dimensional triangular grid with variable node density can be generated with more general tools [2]. For this test case, two other more refined grids were created by refining the coarse grid uniformly by a factor of 2 in the longitudinal direction (grid 3, 2045 nodes) and by a factor of 4 (grid 4, 4085 nodes). Sensitivity tests were conducted to assess the effects of grid refinement, time-step size, bottom friction coefficient, and the coefficient used in the shock capturing dispersion coefficient. The final set of results are shown in Figure 2 for the 3 benchmark cases, and the values for the parameters are shown in Table 3. The model used was the lumped, explicit version of the wave equation so there is a Courant number restriction on the time-step size. The computational resources used are noted in Table 4.

In general, the results depict the shortcomings of shallow water theory in simulating the propagation of solitary waves. In case a, finite amplitude effects lead to a steepening of the wave front. In case b and c, the waves approach the limiting case of a vertical leading edge and a triangular wave shape.

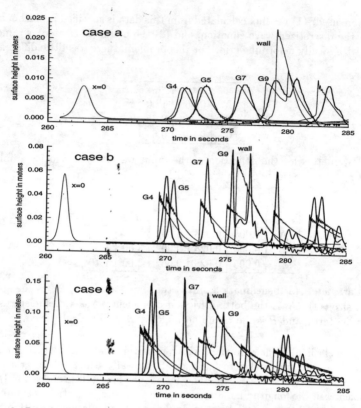

Figure 2: Results for test cases. The heavy lines are the observations and the light lines are the model results. The measurement locations are noted above each waveform.

6. Results: Sensitivity Tests for Problem 3

The sensitivity tests included an examination of the effects of grid refinement, time-step size, bottom friction coefficient C_D, and dissipation coefficient C_l. These results are plotted in the following figures and the parameters used in each simulation is listed in the corresponding table. The conditions for test case C are used because these generate the wave that is the most demanding to simulate. The base grid used in these tests is grid 3, the medium resolution grid.

The grid refinement tests (Figure 3, Table 5) indicate the importance of adequate grid refinement. In the coarse grid, the wave front does not steepen sufficiently and the waves of small-wavelength following the crest are pushed to longer wavelengths than the more refined grids. The source of these waves is both a Gibb's effect caused

grid	nodes	steps	time	memory
2	1025	5000	0000	0.56
3	2045	5000	0000	0.78
4	4085	5000	0000	1.20

Table 4: Computer resources for benchmark 3. Grid refers to the 3 levels of refinement noted in the text, nodes is the total number of nodes in the 2-dimensional grid, steps is the number of time steps, time is total run time in seconds, and memory is the total memory in megabytes.

by the cutoff wavelength of the grid and a nonlinear generation by the finite amplitude and advection terms. In comparison with results using the unlumped version of the model (which requires a matrix solution), it was found that the short waves following the wave crests are to a large degree caused by the poor phase characteristics of the lumped model. In the unlumped model, these waves are virtually absent. This behavior is explained by the dispersion relation for each model: the lumped model has phase lag at short wavelength, while the unlumped version has phase lead.

test	grid	Δx_{wm}	Δx_{wall}	Δt	C_D	C_l
1	2	0.122	0.075	0.01	0.0	2.
2	3	0.061	0.036	0.01	0.0	2.
3	4	0.031	0.018	0.01	0.0	2.

Table 5: Parameters used in the grid refinement tests. Δx_{wm} is the grid size (m) at the wavemaker end, Δx_{wall} is the grid size (m) at the wall end, Δt is the time step (sec), C_D is the bottom friction coefficient, and C_l is a coefficient used in the dispersion term

The time-step size tests indicate that the wave arrival times vary somewhat, but not the overall shape of the wave as the time-step size is reduced from 0.01 to 0.002 seconds. In general, the arrival time becomes later as the time-step is refined. A small time step is required to resolve the steep rise at the leading edge of the wave.

Similar to the results of the time-step size tests, a variation in bottom coefficient causes a small variation in wave arrival times, but not the overall shape of the wave.

An increase in dissipation coefficient C_l (Figure 4, Table 6) causes a significant reduction in the short-wave signal near the wave crest, but otherwise does not affect the wave shape. This provides confirmation that the dissipation term follows the design criteria reasonably well.

7. Some Conclusions

The wave equation formulation of the shallow water equation with a finite element approximation in space and a finite difference approximation in time has been applied to the wave propagation problem where finite amplitude and advective effects are

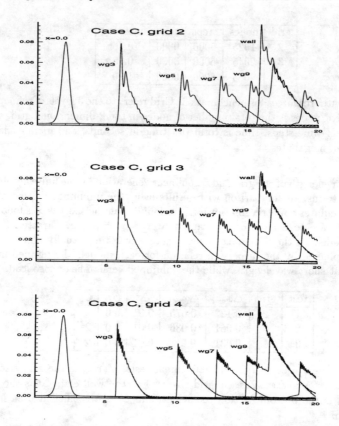

Figure 3: Sensitivity tests for grid refinement. The coarsest grid is at the top (test 1), and the most refined grid is at the bottom (test 3).

test	grid	$\triangle x_{wm}$	$\triangle x_{wall}$	$\triangle t$	C_D	C_l
1	3	0.061	0.036	0.01	0.0	1.
2	3	0.061	0.036	0.01	0.0	2.
3	3	0.061	0.036	0.01	0.0	3.

Table 6: Parameters used in C_l tests. $\triangle x_{wm}$ is the grid size (m) at the wavemaker end, $\triangle x_{wall}$ is the grid size (m) at the wall end, $\triangle t$ is the time step (sec), C_D is the bottom friction coefficient, and C_l is a coefficient used in the dispersion term

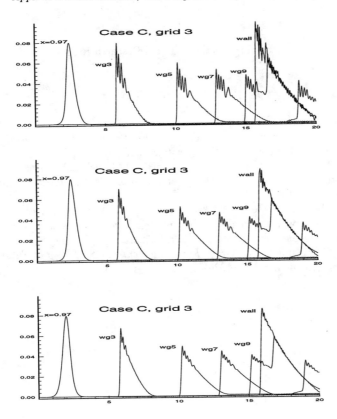

Figure 4: Sensitivity tests for C_l. The smallest value for C_l is at the top, and the largest value at the bottom.

important. The model is ideally suited to wave propagation at the ocean basin and coastal scales where complex topography can be accomodated easily. Within the constraints of shallow water theory, the model simulates the propagation of a solitary wave in the benchmark problem without numerical difficulties. The form of the dissipation term used in this study was found to perform well in this regard. The results showed the importance of adequate grid refinement and the necessity of some form of dissipation for the short-wavelength waves that are generated at the wave crest. The form of the dissipation term used in this study was found to perform well in this regard. Otherwise, the simuations are not very sensitive to any of the coefficients.

8. Acknowledgements

This work was supported through the National Research Program of U. S. Geological Survey.

9. References

1. Becker, E., G. F. Carey, and J. T. Oden, *Finite Elements: An Introduction* (Prentice-Hall, Englewood Cliffs, New Jersey, 1981).
2. Henry, R.F. and R. A. Walters, A geometrically-based, automatic generator for irregular triangular networks, *Communications in Numerical Methods in Engineering* **9** (1993) 555.
3. Kolar, R.L., J.J. Westerink, M.E. Catekin, and C.A. Blain, Aspects of non-linear simulation using shallow-water models based on the wave continuity equation, *Computers and Fluids* **23** (1994) 523.
4. Pinder, G. F. and W. G. Gray, *Finite Elements in Surface and Subsurface Hydrology* (Academic Press,1976).

NUMERICAL MODELLING OF SOLITARY WAVE PROPAGATION AND BREAKING ON A BEACH AND RUNUP ON A VERTICAL WALL

G. Watson,[a] T.C.D. Barnes[b] and D.H. Peregrine[c]

1. Introduction

There are still many kinds of laboratory wave flume experiments whose results cannot be accurately predicted. The difficulties associated with modelling one such case, in which a solitary wave travels along a flume onto a beach, breaks, and then reflects off a vertical wall in shallow water, are discussed here.

Benchmark Problem #3 of the International Workshop on Long-Wave Runup Models was tackled using a variety of numerical and analytical approaches. Other papers in this volume give details of the experiment and the various models. Three test solitary waves were used: Case A was a small wave which did not become steep enough to break; Case B was a larger wave which broke just before hitting the wall, and case C was the largest, which broke somewhere between points 7 and 8.

Models were run using only the flume profile and the paddle displacement time-series, and results were presented at the workshop without prior knowledge of the experimental data (including observations of wave breaking). None of the models were able to predict satisfactorily all of the results, i.e. wave profiles at each gauge, and maximum runup at the wall. It is informative to consider under what circumstances the models are good, and when and why they break down.

For unbroken waves, the best results were obtained using fully-nonlinear potential theory models. The onset of breaking was predicted reasonably well, but computations had to be terminated after this occurred.

A number of participants used the nonlinear shallow-water equations (SWE), as used in various tsunami and other coastal models. Discontinuities in the solution, at which mass and momentum are conserved but energy is dissipated as in a hydraulic jump, can represent breaking waves or bores. This approach, which ignores the detailed structure of the bore, is often a reasonable approximation, especially in the swash zone. The breaking region in a real bore is rarely more than three times the depth on the higher side of the bore, so it is small compared with the wavelength which in many cases is much more than the depth (e.g. 20 times).

However, in this test the water in most of the flume was too deep for the shallow-water approximation to be good. The nondispersive SWE predict waves which steepen and break too quickly and as a result the wave shapes are wrong, and except for small non-breaking waves the runup is also under-predicted. In shallower water the

[a]Disaster Prev. Res. Inst. Kyoto Univ., Uji, Kyoto 611, Japan (gary@rs01.dpri.kyoto-u.ac.jp).
[b]Maths Dept. Bristol Univ., Bristol BS8 1TW, U.K. (tim.barnes@bristol.ac.uk)
[c]Maths Dept. Bristol Univ., Bristol BS8 1TW, U.K. (d.h.peregrine@bristol.ac.uk)

equations are more accurate so, as might be expected, in cases where the problem was initialised on the beach (and the propagation distance was less), better results were obtained.

Others used Boussinesq equations, which are valid in deeper water since they include some dispersive terms. These performed better than SWE until the waves broke, successfully producing solitary waves which travelled along the flume with little change of form. The solitary wave is the most typical finite-amplitude solution of the Boussinesq equations, with its propagation being characterised by a balance between the steepening of the front of the wave in shallow water and the dispersive spreading which characterises linear waves in deeper water.

As the wave travels into shallower water, the steepening effects continue to be partly balanced by the excess pressure arising from the increasing vertical accelerations at the front of the wave. Although the approximations used in deriving Boussinesq's equations break down well before the general character of the motion changes, there seem to be a number of examples where Boussinesq's equations, when extended by the inclusion of further dispersive terms, give adequate solutions in these circumstances until the waves have almost reached breaking [1-4]. Satisfactory results were thus obtained for case A, in which the wave was small and did not break. However, the equations are only valid for small waves; in a large wave, the nonlinear terms are not strong enough to overcome the dispersive terms and the wave cannot steepen into an energy-dissipating bore. Instead it continues to grow to an unrealistic height when in reality it would break. Thus, in the case of breaking waves the Boussinesq equations were also not able to predict the runup accurately.

To represent breaking waves, Boussinesq's equations may be modified, for example using either empirical dissipation terms, or more recently by inclusion of a "surface roller" [5]. However, such models were not tested on this workshop problem.

This paper presents results from a composite model consisting of a Boussinesq model in the deeper water joined to a SWE model in the shallower water. Such a model has been compared with experiments [6], showing that for waves on a beach of gentle slope the overall results were satisfactory. However, examination of the details near breaking and broken waves revealed discrepancies of amplitude and phase. Originally, this model was used to investigate the nonlinear transfer of energy from wave groups into low-frequency (infragravity wave) motions [7,8]. For such investigations the details of wave shape are not so important and good results were obtained. The low-frequency results were found to be only weakly dependent on the position of the join, whereas in this problem the individual wave profiles and heights after breaking were strongly dependent on this, as discussed below.

Although it would be more accurate to use a fully-nonlinear model rather than the Boussinesq Equations for non-breaking waves, then Boussinesq equations with improved dispersion and rollers for waves just after breaking, and finally SWE in the swash zone, such a model has not yet been assembled. It is informative to study the extent of the present model's usefulness compared with the pure Boussinesq or pure SWE models. Further, apart from the position of the join, there are no adjustable empirical parameters such as those in the enhanced Boussinesq models and it is instructive to judge the accuracy of this more fundamental approach.

2. Composite Numerical Model

2.1. Mathematical Model and Numerical Methods

In the surf zone, the nonlinear shallow-water equations for the conservation of mass and momentum, were used:

$$d_t + (ud)_x = 0 \tag{1}$$

$$u_t + uu_x + g\eta_x = 0 \tag{2}$$

where $\eta = d - h$ is the surface elevation, d is the total water depth, h the undisturbed water depth, u the flow velocity and g is gravity.

The solution can steepen to form discontinuities or bores, and various numerical schemes have been developed to handle these automatically, efficiently and accurately, with minimal numerical dissipation. A variant of a popular scheme, the Weighted Average Flux (WAF) method, is used here to solve the equations. The original constant-depth version was modified for use on a beach by incorporating sloping bottom terms, friction, and a moving shoreline [7]. For this problem only the sloping bottom terms are required. Runs using a simple quadratic friction term gave results which were negligibly different from those shown here.

In deeper water, the Boussinesq equations are appropriate because they include dispersive terms. A Boussinesq model essentially the same as that due to Peregrine [9] was used:

$$d_t + (ud)_x = 0 \tag{3}$$

$$u_t + uu_x + g\eta_x = \tfrac{1}{2}h(hu_t)_{xx} - \tfrac{1}{6}h^2 u_{xxt} \tag{4}$$

The improved dispersive terms mentioned in the introduction have not yet been included in this model.

The models were matched at the join using a characteristic boundary condition. In the SWE, the wave signals propagating in each direction are given by the Riemann invariants $R^+ = 2c + u$ and $R^- = 2c - u$, where $c = \sqrt{gd}$. This is also approximately true for the Boussinesq equations, and we use the same condition for them, which seems to work well in most conditions. Thus, an almost non-reflecting join is made by taking R^+ from the last point of the Boussinesq section and feeding it into the first point of the SWE section, and taking R^- from the first point of the SWE section and feeding it into the last point of the Boussinesq section. A grid spacing of 2 cm was used throughout.

The wavemaker boundary condition used the derivative of the measured time series of paddle position in order to set the velocity at the paddle. In Case A the data required some smoothing. At each timestep, the lower grid limit was moved to the grid point closest to the moving paddle. At this boundary point, the characteristic form of the SWE was then used to estimate surface elevation. Previous attempts used a simpler condition in which the paddle velocity was always applied at the same grid point, i.e. that closest to the mean paddle position. However, this resulted in waves which were too large, especially for case C in which the paddle moved through 20 grid points. This is because the effect of this approximation was to introduce a greater net mass into the wave.

The wall boundary condition was made reflective by forcing the solution at two external grid points (beyond the wall) to mirror the solution immediately inside.

3. Results for Benchmark Problem #3, Cases A, B and C

3.1. Case A

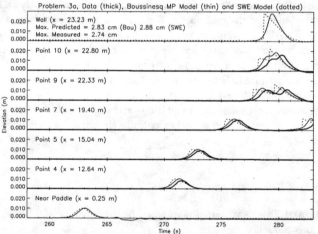

Figure 1: Case A; Data (Thick), Boussinesq (Thin), SWE (Dotted)

Case A was a small wave which did not become steep enough to break. To illustrate the main difference between the Boussinesq and SWE models, results from each of these models alone are shown in Figure 1, together with the measured data. Results are similar near the paddle, but there are no data here against which to check the height and width of the solitary wave. By the time the wave reaches Point 4, the SWE wave has steepened in the front face, whereas the Boussinesq result has kept its form and is still a symmetrical solitary wave. The Boussinesq result follows the data quite well except for the errors in amplitude and propagation speed noted below.

The SWE result continues to steepen and by the time it reaches point 9 its shape is almost triangular, as is typical for SWE models. This is not in agreement with the data, but the wave height is about right and the predicted runup is almost identical to that in the Boussinesq case.

In both cases the amplitude is slightly higher than that of the measured wave. However, this is a small wave and the error is not much greater than the experimental error in the probe data, which is of order ±1 mm. There is also a timing error, with the predicted wave travelling about 3% too fast. This is probably due to the dispersive terms being insufficiently accurate in these basic Boussinesq equations. The runup is overpredicted by about 5%. This result is considered to be reasonably good, given the approximations in the equations and the length of the wave flume.

3.2. Case B

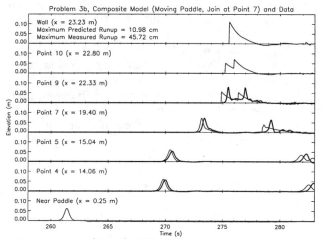

Figure 2: Case B; Data (Thick), Composite Model, Join at Point 7 (Thin)

Case B was a larger wave which broke just before hitting the wall. The data are shown as the solid line in Figure 2, but unfortunately those from Point 10 were clipped and are not shown.

With a pure SWE model (not shown), the wave steepens and becomes bore-like within the first 10 m. After this it gradually loses amplitude so that by the time Point 7 is reached it only has about half of the observed height. The predicted runup is only 7.0 cm, compared with the observed value of 45.7 cm.

With a composite model, the most notable thing is that the results are strongly dependent on the position of the join. As the join is moved towards the wall, the shoaling effect of the Boussinesq equations continues to amplify the waves to larger values before the SWE start to dissipate wave energy. The predicted runup value at the wall thus increases. In the case of a pure Boussinesq model (not shown), the runup value at the wall was 68.2 cm, greater than that which was observed.

The main question is thus how to decide where to join the models. It is suggested that the most justifiable strategy is to put the join at the break point, since in both the real flume and in the model, this is where the waves stop growing and begin to lose amplitude. With no prior knowledge of the breaking point, the common criterion that the wave breaks when its height becomes some fraction of the water depth (e.g. $a = 0.8h$) suggests itself. However in this case the wave was observed to break just before hitting the wall, where the height was at least $1.2h$, so this criterion is clearly inadequate. For this reason, a fully nonlinear model is recommended order to predict the breakpoint.

Figure 2 shows the result for the join at Point 7. This is chosen because in this case the results at Points 7 and 9 were closer to the data than with other locations of the join. Point 7 does happen to correspond approximately to $a = 0.8h$, but the real wave broke in much shallower water. The wave amplitude is well predicted in the Boussinesq region between the paddle and Point 7. There is a similar error in wave speed to that in Case A. However, the agreement after Point 7 cannot be described as very good. The predicted wave at Point 9 is smaller than was measured and is less peaked. In this case the SWE do not accurately represent the early stages of breaking.

The runup in this case was underpredicted by 24%. However, the observed wave was reported as causing a particularly high splash of water which occurred because it broke close to the wall. Since the model does not attempt to include such splashes, runup comparisons are not meaningful in this case.

3.3. Case C

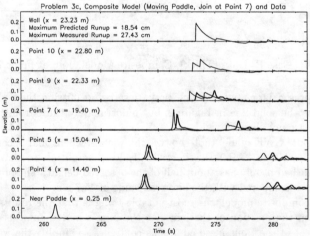

Figure 3: Case C; Data (Thick), Composite Model, Join at Point 7 (Thin)

Case C was the largest wave, which broke somewhere between Points 7 and 8. Results with the join at Point 7, close to where the wave was observed to break, are shown in Figure 3. The wave amplitude is 25% too large at Point 7, showing that the wave in this case is too large for the equations to be precise. The speed error is similar to Cases A and B.

As in Case B, the amplitude reduces too rapidly just after breaking, so that it is fortuitously correct at Point 9. The wave shape however is quite good. The runup is underpredicted by 32%. As in Case B, the predicted runup depends on the position of the join. The splash was not reported as being as large as in Case B but may still have contributed to some of this disagreement.

4. Conclusions

The Boussinesq equations worked reasonably well for non-breaking waves except for an overprediction of amplitude for the largest wave. There was also a small error in wave speed which could probably be reduced by including more accurate dispersive terms. This model also cannot predict the occurrence of breaking. For these reasons, a fully-nonlinear potential flow model is recommended for the modelling of wave evolution up until breaking.

For non-breaking waves, the SWE are inappropriate in the main part of the flume but although the predicted waves are too steep, the amplitude is approximately correct. The SWE reduce the wave amplitude too quickly just after breaking and do not model the wave shape well in this region. Although other work has shown SWE to be good in the swash zone, there were no data there in this experiment. For breaking waves, the predicted runup depended on the position of the model join.

In summary, some aspects of the results are quite well reproduced by this model. However, it cannot accurately predict either the onset or early stages of breaking and this forces us to conclude that it is not good enough for this problem.

5. References

1. P.A. Madsen et al., *Coastal Engineering* **15** (1991) 371-388.
2. O. Nwogu, *J. Wat. Port Coast. Ocean Eng. ASCE* **119** (1993) 618-638.
3. G. Wei et al., *J. Fluid Mech.* **294** (1995) 71-92.
4. M.W. Dingemans, *Water wave propagation over uneven bottoms*, (World Scientific, Singapore, 1996).
5. P.A. Madsen et al., *Proc. 24th Int. Conf. Coastal Eng., ASCE* (1994) 399-411.
6. T.C.D. Barnes et al., *Proc. Int. Symp. Waves - Physical and Numerical Modelling, Vancouver, Aug. 1994*, 280-286.
7. G. Watson and D.H. Peregrine, *Proc. 23rd Int. Conf. Coastal Eng., ASCE* (1992), 818-831.
8. G. Watson et al., *Proc. 24th Int. Conf. Coastal Eng., ASCE* (1994) 776-790.
9. D.H. Peregrine, *J. Fluid Mech.* **27** (1967) 815-827.

Runup of Solitary Waves on a Vertical Wall

Sigurdur M. Gardarsson

University of Washington

Department of Civil Engineering, Box 352700

Seattle, WA 98195-2700, USA

E-mail: smg@u.washington.edu

Abstract

A numerical method, called the Random Choice Method, is used to solve benchmark problem 3. It is based on the the shallow water wave equations. This method is capable of preserving a shock, i.e abrupt spatial changes in the flow conditions. This means that wave breaking can be modeled without any special treatment for numerical instabilities at the bore front. The method is unconditionally stable provided that the Courant-Friedrichs-Lewy condition is fulfilled. The solution is advanced in time by a series of operations which includes solving a Riemann problem at each gridpoint at each timestep. The Riemann problem is computationally expensive so considerable efforts were made on speeding up the solution procedure. This scheme was originally developed by Chorin[1] based on the theory by Glimm[4] to solve the Euler equations.

1. The Random Choice Method

For simulation in one spatial dimension the Random Choice Method divides the fluid domain into n intervals with $n+1$ gridpoints. At a time t_j the Riemann problem is solved at each pair of gridpoints i and $i+1$ (i is an integer). The Riemann problem of the shallow water waves is also called 'the dam break problem', where a fictitious discontinuity is located initially at the midpoint between the gridpoints i and $i+1$ and the problem is solved for the flow condition at $\Delta t/2$ after the discontinuity is removed. Figure 1 shows an example for a part of the calculation space at time $t = t_j$. The notation $w_{i,j}(x, t)$ will be used to denote the solution of the Riemann problem at gridpoint i for time t_j. From the solution $w_{i,j}(x, \Delta t/2)$ only one point between $x = i\Delta x$ and $x = (i + 1)\Delta x$ is sampled randomly, say $x = x^*$, and the sampled value $w_{i,j}(x = x^*, \Delta t/2)$ is assigned to the gridpoint $i + \frac{1}{2}$. This process is repeated for the next half time step to obtain a solution for gridpoint i and time $t = t_{j+1}$. The procedure for the sampling is based on a uniformly distributed pseudo-random variable $\theta_j \in [-\frac{1}{2}, \frac{1}{2}]$ which is generated at each half time step. The method used to generate the random number θ_j is described in detail by Chorin[2].

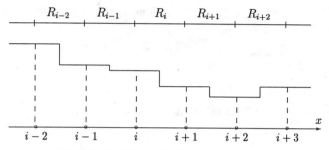

Fig. 1. Part of the calculation space at time t_j.

1.1. *The Shallow water wave equations*

The simulation is based on solving the Riemann problem for the shallow water wave equations:

$$q_t + f(q)_x = 0, \tag{1}$$

or

$$q_t + A(q)q_x = 0, \tag{2}$$

where

$$q = \begin{bmatrix} h \\ m \end{bmatrix}, \quad f(q) = \begin{bmatrix} m \\ \frac{m^2}{h} + \frac{1}{2}gh^2 \end{bmatrix}, \quad m = hu. \tag{3}$$

Here $h = h(x,t)$ is the water depth, $u = u(x,t)$ is the water velocity, g is the earth gravity, and $A(q)$ is the Jacobian matrix:

$$A(q) = f'(q) = \begin{bmatrix} 0 & 1 \\ -\frac{m^2}{h^2} + gh & \frac{2m}{h} \end{bmatrix}. \tag{4}$$

This system is nonlinear because $f'(q)$ is not a constant matrix but a function of the dependent variable q. The eigenvalues of the Jacobian matrix are

$$\lambda_1(q) = m/h - \sqrt{gh}, \qquad \lambda_2(q) = m/h + \sqrt{gh}, \tag{5}$$

with the corresponding eigenvectors

$$r_1(q) = \begin{bmatrix} 1 \\ m/h - \sqrt{gh} \end{bmatrix}, \quad r_2(q) = \begin{bmatrix} 1 \\ m/h + \sqrt{gh} \end{bmatrix}. \tag{6}$$

An approximate Riemann solver is used to solve the Riemann problem. This solver is based on a scheme suggested by Roe[5]. The idea is to rewrite the above nonlinear system (2) as a linear system:

$$\hat{q}_t + \hat{A}(q_l, q_r)\hat{q}_x = 0, \tag{7}$$

where $\hat{A}(q_l, q_r)$ is a constant matrix. This system is much easier to solve.

To determine $\hat{A}(q_l, q_r)$ in a reasonable way, Roe[5] suggested that the following conditions should be imposed on $\hat{A}$:

a) $\hat{A}(q_l, q_r)(q_r - q_l) = f(q_r) - f(q_l)$: this guarantees that the solution will be conservative.

b) $\hat{A}(q_l, q_r)$ is diagonalizable with real eigenvalues: this guarantees that the problem is hyperbolic and solvable, and

c) $\hat{A}(q_l, q_r) \to f'(\bar{q})$ smoothly as $q_l, q_r \to \bar{q}$.

The matrix $\hat{A}(q_l, q_r)$ that fulfills these three conditions is found to be:

$$\hat{A}(q_l, q_r) = \begin{bmatrix} 0 & 1 \\ \frac{1}{2}g(h_l + h_r) - \bar{v}^2 & 2\bar{v} \end{bmatrix}, \quad \bar{v} = \frac{h_l^{1/2}u_l + h_r^{1/2}u_r}{h_l^{1/2} + h_r^{1/2}}. \tag{8}$$

1.2. The Strang splitting for a sloping bottom

For the sloping part of the channel the shallow water equations needs to be modified in the following way:

$$h_t + m_x = 0, \tag{9}$$

$$m_t + \left(\frac{m^2}{h} + \frac{1}{2}gh\right)_x = g\alpha h, \tag{10}$$

where α is the bottom slope from the horizontal. In this case there is a source term on the right hand side of equation (10). To solve this we adopt the splitting technique introduced by Strang[3]. The homogeneous shallow water wave equations is first solved using the Random Choice Method at time t_j. That yields $h^*(x_i, t_j)$ and $m^*(x_i, t_j)$ where $i = 1, 2, ..., N$, where N is the total number of gridpoints. Then to obtain h_i^{j+1} and m_i^{j+1}, the following equations are solved:

$$h_t = 0, \tag{11}$$

$$m_t = g\alpha h. \tag{12}$$

Equation (11) has the solution $h_i^{j+1} = h_i^*$ and equation (12) can be integrated

$$\int_{m^*}^{m_i^{j+1}} dm = \int_0^{\Delta t} g\alpha h_i^{j+1} dt, \tag{13}$$

yielding,

$$m_i^{j+1} = m^* + g\alpha h_i^{j+1}\Delta t \tag{14}$$

since h is held constant at this stage of the splitting procedure.

1.3. The Courant-Friedrichs-Lewy condition

The solution of the Riemann problem for the shallow water wave equations consists of one of four combinations of shocks and rarefactions, (see e.g. Stoker[6]):

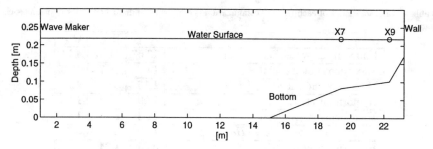

Fig. 2.   A schematic view of the flume.

1. $m_l/h_l - \sqrt{gh_l} \geq m_m/h_m - \sqrt{gh_m}$: Left wave is a shock,

2. $m_l/h_l - \sqrt{gh_l} < m_m/h_m - \sqrt{gh_m}$: Left wave is a rarefaction,

3. $m_m/h_m + \sqrt{gh_l} \geq m_r/h_r + \sqrt{gh_r}$: Right wave is a shock,

4. $m_m/h_m + \sqrt{gh_m} < m_r/h_r + \sqrt{gh_r}$: Right wave is a rarefaction,

where the subscript m denotes the constant middle state. The Random Choice Method is stable if there is no interaction between adjoining Riemann problems for each time step. This means that waves cannot reach a boundary of a cell. By using the above characterization of shock and rarefaction following stability can be established:

$$\frac{\Delta x}{\Delta t} \geq \begin{cases} \max\left(|s_l|, |s_r|\right), & \text{cond. 1 and 3 combined,} \\ \max\left(|s_l|, |m_r/h_r + \sqrt{gh_r}|\right), & \text{cond. 1 and 4 combined,} \\ \max\left(|m_l/h_l - \sqrt{gh_l}|, |s_r|\right), & \text{cond. 2 and 3 combined,} \\ \max\left(|m_l/h_l - \sqrt{gh_l}|, |m_r/h_r + \sqrt{gh_r}|\right), & \text{cond. 2 and 4 combined,} \end{cases} \quad (15)$$

where s_l and s_r are the left-running and the right-running shock speeds, respectively.

2. Simulation

Benchmark problem 3 is to simulate a solitary wave run-up on a vertical wall in a 23.2 m long by 45 cm wide flume. The bathymetry consists of a 15.04 m long flat bottom extended from a wave maker and three sections of sloping bottom (1:53, 1:150, 1:13) ending at a vertical wall. The water depth in the flat section is 21.8 cm. Wave gages are located where the slope changes from 1:53 to 1:150 (X_7) and from 1:150 to 1:13 (X_9). A sketch of the flume is shown in Figure 2.

The computational domain consist of 4000 gridpoints which requires $dx = 0.0058075$ m. For case A the maximum wavespeed a was found to be around $a = 2$ m/s and for B and C around $a = 3$ m/s. Hence, equation (15) determines the timestep size as $dt = dx/a$. The simulations were made for approximately 30 seconds in each case.

Table 1. Characteristics of each wave. Normalization is done with undisturbed water depth at each location.

	A	B	C
Undisturbed water depth [cm]	13.57	11.62	4.74
Max. depth at X_7 [cm]	14.7	16.7	17.8
Max. wave height at X_7 [cm]	1.16	3.1	4.2
H_7 normalized [-]	1.09	1.23	1.31
Max. depth at X_9 [cm]	12.8	14.5	15.4
Max. wave height at X_9 (H_9) [cm]	1.19	2.8	3.8
H_9 normalized [-]	0.65	1.24	1.33
Max. depth at the wall [cm]	7.6	11.2	13.4
Max. wave height at the wall (H_W) [cm]	2.9	6.4	8.6
H_W / H_9 [-]	2.39	2.28	2.26
Experimental result for H_W [cm]	2.7	45.7	27.4

The simulation was also performed with the computational domain consisting of 2000 gridpoints which essentially gave the same result.

The paddle motion is fitted with a curve of the form $x(t) = a(\tanh(bt + c) + 1)$, where a, b, c are constants.

Boundary conditions are mirror boundary conditions at the flume end wall. The boundary conditions are modified for the velocity at the wave paddle $u_{p+1} = -u_p + 2\dot{x}(t)$, where the paddle is located between gridpoints p and $p + 1$ and $\dot{x}(t)$ is the velocity of the paddle.

3. Results and Discussion

In Table 1 the water depths at locations X_7, X_9 and the run-up at the wall are listed. The time series of the free surface displacement is plotted for the points $X = X_7 = 19.40$ m and $X = X_9 = 22.33$ m in Figure 3. Note that the time origin is at the start of the time series which was given for the wave paddle motion. In Case A, it is observed that the solitary wave in the laboratory is not broken at X_7 nor at X_9. The simulated wave is not broken when it passes point X_7 because the wave rise is not sharp but gradual. It is close to be fully broken at point X_9 because the wave rise is sharper.

For Cases B and C the wave is completely broken at X_7 and in fact, by viewing simulation of the wave propagation, the wave breaks shortly after it starts to propagate for Case B and breaks almost immediately for Case C. Therefore it is expected that the waveheight is much lower for the simulation than the experiment because as soon as the wave breaks in the simulation the waveheight decreases as it continues to travel due to energy dissipation at the wave front.

For the simulation the reflected wave height is in all cases larger at X_9 than the incoming wave height. At X_7 the reflection wave is smaller for both Case B and C,

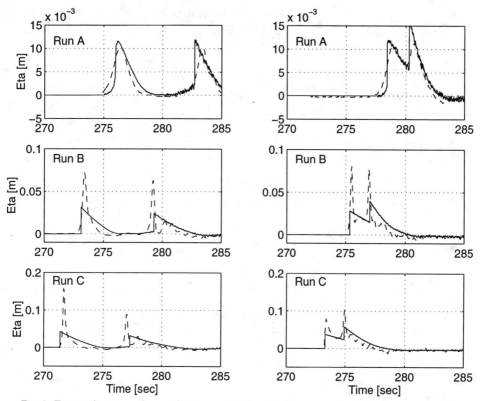

Fig. 3. Temporal water-surface profiles for run A, B and C. Figures on the left are at wavegage X_7 and figures on the right are for wavegage X_9 where (- -) denotes experiments and (-) simulations. Note that the first peak in each figure is the incoming wave and the second peak is the reflected wave.

but for Case A the reflected wave is of similar height as that of the incoming wave.

In Table 1, comparison between wave run-up at the end wall (H_w) for the experiment and simulation is made. For Case A the run-up values are in fairly good agreement due to the fact that the wave in the simulation breaks only short time before hitting the wall. For Case B simulation underestimates greatly the run-up height because the simulated wave breaks well before it hits the wall. For Case C the agreement is slightly better than in Case B but that is actually due to the fact that the solitary wave in the experiment broke before it hits the wall although the discrepancy is still significant.

Snapshots of the computed wave profile are shown in Figure 4. The different characteristics between Case A and Cases B and C regarding different wave-breaking

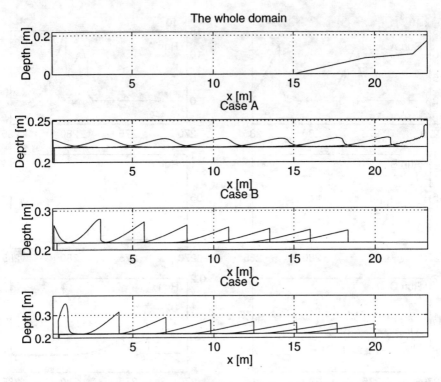

Fig. 4. Snapshots of water-surface profiles in the computational domain to demonstrate the different behavior of breaking and non-breaking waves.

pattern can be observed. The run-up height is strongly correlated to the wave breaking location because as soon as the wave breaks it starts to decrease in height but while it remains solitary wave it sustains its original height. Hence, the run-up height for the simulation is a function of the wavepaddle location which is not true for solitary waves. This is to be expected because shallow water wave theory predicts that all waves eventually break given enough travel time. Therefore it turns out that only in case A the simulation is close to predicting the correct run-up value.

As an example of the run time, case A with 2000 gridpoints took about 15 min on Sun Sparc 2. The executable file is 213 kbyte. The main memory requirement are 3 double precision arrays of length N, where N is the number of gridpoints in the computational domain.

4. References

1. A.J. Chorin. Random choice solution of hyperbolic systems. *Journal of Computational Physics*, 22:517–533, 1976.
2. A.J. Chorin. Random choice methods with applications to reacting gas flow. *Journal of Computational Physics*, 25:253–272, 1977.
3. Strang G. On the construction and comparison of difference schemes. *SIAM J. Num. Anal.*, 5:506–517, 1968.
4. J. Glimm. Solution in the large for nonlinear hyperbolic systems of equations. *Communications on pure and Applied Mathematics*, 18:697–715, 1965.
5. P.L. Roe. Approximate Riemann solvers, parameter vectors, and difference schemes. *Journal of Computational Physics*, 43:357–372, 1981.
6. J.J. Stoker. The formation of breakers and bores. *Communications on Applied Mathematics*, 1:1–87, 1948.

Calculation of Runup using a Local Polynomial Approximation Model:

Discussion for Benchmark Problem 3

Andrew Kennedy

Department of Mechanical Engineering, Monash University

Clayton, Victoria, Australia, 3168

Summary

Solutions were attempted for cases A and B using a potential flow model. Case C was not attempted, as the wave would have broken before reaching the wall. Runup was predicted well for case A, although agreement with recorded wave gauge surface elevations is not as good. Runup is overpredicted in case B, which is partly due to the inherent neglect of viscous effects in the potential flow assumption.

Brief Description of Numerical Model

The model presented here calculates the evolution of nonbreaking waves over variable topography for one dimension in plan. Potential flow is assumed. The overall domain is divided into subdomains where the velocity potential is locally represented by a polynomial analytically satisfying Laplace's equation. At each time step, Laplace's equation is solved for known boundary conditions, and the system is advanced to the next time step using the free surface evolution equations. Due to the bandedness of the matrix equations, the computational expense for each time step is directly proportional to the number of subdomains. This makes the method quite efficient for large problems. A somewhat more detailed description of the model including typical accuracies and examples may be found in ([1]).

Solution of Laplace's Equation

The velocity potential satisfying Laplace's equation

$$\phi_K(x_K, y, t) = \left[A_{0R} + \text{Re}\left(\left(\sum_{j=1}^{n-1} (A_{jR} + iA_{jI})(x_K + iy)^j \right) + (x_K + iy)^n \times \begin{cases} A_{nR}, & n \text{ odd} \\ iA_{nI}, & n \text{ even} \end{cases} \right) \right]_K ,(1)$$

has $2n$ independent time-varying A coefficients, where n is an integer ≥ 2 which defines the level of approximation. It is defined only in subdomain k, as shown in Figure 1, while similar functions are valid in the other regions. Depending on the value of n, this velocity potential could conceivably provide a good representation of the actual flow for a wave in finite depth. For all computations presented here, a ninth degree polynomial (*i.e.* $n=9$) was used during wave generation, and a seventh degree polynomial for all subsequent times.

Although (1) is the most straightforward representation of the velocity potential, an equivalent form is preferred, for reasons which will shortly become clear. The new form will be

$$\phi_K(x_K, y, t) = \left[\sum_{j=0}^{n-1} (B_{jL}\phi_{jL}(x_K, y) + B_{jR}\phi_{jR}(x_K, y)) \right]_K , \tag{2}$$

306

where B_{jL} and B_{jR} are independent coefficients which may vary in time. ϕ_{jL} and ϕ_{jR} are independent subsets of ϕ_K defined such that

$$\phi_{jL} = \begin{cases} y^j, & x_K = -\Delta x_K/2 \\ 0, & x_K = \Delta x_K/2 \end{cases}$$

$$\phi_{jR} = \begin{cases} 0, & x_K = -\Delta x_K/2 \\ y^j, & x_K = \Delta x_K/2 \end{cases}$$

The velocity potential may now easily be made continuous across the interior boundary between subdomains k and $k+1$ by setting $B_{j,K} = \left(B_{jR}\right)_K = \left(B_{jL}\right)_{K+1}$. When continuity is imposed at all interior boundaries, the number of independent coefficients is almost halved, which greatly increases computational efficiency and justifies the change in basis functions.

Now that the form of the velocity potential has been established, the boundary value problem may be considered. Boundary conditions are

$\phi = \phi_s(x)$ on the free surface,

$\partial \phi / \partial n = 0$ on the bed and right wall,

$\partial \phi / \partial n = f(t)$ on the wavemaker.

Based on the discontinuity in horizontal velocity that will exist across the boundary between subdomains, there is an additional condition

$(\partial \phi / \partial x)_+ = (\partial \phi / \partial x)_-$ on interior boundaries,

where the positive and negative signs indicate the right and left sides of the boundary, respectively.

An obvious choice would be a weighted residual solution, which, since Laplace's equation is analytically satisfied, would only have to be evaluated on the boundaries of the computational subdomains. This was considered, but was found to require too many function evaluations and multiplications.

Instead, a collocation method was used. This method may be easily explained if the velocity potential is thought of as having had n degrees of freedom at each interior boundary, and at each of the left and right global boundaries. At interior boundaries, one constraint set the velocity potential at the free surface to the known value, and another imposed the bottom boundary condition at the bed. The remaining $n-2$ constraints imposed continuity of horizontal velocity across the boundary at the Chebyshev points for $N=n-2$, taking the free surface and bed as limits. At the left and right global boundaries, the free surface and the bed were treated in a similar manner, but to fill the remaining constraints, instead of continuity, horizontal velocities were set to known values at the Chebyshev points. (There was one exception to this. During wavemaking, weighted residual equations were obtained for the set of functions closest to the wavemaker. The other equations were unchanged.)

The resulting set of linear equations was solved using a banded matrix solver, which has a computational expense of solution that increases only linearly with the number of variables. As noted earlier, this becomes highly desirable when dealing with large domains.

Dispersion Characteristics

As mesh lengths go to zero, analytical expressions may be found for the local polynomial approximation to the phase speed of a small amplitude wave over a level bed. While these expressions and their derivation will not be given here, a comparison between the exact small amplitude phase speed and present results with $n=7$ (*i.e.* a seventh degree interpolating polynomial) is shown in Figure 2. As dispersion relationships change somewhat with the choice of collocation points, results are given for both the Chebyshev points (used for all benchmark problems), and the Gauss-Legendre quadrature points (which, at this level of approximation, also conserve overall flow and three weighted moments of flow between subdomains), and an empirical set of points chosen to provide good agreement over this range.

All sets of points give extremely good results, with errors of less than one part in ten thousand at the deep water limit of $kd=\pi$, and remaining less than one percent at three times the deep water limit. The empirical points, in particular, give good high wavenumber performance, although accuracy is sacrificed slightly for longer waves. It must be noted, however, that "tuning" for linear performance over a level bed is not necessarily the best manner in which to choose a set of collocation points, as it says little about nonlinear performance over variable topography.

Representation of the Free Surface and Bed

While the wave generator was in motion, surface slopes were calculated using second order central differences, except at the wave generator, where a first order forward difference was used. After the generator had finished moving, a 5th degree polynomial spline was used, with odd spatial derivatives set to zero at the global boundaries. The bottom topography was as defined, with slightly rounded corners.

Time Stepping

Following the solution of Laplace's equation, the free surface elevations and velocity potentials at the interior and exterior boundaries were advanced in time using the kinematic and dynamic free surface boundary conditions. While waves were being generated, a second order leapfrog technique was used, and a third order Adams-Bashforth technique was employed after this.

Results and Discussion

This model provided an accurate potential flow solution for case A. Maximum runup was calculated to be 2.75 cm. During wave generation, the volume displaced was conserved to 5 parts in 10^7 and, after, energy was conserved to 2 parts in 10^4. Total CPU time for the entire problem was about 13 hours on a 80486 DX2-66 based PC, although this could likely be improved. Figure 3(a) shows a sequence of surface profiles as the wave shoals, reflects and propagates back down the slope.

Figure 4(a) gives computational and experimental time traces of surface elevation for case A at points 7 and 9. Due to the small amplitude of the incident wave, a vertical offset in experimental results was non-negligible, and a correction was applied to bring the still water level closer to zero.

Agreement between computed and experimental values is moderate. Qualitative features are in complete agreement, but computed maximum surface elevations are significantly higher than measured values. As well, there is a significant phase lag, and the speed of the computational wave appears to be slightly greater. The major source of error seems to arise from the discrepancy in heights between the computed and measured wave as it propagated along the level bed before the slope: the computed wave had a height to depth ratio of 0.0475, while the measured value at probe 4 was 0.038. Part of this difference may come from digitisation error of the wave paddle time trace, but this is not likely to be more than several percent. Since the desired height of the wave when programming the wave paddle was 0.05, the measured value seems even more odd. However, as seen in Table 1, computed and measured maximum runups are almost identical, which only serves to increase the confusion.

All of this leads to Figure 4(b), which gives experimental data from case A and the computed time traces for a solitary wave with initial height to depth ratio of 0.038, as was measured. Agreement with experiment, although not perfect, is greatly improved. However, the computed maximum runup of 2.15 cm is significantly less than the measured value of 2.74 cm.

The runup for case B, in contrast, could not be calculated directly using the present method. Computations proceeded accurately until the wave reached the vertical wall. Here, despite shrinking the computational domain and greatly increasing the number of subdomains, accuracy quickly deteriorated, and the method became unstable and blew up. Figure 3(b) shows the final accurate surface profile, where energy was still conserved to better than 3 parts in 10^3. The initial development of a vertical jet at the wall can clearly be seen. By using the jet elevation and vertical velocity at the wall, it was possible to get a lower bound on the inviscid runup by assuming that the surface was in free fall. This gave a minimum runup of 77.7 cm. As high as this is, the full potential flow solution would almost certainly have given a significantly higher value, as the jet was not yet fully developed. Frictional effects would likely have been important in limiting runup, but cannot be assessed with this model.

Traces of surface elevation for points 7 and 9 are shown in Figure 4(c). Again, computed and experimental results show qualitatively very similar features. Looking at point 7, a small phase lag is visible but, without this, agreement would be good. A phase lag is also visible at point 9 but, here, the predicted amplitude of the wave is somewhat greater than the measured value. Table 1 shows predicted runup for case B to be much greater than the measured value, which would appear to confirm the importance of viscous effects in limiting runup for near-breaking waves.

Case C was not attempted, as it was obvious that the wave would break well before the wall and runup could not be estimated with a potential flow model.

Looking back, it seems that mixed results were obtained. Accurate potential flow solutions should give very good estimates of solitary wave evolution over varying topography, but were not entirely impressive for either of cases A or B. Runup was very well predicted for case A, but it is not known what significance this should be given due to the poor time seies prediction. A large part of the problem is due to the unusually large discrepancy between the desired and measured incident wave heights. Runup was predicted more poorly for case B, which can partly be ascribed to the failure of potential flow to account for viscous effects which, given the strength of the predicted jet, would have been very important.

Attention must now be focused on the model itself. While the potential flow solution for case A was quite accurate, this degree of exactness was not likely necessary. Any of the various Boussinesq equations ([2], [3]) would provide a similar answer for this problem with less computational cost. For case B, the problem is somewhat different. A method capable of dealing with highly nonlinear waves must be used to calculate wave evolution as it moves towards the wall, but the assumption of potential flow inherent in almost all high accuracy methods breaks down when calculating the extreme runup. To accurately calculate runup for this problem, viscous effects must be included in some way, while maintaining an extremely fine resolution.

It seems that the real strength of the present method in dealing with this type of runup problem would be for waves intermediate in height between those of cases A and B. In this range accurate potential flow solutions may be obtained with a computational cost per time step that increases only linearly with the number of subdomains. This makes the method highly suitable for nonlinear wave propagation and runup problems with a large domain. While only runup problems on a vertical wall have been presented here, calculation of runup on a sloping beach should also be possible by using an accordion-like mesh, as was done while waves were generated.

Finally, it should be noted that with a fixed domain, such as was found here after the wavemaker stopped, great increases in computational efficency may be had with some loss of accuracy. By converting the local polynomial approximation method into a free surface expansion technique, the matrix equation for solving Laplace's equation must only be set up and decomposed once at the beginning of the computations, with a consequent increase in speed. Additionally, with a proper choice of basis functions, the number of unknowns at each interior boundary may be reduced to one and the bandwidth of the matrix equations may also be reduced. All of this is presently being investigated and will be reported on in the future.

References

1. Kennedy, A.B., and Fenton, J.D. (1995). Simulation of the propagation of surface gravity waves using local polynomial approximation. *Proc. 12th Aust. Coastal and Ocean Eng. Conf.*, Melbourne, Australia, pp. 287-292.
2. Peregrine, D.H., (1967). Long waves on a beach. *J. Fluid Mech.*, 27(4):815-827.
3. Nwogu, O., (1993). Alternative form of Boussinesq equations for nearshore wave propagation. *J. Waterway, Port, Coastal and Ocean Eng.*, 119(6):618-638.

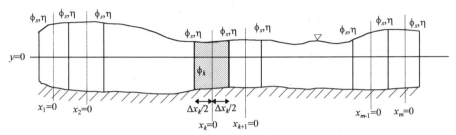

Figure 1. Definition sketch for wave motion

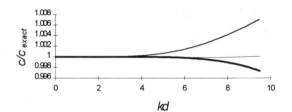

Figure 2. Local polynomial approximation to phase speed for $n=7$. Thin line uses empirical collocation points; medium line uses Chebyshev points; thick line uses Gauss-Legendre points.

	Runup (cm)	
	Case A	Case B
Computed	2.75	77.7
Measured	2.74	45.72

Table 1: Computed and measured maximum runup

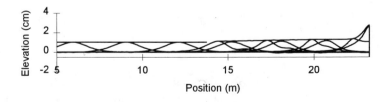

(a) Case A

Figure 3. Computed Water Surface Profiles

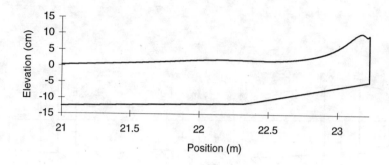

(b) Case B

Figure 3. (cont.) Computed Water Surface Profiles

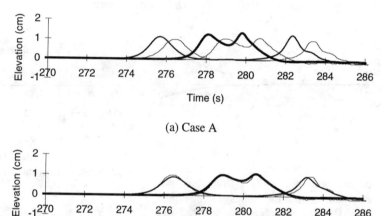

(a) Case A

(b) Case A data; Computational results for wave with *H/d*=0.038

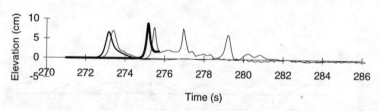

(c) Case B

Figure 4. Surface Elevation Time Traces. Thick lines are computations at point 9; medium lines are computations at point 7; thin lines are all measured data.

Numerical Simulations of Solitary Wave Generation and Runup on a Canal Compared with Experiment

A. F. Teles da Silva
Departamento de Engenharia Mecânica
Universidade Federal do Rio de Janeiro, Brasil

J. T. Aquije Chacaltana
Departamento de Engenharia Mecânica
Universidade Federal do Rio de Janeir, Brasil

D. H. Peregrine
School of Mathematics, Bristol University, UK

1. Introduction

In the present work we simulate, numerically, flume experiments conducted at the U.S. Army Engineer Waterways Experiment Station, at Vicksburg, MS. The flume, shown in figure 1, is 23.23 m long and 0.45 m wide; the undisturbed water depth at the flat portions of the flume is 0.218 m and at its left hand side there is a piston type wave-maker. Ten wave gages are placed along the canal, in our computations we take readings of gages 5, 6, 7, 9 and 10; gages 5, 7 and 9 are respectively placed at the exact beginning of each of the three slopes; gages 6 and 10 are placed respectively at the middle of the first and last slope. We also take readings of the runup at the wall.

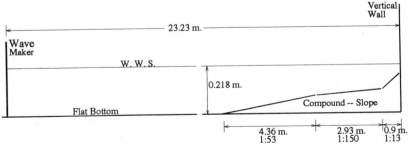

Fig. 1. Sketch of the flume, with the three sections of compound-slope

Three waves, namely cases A, B and C, are produced by the wave-maker with motions graphed in figure 2. These waves are to be produced in the flume and allowed to propagate along the canal, to reflect at the wall, at the right end, and move towards the paddle. Gage readinds for the numeric simulation are to be compared with actual readings from experiment.

2. Numerical method

The numerical simulations are computed with a Boundary Integral code that is a modification and extension of the code described by Tanaka, Lewy, Peregrine and Dold (1987) which is based on the Cauchy integral formula for the complex velocity $q(z) = u - iv$. This code, called "chyis", is described in Teles da Silva and Peregrine (1990); "chyis" computes the completely nonlinear potential flow evolution of an initial perturbation set in an infinitely long layer of water; this layer is allowed to have constant different velocities and depths at both extremeties, far upstream and far downstream, and any arbitrary smooth bottom topography, non surface piercing, at places far from the extremeties, $-\infty, +\infty$; see figure 3. Initial conditions, for the evolution, are the position of the free surface and the potential of velocities at the free surface. This code is extremely efficient and precise; mainly because integration is performed by the trapezoidal rule which under certain conditions is very accurate and the solution of the linear equation system is obtained by a Gauss-Seidel iterative procedure within a few, six to eight, iterations. Computations were made on a Sun Sparc Station 2 and a simulation for case A for 15 seconds in physical units uses about 1453 seconds of CPU time.

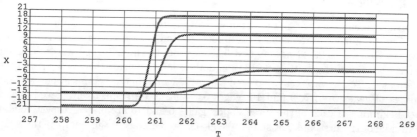

Fig. 2. The three paddle motions for cases A, B and C waves in order of increasing amplitude

Fig. 3. Sketch of the computational setting for "chyis"

Stringent requirements for smoothness is one disadvantage of the numerical scheme; the parameterization of functions and contours have to be twice differentiable and the way to deal with sharp corners is through the use of images or conformal mapping.

For each simulation the computations have been divided in two parts. In the first part waves are obtained from the recorded paddle motions and allowed to develop until crests travel to the midpoint of the flat section of the canal; this length has been chosen in order to avoid any significant interaction of the wave with the slope which is a necessary condition for the splitting of the computation, in two parts, to produce reliable results. In the second part these waves are set on the canal, over the flat bed, at a distance to the slope, corresponding to 50% of the length of the flat portion of the flume. These two parts are described in the next two sections, sections 3 and 4, and in section 5 results are presented.

In the next sections $h' = 21.8cm$ is the undisturbed depth of the canal, η' is the free surface elevation above the undisturbed level, g' is the acceleration of gravity, t' is time, x' and y' are the horizontal and vertical coordinates with origin at the undisturbed water level at the wall side of the canal, u' and v' are the horizontal and vertical velocity components and ϕ' is the velocity potential. Nondimensional quantities, with the same notation without primes, are obtained by re scaling lengths by h' and accelerations by g'; time is, accordingly, scaled by $\sqrt{\frac{h'}{g'}} \simeq 0.1491s$.

3. Solitary wave generation

In order to obtain the solitary waves for the three paddle movements, cases A, B and C, we use three different approaches:
i) Shallow water solutions,
ii) Integration of the horizontal particle velocities for the solitary wave exact solution of the Kortweg-De Vries equation,
iii) Simulation of the wave maker movement.
The comparison of the different results validate the techniques as waves generated by approaches i), ii) and iii) are found not to differ very significantly. Except for case C, where the wave obtained by the shallow water technique steepens and breaks too quickly. The digitized data of the paddle motions for cases A, B and C have been numerically smoothed and time differentiated. In this way data for the paddle velocities have been produced. A brief description of each of the three methods is given below but solutions are going to be presented only for the simulation of the wave maker motion, iii), that produced the better results for comparison with experiments.

3.1. Shallow water solutions

Numerical differentiation of the data for the paddle motions provides paddle velocities $\frac{dx}{dt}$ at each position; these velocities are the particle horizontal velocities set in the water by the prescribed paddle movement. If we take the elevation η of the free surface, above the undisturbed level, corresponding to the speed of paddle motion to be approximated by a long shallow water of elevation, cf Lamb, 1932, §187, then we have the following equation for η in terms of u:

$$u' = 2\sqrt{g'h'} \left(\sqrt{1 + \frac{\eta'}{h'}} - 1 \right)$$

from the values of η and u we calculate v as:

$$v' = -(h' + \eta')(\frac{\partial u'}{\partial x'})$$

With these quantities, we compute the position of the free surface, η, and the velocity potential at the free surface that are used as initial conditions. Computing these conditions for cases A, B and C we find approximations of the waves generated by the paddle motions, to be set at the wave maker side of the flume.

3.2. Integration of the solitary wave exact solution for the Kortweg De Vries equation

An exact small amplitude solitary wave solution, Whitham, 1974, is given by

$$\eta' = \eta_0' sech^2 \sqrt{\frac{3\eta_0'}{4h'^3}}(x' - U't')$$

$$U' = \sqrt{g'h'}(1 + \frac{1}{2}\frac{\eta_0'}{h'})$$

$$u' = \sqrt{\frac{g'}{h'}}\eta_0' sech^2[\sqrt{\frac{3\eta_0'}{4h'^3}}(x' - U't')]$$

now considering the particle velocity u' to be $u' = \frac{dx'}{dt'}$ we form a first order ordinary differential equation for the paddle motion $x(t)$ depending on the parameter η_0; numerical solutions with a fourth order Runge-Kutta method produce a good agreement with the digitized data for cases A, B and C for values of η_0 respectively equal to 0.0440, 0.2378 and 0.4882. In order to obtain solitary waves with these amplitudes we do not use $sech^2$ approximations, which would be precise for the case A wave only; but we use, instead, nonlinear exact numeric solutions for steep solitary waves produced by a code named "stchy" described by Teles da Silva and Peregrine, 1988.

3.3. Simulation of the paddle movements

The numerical code "chyis", mentioned above, which follows the transient potential flow evolution of an initial perturbation allows for different uniform velocities and depths at each extremety. If we set, as an initial condition, a flat surface over a flat bed, with uniform speeds U and $-U$ respectively at the left, $(-\infty)$, and at the right, $(+\infty)$, it works out as an image method; it is the same as if on both sides water runs against a rigid wall. As the motion is inviscid, the flow, to the right or to the left, can be seen as the flow produced by the impulsive movement, at speed $-U$ or $+U$ respectively, of a paddle viewed from a frame of reference moving with the paddle. In this way the impulsive movement of a paddle at constant speed U could be modelled exactly.

In order to model the paddle movements, with variable speeds, given in figure 1, the velocities, $+U$ and $-U$, at both ends, for the "chyis" code, should be functions of time. As U varies with time the frame of reference is no longer inertial but has an acceleration $\frac{dU}{dt}$; in this frame of reference we have to solve the potential flow motion, for the potential ϕ, which is governed by: $\nabla^2\phi = 0$ with conditions of non penetration, $\phi_n = 0$, at the flat bed and the paddle; far from the paddle the water speed is uniform and equal to $(-U(t), 0)$; at the free surface, parameterized by $\mathbf{R}(\xi) = (x(\xi), y(\xi))$, ϕ must obey: $\phi_t + \frac{1}{2}(\nabla\phi)^2 + \frac{dU}{dt}x + y = 0$ and $\frac{D\mathbf{R}}{Dt} = \nabla\phi$ where $\frac{D}{Dt} = \frac{\partial}{\partial t} + \nabla\phi.\nabla$. The implementation of these boundary conditions represented a very minor modification to "chyis" and results appear very good, they are in fact the exact nonlinear potential flow solution for the motion of a piston type wave maker.

4. Simulation of runup over the compound slope and wall

The wall was set by the use of images, reflecting the flow about its position. The compound slope is made of straight segments joined by small smooth arcs in order to meet the smoothness requirements of "chyis". The waves obtained as described in section 2 are set over the flat portion of the bed before the slope and allowed to propagate towards the wall, to reflect there and move back towards the paddle. The simulation stops as waves overturn to break; the case B wave breaks just before its crest arrives at the wall and the case C wave breaks just before its crest arrives at the second portion of the compound slope.

5. Results

For many reasons the results obtained by the simulation of the paddle motions are the best for comparison with the experimental data. The two other techniques, i) and ii), produce ready made waves that should be placed somewhere; the gage readings for the experimental data could help finding a good initial position. But if the paddle motion is simulated, technique iii), we just place the paddle at the prescribed initial paddle position and move it according to the digitised data; without any interference with the position of the crest. Gage readings of surface elevation, for the case A wave, for numerical simulation and experiment are compared graphicaly in figure 4. In both simulation and experiment the lefmost peak represents the wave coming from the wave maker and the rightmost peak represents the wave reflected at the wall. Figure 5 gives the case A wave runup against the wall. The computational results represent the completely nonlinear solution with no viscosity and are consistently higher in amplitude than the experimental ones. Differences in phase are remarkably small, even for the reflected waves, measurable in tenths of seconds. The generated wave at the end of the first part of the computation, the generation stage, attained a nondimensional amplitude of 0.0473 which is very close to the amplitude of the speed of the paddle which is 0.0462; the Kortweg & De Vries equation predicts that they should be equal and they would be actually equal in an asymptotic time span (no less than 300 nondimensional time units, for a length at the least three times bigger than the whole length of the flume). Strangely the maximum runup at the wall, found to be 0.1251 in nondimensional units, is slightly lower than the experimental value of 2.74 cm or 0.1257 in nondimensional units; we have no explanation for this mismatch.

Figure 6 show readings for gages 5, 7 and 9 for both computation and experiment for the case B wave. Gage 10 is not included because the wave, in both experiment and computation, breaks as it passes this gage. In the experiment wave breaking occurs at time approximately 119.1 whilst in the computations breaking occurs at time 119.8 when the crest is at a nondimensional distance of 0.31, 6.75 cm, from the wall; note that at this point the wave is already interacting with its own reflection. This set of experimental results is, surprisingly, in better agreement with computations than the previous one.

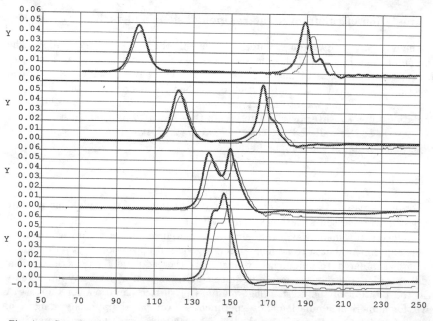

Fig. 4.   Case A wave: from top to bottom gage readings for stations 5, 7, 9 and 10; the thick lines represent surface elevations for the simulation and the thin lines surface elevations for the experiment.

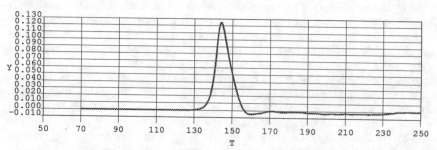

Fig. 5. Runup against the wall, (numerical simulation).

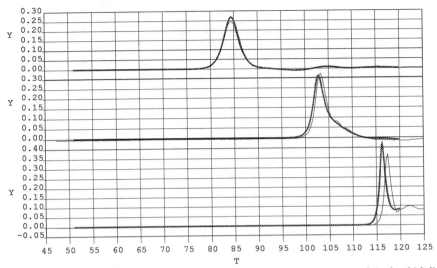

Fig. 6. Case B wave: from top to bottom gage readings for stations 5, 7 and 9; the thick lines represent surface elevations for the simulation and the thin lines surface elevations for the experiment.

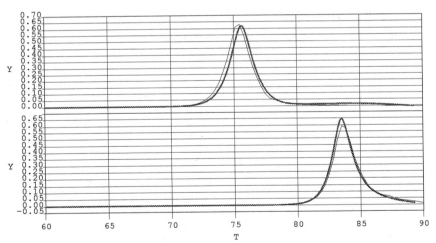

Fig. 7. Case C wave: from top to bottom gage readings for stations 5 and 6; the thick lines represent surface elevations for the simulation and the thin lines surface elevations for the experiment.

A good agreement between experiment and computation occurs, also, in Figure 7 where readings for gages 5 and 6 are shown for the case C wave. In the experiment the case C wave breaks somewhere between gages 7 and 8, respectively at distances of 17.57 and 10.85 from the wall, whilst computations find breaking when the crest is at a distance of 20.13 to the wall.

6. Conclusions

For the authors, the greater achievement in this work has been the successfull computational implementation of an exact nonlinear inviscid flume with a piston type wave maker. Top of figures 6 and 7, cases B and C waves, show how close computational and experimental results are; the smaller differences are mostly due to the effects of viscosity. Case A wave shows a good qualitative agreement, (figure 4); although, being a very small wave and so less influenced by viscosity, we should expect it to be the best match between experiment and computation. The paddle data, figure 2, for the case A wave, when numerically differentiated in time produces a speed for the paddle that matches the expression for u given by the Kortweg & De Vries equation; this equation predicts that the nondimensional wave amplitude is the same as the velocity amplitude; so we should produce a wave that after a long time settles its amplitude to the amplitude of the velocities of the paddle motion, and our case A wave actualy settles to 0.0462 after moving for more than 300 depths of the canal.

Data is provided for a gage 4, positioned near middle of the flume over its flat section of bed. This gage would have been ideal to check the match between experiment and computation for the wave generation; but we allways found a poor agreement for this gage alone not for the others further down the flume.

7. References

1. H. Lamb. *Hydrodynamics*, (Dover Pub., New York, 1932).
2. G. H. Keulegan, *J. Res. Nat. Bur. Standards*, **40**, (1948), p. 487.
3. G. B.Whitham, *Linear and Nonlinear Waves*, (John Wyley & Sons, New York, 1974).
4. M. Tanaka, M. Lewy, D. H., Peregrine, & J. W. Dold, *J. Fluid Mech.* **185**, (1987), p.235.
5. A. F. Teles da Silva & D.H. Peregrine, *J. Fluid Mech.* **195**, (1988) p.281.
6. M. J. Cooker, D. H. Peregrine, C. Vidal & J. W. Dold, *J. Fluid Mech.* **215**, (1990), p. 1.
7. A. F. Teles da Silva & D.H. Peregrine, *Eng. Anal. with Bound. Elem.* vol **7** n**4**, (1990), p. 214.

Three-Dimensional Analysis of Long Wave Runup on a Conical Island by using the MAC Method

Kenji Masamura and Koji Fujima
Department of Civil Engineering
National Defense Academy of Japan

1. Outline

In order to reproduce the runup process of solitary wave on an island accurately, both the nonlinearity and the dispersion effect should be taken into account in the numerical simulations. In this study, some improvements are made on the MAC method to conduct a numerical simulation of tsunami around a conical island, and the results are compared with the experimental ones.

2. 3-D MAC method

2.1. Assumptions

The MAC method is an Eulerian finite-difference method for computing the dynamics of incompressible fluids with multiple free boundaries. However, in the present computation, viscous and bottom friction forces and overturning of wave are ignored.

The marker particles are not used in this computation, but the height function η is used in order to determine the surface elevation. η denotes the water surface elevation from the initial water level.

2.2. Mesh configuration

The computation is conducted using an Eulerian fixed mesh of rectangular parallelepiped cells. Pressure is computed at the center of cell, and velocities are on the boundaries of cell. The conical island is considered to be the complex of obstacle cells in which water particle can not pass through the boundaries (Figure 1). The cells whose center points are below the surface of conical island are treated as the obstacle cells. Therefore, this numerical method seems an extension of the usual method of numerical simulations of tsunamis using the leap-frog scheme.

The variable mesh is used to decrease the memory size. Figure 2 shows the grid size used in the computations. Both Δx and Δy are 10cm near the island model, and

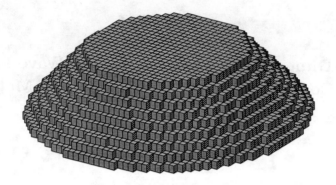

Figure 1: The complex of obstacle cells above $z = $ -2cm

are 30 ~ 40cm far from the island. Δz is 2.5cm. Totally, grid number is 148 × 155 × 21.

2.3. Initial and boundary conditions

(1)Initial conditions

At the initial time step, water elevation and velocities are zero everywhere, and pressure values are assumed to be the hydrostatic pressure.

(2)Bottom and wall boundaries

The bottom and three walls where the wave maker does not locate are treated as free-slip wall. Thus velocity component normal to wall is zero, and components parallel to wall should have no normal gradients.

(3)Wave-paddle boundary

Trajectory curves of wave-paddle are used in the present computations to determine the boundary condition at the wave maker. By fitting the given curves to curves of *tanh*-function, the position of wave paddle x_p is approximated by the functions listed in Table 1. The velocity at the wave paddle can be estimated by the differentiation of x_p. However, displacement of wave paddle, which is ± 15cm in Case C, is ignored. Thus, the estimated velocity is always given at $x = 0$.

(4)Free surface boundaries

Since the pressure at the free surface should be zero, pressure $p_{i,j,k}$ in the cells containing a free surface are set equal to the value obtained by a linear interpolation of the surface pressure (=zero) and the pressure inside the fluid ($p_{i,j,k-1}$).

(5)Wave tip boundaries

The wave tip on the slope of conical island always agree with the mesh boundary. Near the wave tip, surface cell is mostly located on obstacle cell. The pressure in surface cells located on obstacle cell are estimated by the assumption of hydrostatic pressure distribution. Therefore, the present computation is equivalent to the computation based on the shallow water theory neglecting friction force at the wave tip cell.

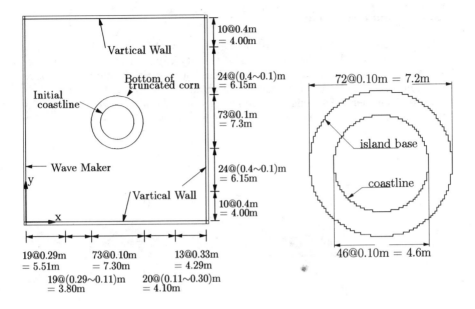

Figure 2: Model island (Plain View)

By the way, the wave maker starts to move after $t=22$s. Then, at the initial time step, t is set to be 22s.

All computations were performed by using DEC 3000/400. Table 2 shows the list of CPU time and machine requirement.

Table 1: Approximate curves of x_p and $u_{1,j,k}$

	x_p	$u_{1,j,k}$
Case A	$8.01\tanh(1.06(t-22)-3.74)$	$8.49\,\mathrm{sech}^2(1.06(t-22)-3.74)$
Case B	$11.31\tanh(1.48(t-22)-3.46)$	$16.74\,\mathrm{sech}^2(1.48(t-22)-3.46)$
Case C	$14.76\tanh(2.19(t-22)-3.69)$	$32.32\,\mathrm{sech}^2(2.19(t-22)-3.69)$

Table 2: CPU time and machine requirement

	case A	case B	case C
CPU time	32.1hrs.	37.8hrs.	42.8hrs.
	$(t=22\sim39\mathrm{s})$	$(t=22\sim39\mathrm{s})$	$(t=22\sim39\mathrm{s})$
memory	22MB	22MB	22MB
storage	43MB/step	43MB/step	43MB/step

3. Results and discussions

Figure 4 shows the side view of computed velocity vector at y=13.8m in Case C. Figures 5 shows the plane view of computed velocity vector at free surface in Case C. It is obvious that phenomena can be reproduced symmetrically and runup process can be computed stably. However, the water which inundates the dry step tends to remain too long there. This might be caused by the fact that the back-flow velocity could not be accelerated suitably by using the upwinding method in the evaluation of advection terms.

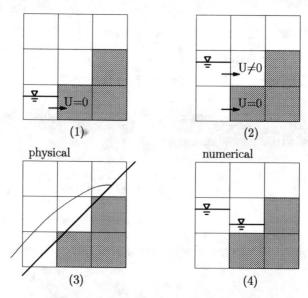

Figure 3: Evaluation of wave profile near the wave tip

Figures 6, 7 and 8 show the time histories of water elevation at several points in cases A, B and C. The present computations reproduce the variation of water elevation satisfactorily in all cases.

Runup height distribution of cases A, B and C are shown in Figures 9, 10 and 11. There are abrupt jumps at some points. These sudden jumps on runup height distribution appear on the region where the arrangement of obstacle cells is complex. In the regions where the arrangement of cells is relatively simple, for example $\theta = 0°$ and $90°$, a smooth distribution is obtained. Of course, if the smaller grid size would be used in the computations, amplitude of sudden jumps would become small. However, in Case A, the computed runup height distribution agrees well with the measured one. In Case B, the computed values of runup height agree well with the measured ones except for the points at the lee of the island. It should be noted that the measured water elevation at x=15.56m, y=13.8m is reproduced satisfactorily in Case B (see Figure 7). Only runup height at the lee of the island is underestimated. In Case C, the simulation underestimates the runup height at the front and lee of the island. However, at the front of the island, measured water elevations at (9.36m, 13.8m) and (10.36, 13.8m) are reproduced satisfactorily. It shows that the volume of runup water is estimated accurately, but the only runup height is underestimated. This underestimation of runup height might be caused by the following:

In the present computation, the runup process is computed by the procedure as described follows.
1. The water elevation in wave tip cell is increasing, although the position of wave tip is stopped by the vertical walls. (Figure 3(1))
2. If the water elevation exceeds the height of obstacle cell, water volume starts flooding to the neighbor cell.(Figure 3(2))
Therefore, if the volume of runup water is less than $\Delta x \times \Delta y \times \Delta z$, computed runup height cannot exceed Δz. Namely, when the real thickness of runup water is smaller than Δz as show in Figure 3(3), profile of runup water is evaluated as shown in Figure 3(4). Then, runup height becomes underestimated. It is considered that the similar phenomena are occurred at the front and lee of the island in Case C and the lee of the island in Case B.

4. Conclusions

Three-dimensional numerical analysis method is developed based on the concept of the MAC method. Both the nonlinearity and the dispersion effect can be taken into account in the computations by using the present method. Further, since the present method seems an extension of the usual method of numerical simulation of tsunamis, it might be useful for the practical problems.

Through the comparisons of the numerical results with the experimental results, the validity of the numerical simulations are verified. However, following matters are pointed out in this paper.
1. Runup height distribution shows abrupt jumps at some points.
2. Runup height is underestimated when the thickness of runup water is less than Δz.
3. Runup water seems to stay in an inundation area too long.
These matters might be improved by using the fine mesh.

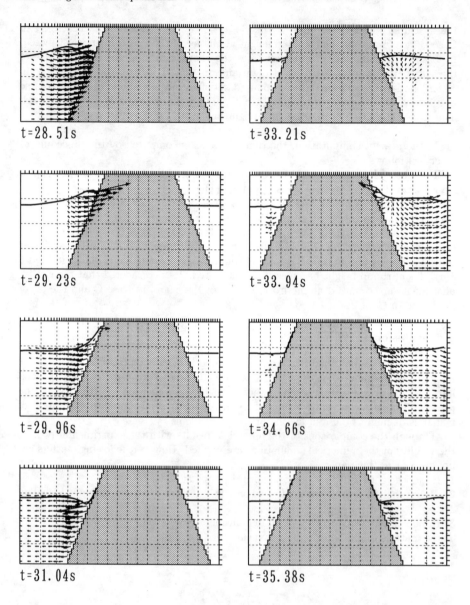

t=28.51s   t=33.21s

t=29.23s   t=33.94s

t=29.96s   t=34.66s

t=31.04s   t=35.38s

Figure 4: The side view of velocity vector at $y=13.8$m in Case C

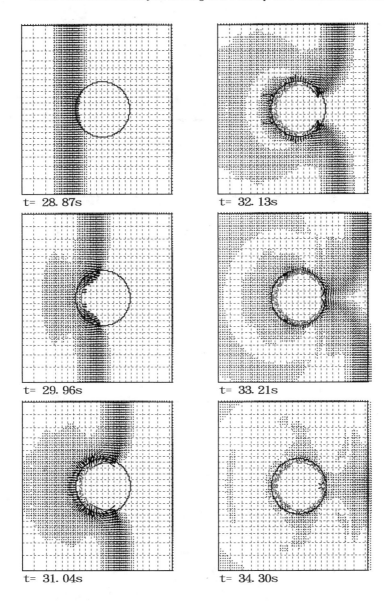

t= 28. 87s   t= 32. 13s

t= 29. 96s   t= 33. 21s

t= 31. 04s   t= 34. 30s

Figure 5: The plan view of velocity vector in Case C

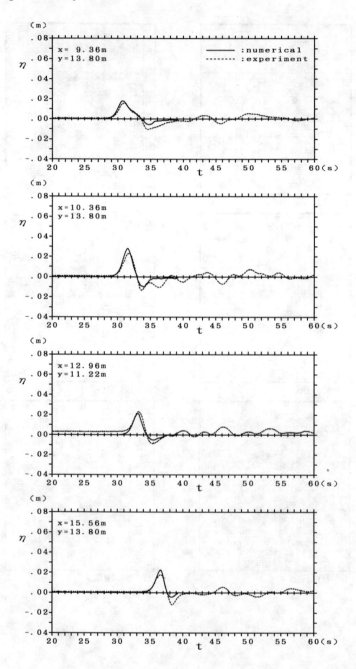

Figure 6: Time history of water elevation in case A

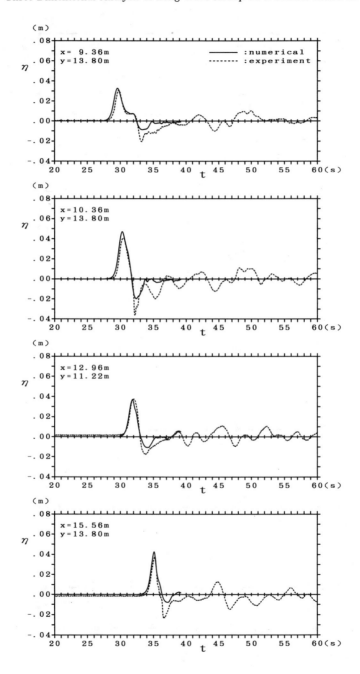

Figure 7: Time history of water elevation in case B

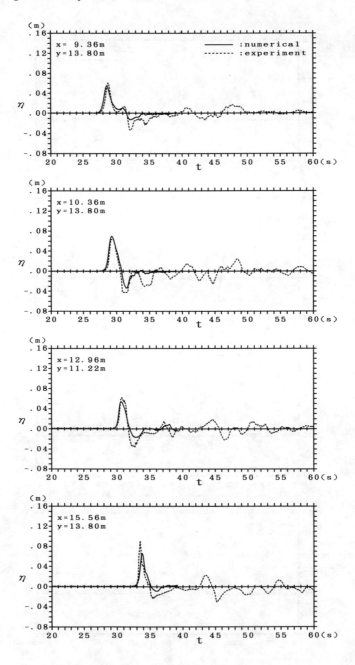

Figure 8: Time history of water elevation in case C

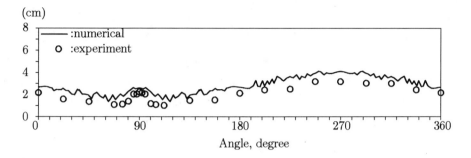

Figure 9: Runup height distribution in case A

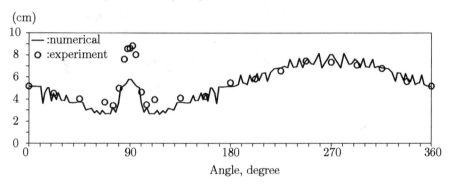

Figure 10: Runup height distribution in case B

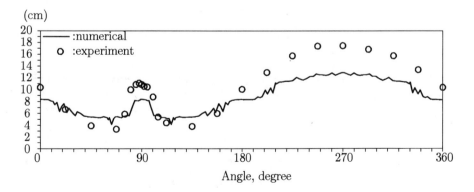

Figure 11: Runup height distribution in case C

TSUNAMI RUNUP IN A SLOPING CHANNEL

E. Pelinovsky [1], O. Kozyrev [2], E. Troshina [3]

Abstract

In this paper we investigate the problem of 2D tsunami runup in a channel of arbitrary cross-section (meaning a parabolic family of functions) and we simulate tsunami waves climbing the beach. We use both analytical and numerical approaches, based on the assumption that the hydraulic approach is valid. The nonlinear set of shallow water equations for waves in an open channel is transformed using the hodograph approach into one linear differential equation with the coefficients depending on the specific geometry of the bay.

Computations are carried out through finite-difference numerical integration of the 2D nonlinear shallow water equations. The maximum of the runup appears to be not sensitive to non-linear effects. The wave parameters for breaking appear to be independent of the cross-section of the bay.

1. Boundary Value Problem

The combination of analytical approaches and numerical simulation is probably the most reasonable way to study the behavior of tsunamis in shallow water and the climbing of the waves onto the dry land. The tsunami problem basically consists of:

(1) choosing appropriate physical and mathematical models for the generation process, the open ocean propagation stage and the runup on a shore;

(2) taking into account that tsunami waves are enhanced in bays;

(3) constructing suitable algorithms (analytical and numerical) for solution of corresponding sets of differential equations.

The main goal of the present work is to study the runup characteristics of a tsunami wave climbing a channel and to test the numerical scheme.

For the simulation we use two-dimensional nonlinear shallow water equations:

$$\frac{\partial u}{\partial t} + u\frac{\partial u}{\partial x} + v\frac{\partial u}{\partial y} = -g\frac{\partial \xi}{\partial x},$$

---

[1] Institute of Applied Physics, Nizhny Novgorod, Russian Academy of Science
[2] State Technical University, Nizhny Novgorod, Russia
[3] Institute of Marine Science, University of Alaska Fairbanks

$$\frac{\partial v}{\partial t} + v\frac{\partial v}{\partial y} + u\frac{\partial v}{\partial x} = -g\frac{\partial \xi}{\partial y},$$

$$\frac{\partial \xi}{\partial t} = -\frac{\partial}{\partial x}(Du) - \frac{\partial}{\partial y}(Dv), \tag{1.1}$$

where x, y are the horizontal coordinates, t is time, $u(t, x, y), v(t, x, y)$ are particle velocities in the x and y directions, ξ is the vertical displacement of water surface above the undisturbed water level, g is the gravitational acceleration, $D = h(x, y) + \xi(t, x, y)$ is the total water depth (h is the water depth).

In order to simulate a tsunami wave climbing an indented beach we assume that the cross-section of the bay is described by an arbitrary parabolic function.

2. Analytical Approach

In this section we consider an analytical solution for the runup maximum of a wave climbing a sloping bay.

Let us assume for simplicity that we have the bay with a parabolic cross-section:

$$h(x, y) = -\alpha x - a \cdot |y|^m, \tag{2.1}$$

where α is the slope of the beach, a and m are constants ($m > 0$; if $m \to \infty$, we have a sloping rectangular channel with slope α).

For that case our hyperbolic differential set of equations has Riemann invariants as follows: $V_{\pm} = u \pm 2\sqrt{\frac{m+1}{m}gD} + g\alpha t$. This allows us to apply *Carrier and Greenspan* transformation to represent the initial set of shallow water equations as a second order linear differential equation [1, 2]:

$$\frac{\partial^2 \Phi}{\partial \lambda^2} - \frac{\partial^2 \Phi}{\partial \sigma^2} - \frac{m+2}{m}\frac{1}{\sigma}\frac{\partial \Phi}{\partial \sigma} = 0. \tag{2.2}$$

All physical parameters can be obtained through Φ. A moving boundary between wet and dry domains becomes a fixed one ($\sigma = 0$) in new coordinates (λ, σ).

An especially simple solution of the equation has a form that corresponds to a harmonic wave

$$\Phi = A \cdot \frac{J_{1/m}(l\sigma)}{\sigma^{1/m}}\sin(l\sigma), \tag{2.3}$$

where $J_{1/m}$ is a Bessel function, l and A are constants.

Then, using asymptotic expansions for Bessel functions and approximate forms of *Carrier and Greenspan* transformations, we can construct formulas for wave field representation and for values of maximum penetration height [2]:

$$\frac{\xi_R}{\xi_0} = \frac{2\sqrt{\pi}}{\Gamma\left(1 + \frac{1}{m}\right)} \cdot \left(\frac{2\pi L_R}{\lambda}\right)^{\left(\frac{1}{2} + \frac{1}{m}\right)}, \tag{2.4}$$

which are valid for $\xi_R \gg \xi_0$ (ξ_R is the maximum penetration height, ξ_0 is the initial wave amplitude); Γ is a gamma-function.

For $m \to \infty$ this formula describes the well-known *Shuto* relation for a sloping beach [3].

The main goal of investigating of runup in a bay with parabolic cross-section lies in corroboration of this *Shuto's* result that there is a linear relationship between maximum runup height and initial wave amplitude.

3. Numerical Approach

In this section we have carried out a numerical investigation of the runup problem within the framework of the general shallow water model. It gives us an opportunity:

(1) to estimate the domains of validity of linear approaches and

(2) to construct and test an efficient numerical model.

The numerical scheme we used is analogous to that of *Kowalik and Murty* [4], see also *Marchuk et al.* [5]. The x - and y - directions are integrated separately. The integration is split in time into two sub-steps. The integration scheme uses a space-staggered grid for approximation of the different variables. In order to avoid the problems of computational instability and connected with them filtering difficulties we have used the first order scheme with additional implementation of averaging procedures.

Let us rewrite the original differential system in dimensionless variables:

$$\tilde{t} = \omega t; \quad \tilde{x} = \alpha \frac{x}{\xi_0}, \quad \tilde{y} = \alpha \frac{y}{\xi_0}, \quad \tilde{\xi} = \frac{\xi}{\xi_0}, \quad \tilde{h} = \frac{h}{\xi_0},$$

$$\tilde{u} = \alpha \frac{u}{\omega \xi_0}, \quad \tilde{v} = \alpha \frac{v}{\omega \xi_0},$$

where ω and ξ_0 are the frequency and the amplitude of the initial wave, respectively.

Let us denote by m the number of the time step and by $*$ the intermediate values of sea-level variation. Then we can use a finite-difference scheme in the form:

$$\frac{u^{m+1} - u^m}{T} + \left[\left(u\frac{\partial u}{\partial x}\right)^m + \left(v\frac{\partial u}{\partial y}\right)^m\right] = -\frac{1}{Br}\left(\frac{\partial \xi}{\partial x}\right)^m;$$

$$\frac{1}{2}\frac{\xi^* - \xi^m}{T/2} = -\frac{\partial}{\partial x}\left[D^m\frac{u^{m+1} + u^m}{2}\right];$$

$$\frac{v^{m+1} - v^m}{T} + \left[\left(v\frac{\partial v}{\partial y}\right)^m + \frac{u^{m+1} + u^m}{2}\left(\frac{\partial v}{\partial x}\right)^m\right] = -\frac{1}{Br}\frac{\partial}{\partial y}\left(\frac{\xi^* + \xi^m}{2}\right);$$

$$\frac{1}{2}\frac{\xi^{m+1} - \xi^*}{T/2} = -\frac{\partial}{\partial y}\left[D^*\frac{v^{m+1} + v^m}{2}\right]. \tag{3.1}$$

Here

$$Br = \frac{\omega^2 \xi_0}{\alpha^2 g};$$

is dimensionless parameter. There are two principal points in our numerical approach:

3.1. Location of the moving wet-dry boundary

In order to locate the wet-dry boundary, the total depth is assumed to be positive at the wet points and equal to zero at the boundary between wet and dry domains ($\xi = -h$). Thus, a negative value of D indicates dry regions. Such an algorithm was described in *Kowalik and Murty* [4].

A linear extrapolation procedure of *Sielecki and Wurtele* [6] of sea level to the first dry point along a wet-dry boundary was used:

$$\xi^*_{jm+1 k} = \xi^m_{jm+1 k}\big|_{ext} - \frac{T}{\Delta h}\left\{(D\cdot u)^m_{jm+1 k}\big|_{ext} - (D\cdot u)^m_{jm k}\right\}$$

$$\xi^{m+1}_{jm+1 k} = \xi^m_{jm+1 k}\big|_{ext} - \frac{T}{\Delta h}\left\{(D\cdot v)^m_{jm+1 k}\big|_{ext} - (D\cdot v)^m_{jm+1 k-1}\right\},$$

where $[\cdot]|_{ext}$ means linear extrapolation from within the region of calculation.

3.2. Upstream   approximation

It is a known fact that upwind numerical schemes are usually stable. Diffusion can be improved by applying *Mader's* [7] procedure for advective terms in the equations of motion, the continuity equation and in the boundary conditions. This procedure is based on a proper choice of the space derivative. This choice depends on the direction of the current.

Thus, advective terms can be treated as follows:

$$if \qquad u < 0 \quad \Rightarrow \quad \frac{u - |u|}{2} \cdot \frac{u_{j+1}^m - u_j^m}{\Delta h}$$

$$if \qquad u > 0 \quad \Rightarrow \quad \frac{u + |u|}{2} \cdot \frac{u_j^m - u_{j-1}^m}{\Delta h}.$$

In order to test the above numerical procedures we have simulated the problem of nonlinear wave runup on a sloping beach. For some initial condition an analytical one-dimensional solution of that problem was obtained by *Carrier and Greenspan* [1].

The distribution of water level along a sloping beach at various dimensionless times ($t = 0.05; 0.1; 0.15; 0.2; 0.25; 0.3; 0.35; 0.4$) is given in Fig. 3.1. The numerical solution was obtained with a 2D-solver with quasi-one-dimensional initial *Carrier and Greenspan* conditions.

Figure 3.1 shows close agreement between the analytical and numerical solutions, suggesting that the numerical solution is fairly accurate.

4. Parabolic Bay

The next step is the comparison of analytical data for maximum penetration height in a parabolic bay ($m = 2$) with numerical data received for an initially monochromatic wave coming from an open boundary. Figure 4.1 shows maximum runup heights computed analytically and numerically for the sloping channel with a parabolic cross-section. Again, there is a good agreement between the analytical and numerical results, suggesting that a linear approach for the calculation of maximum runup heights and breaking criteria in a 2D problem seems to work for the bay with a parabolic cross-section.

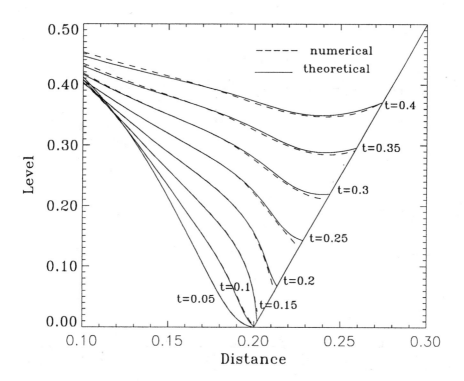

Figure 3.1:

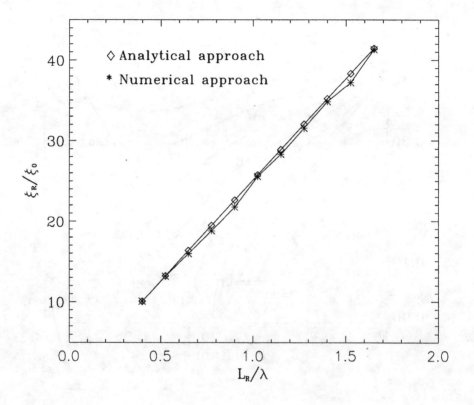

Figure 4.1:

References

[1] G. Carrier and H. Greenspan, *J. Fluid Mech.*, **N. 4.**, (1958), 97.

[2] N. Voltzinger, K. Klevanny and E. Pelinovsky, in *Long-wave dynamics of the coastal zone,* (Leningrad: Gydrometeoizdat, 1989, in Russian).

[3] N. Shuto, *Coast. Eng. Japan*, **V. 10**, (1967), 24.

[4] Z. Kowalik and T.S. Murty, *Marine Geodesy,* **V. 16**, (1993), 87.

[5] A. Marchuk, L. Chubarov and Yu. Shokin, in *Numerical modelling of tsunami waves,* (Novosibirsk: Nauka, 1983; Engl. transl., Los Alamos, 1985).

[6] A. Sielecki and M. Wurtele, *J. Comp. Phys.*, **N. 6**, (1970), 219.

[7] L.C. Mader, in *Numerical modelling of water waves,* (Berkley: University of California Press. 1988).

LONG WAVE RUN-UP MODELLING IN 2D USING FINITE DIFFERENCE GRID WITH SURFACE ROLLER AND MOVING BOUNDARIES

G.S. Prasetya, B. Ontowirjo and Subandono .D[1]

1. Background

Recently, numerous investigations and studies have been carried out in the field of Long wave run-up modelling both for engineering and scientific purposes. This model is developed as a model prototype to simulate the tsunami propagation and run-up on coastal land area. In this model surface roller formations are taken into account and automatic mesh generation are introduced in shoreline boundary to make the model more realistic. Staggered leapfrog finite difference scheme is used to solve the equations numerically.

2. Governing Equation

According to shallow water wave theory, the basic equations used for modelling the Tsunami propagation are continuity equation and the equations of motions :

$$\frac{\partial \zeta}{\partial t} + \frac{\partial P}{\partial x} + \frac{\partial Q}{\partial y} = \frac{\partial Z}{\partial t}$$

$$\frac{\partial P}{\partial t} + \frac{\partial}{\partial x}\left(\frac{P^2}{D}\right) + \frac{\partial}{\partial y}\left(\frac{PQ}{D}\right) + gD\frac{\partial \zeta}{\partial x} + \tau_x + k.Rolx = 0 \tag{1}$$

$$\frac{\partial Q}{\partial t} + \frac{\partial}{\partial x}\left(\frac{PQ}{D}\right) + \frac{\partial}{\partial y}\left(\frac{Q^2}{D}\right) + gD\frac{\partial \zeta}{\partial y} + \tau_y + k.Roly = 0$$

in which Z is the deformation of sea bottom. P,Q, (x,y,t) are the flux densities in x- and y- directions $(m^3/s/m)$ = (uh,vh) ; (u,v) = depth averaged velocities in x- and y- directions and τ_x and τ_y denote the x- and y- component of bottom friction . $D = h + \zeta$ is the total

[1] Tsunami Research Center, LPTP BPP Teknologi, 22nd.fl
Jl.MH.Thamrin no.8. Jakarta 10340 Indonesia

depth, while Rolx, Roly are the additional pressure forces from the roller if surface roller is present during propagation k = 1 or k=0 if the roller doesn't exist.

Bottom Friction

Bottom friction can be described using Chezy (C) number or Manning (M) numbers. In Chezy number the formula can be written as follows :

$$\frac{\tau_x}{\rho} = \frac{g}{C^2.D^2} P\sqrt{P^2 + Q^2} \quad ; \quad \frac{\tau_y}{\rho} = \frac{g}{C^2.D^2} Q\sqrt{P^2 + Q^2} \tag{2}$$

Surface Roller

Surface roller is considered in this momentum equation due to wave breaking phenomena and bore formation. This term are formulated by Fredsoe and Deigaard (1992) as follows :

$$\frac{\partial p_r}{\partial x} = -\rho g\left(\zeta^+ \frac{\partial \zeta}{\partial x} + (D+\zeta)\frac{\partial \zeta^+}{\partial x}\right) = \text{Rolx}$$

$$\frac{\partial p_r}{\partial y} = -\rho g\left(\zeta^+ \frac{\partial \zeta}{\partial y} + (D+\zeta)\frac{\partial \zeta^+}{\partial y}\right) = \text{Roly} \tag{3}$$

where ζ^+ is the local thickness of the roller. This pressure term will modify the dynamics of the wave motion, simulating the interaction between wave and rollers and extract energy from the waves.

3.Numerical Scheme

Both of the linear and nonlinear terms of the shallow water equations are solved using a staggered leap-frog finite difference scheme. Since this scheme is standard numerical scheme, the explanation will be focused on the calculation of the surface roller and the grid arrangement for the moving boundaries on run-up calculation.

3.1.Grid Arrangement

Nested square grid arrangement is chosen to avoid excessive numbers of calculations, while maintaining higher resolution in particular region. A non-uniform grid system will be used with a fine grid in the area of interest and a coarse grid over the rest of the area. To ensure the numerical stability, the Courant number is used and must be less than or equal to 1.

3.2. Calculation of Surface Roller.

According to Schaffer,et al, (1992), Deigard and Fredsoe (1992), the surface roller starts where the water surface has the same slope as the lower boundary of the roller and the breaking will cease when the maximum slope at the wave front becomes smaller than the wave slope, α_o. A simple procedure is used to determine the geometry of the surface roller at each time step in the simulation . The geometrical of the roller can be seen on figure. 1.

The breaker criterion is expressed by the local slope of the water surface $\alpha(t)$. Since breakers often transform rather quickly to the bore-like, an exponential transition is used as proposed by H.A.Schaffer.,et.al as follow :

$$\tan\alpha(t) = \tan\alpha_o + (\tan\alpha_B - \tan\alpha_O)\exp\left(-\ln 2 \frac{t - t_B}{t_{1/2}}\right) \qquad (4)$$

where t_B is the time of incipient breaking and $t_{1/2}$ is the half time for tan $\alpha(t)$. The lower boundary of the roller is assumed to be a straight line with the slope α_o, where $\alpha_o \leq \alpha_{br}$.

To compensate inaccuracies in the determination of the surface roller geometry, this pressure term will be multiplied by calibration factor, Kr . The value of Kr is determined and between 1 and 2

3.3. Calculation of Moving Shoreline

We consider the horizontal distance of shoreline moves using the accelerated movement of wave front with the following formula :

$$S = V.t + 0,5.a.t^2 \qquad (5)$$

where V is the velocity of the leading edge of wave front = u + C , t is time and a is acceleration, u is Eulerian velocity from momentum equation and C is wave celerity at the end of grid at shoreline (at D = 0) and we calculate the water acceleration using the following formula :

$$a = \frac{\partial u}{\partial t} + u\frac{\partial u}{\partial x} \qquad (6)$$

3.4. Grid generation and Boundary condition at Run-up Front

At the shore, boundary moves in the process of wave run-up and drawdown. If the wave moves upwards to the beach, the grid points on the shoreline move upwards and the number of grid size increases. Conversely, if the wave moves downwards, the number of grids decreases.

The horizontal distance of wave movement could be larger than the size of the new grid mesh which has been designed . To solve this problem and to keep the numerical stability during the interval time step on new grid mesh system, we define that if S is ≥ 0.5 Δx then we set to one grid as additional and conversely if S < 0.5 we set to zero (no additional grid). After defining the new grid we proceed to calculate the equation of continuity and momentum at every new grid during one time step. And the calculations are done for the next time step using the same procedure as before.

The initial condition of water elevation on computation of run up on coastal land is equal to the ground level (h) where h in this region are taken negative. Therefore we consider the dried bed if $h + \zeta \leq 0$, and $h + \zeta > 0$ then the cell is submerged.

4.Initial Condition

We set the initial condition of velocity and water elevation to be zero and for actual situation the initial condition for water level are set to tides level which is assumed to be constant during the simulation. For situation where the tsunami source are included in the simulation we use initial condition with dynamic effects of fault motion in reference to Imamura (1995) and Manshina & Smylies (1971). For run-up computation on coastal land the initial water level is equal to the ground level h as mentioned above .

5. Boundary Condition

5.1. Open Boundary

In this model we use the same open boundary as described by Imamura (1995) i.e. : open boundary conditions with characteristics, for forced input and for free transmission.

For free transmission we consider the use of the Sponge Layers (DHI,1993).

5.2. Boundary condition at Run-up Front

Most of boundary conditions at run-up front have been described at section 3.4. Using surface roller condition we check whether the roller is dominant or not. If the surface roller is dominant then bore front must be taken into consideration and wave celerity used in the formula to calculate S (horizontal distance of wave travel) must be changed to bore velocity which is also suggested by Kirkgoz, M.S.(1983).

6. Computational Condition

Bathymetry used in this computation almost same with Benchmark problem #3. This bathymetry was digitized from one area near Bandealit in east Java. The bathymetry was divided into two region, first with Δx = 50 m (300 x 500 m) and second with Δx =

10 m (15 x 500). Sinusoidal waves are input with height (H)= 4 m and period (T)= 3 minutes and for surface roller $\alpha_o = 10°$ and $\alpha_{br} = 25°$ and Kr = 1.5.

We used PC i486 DX 50 Mhz. For simulating 50 minutes event in nature for second area (15 x 500 mesh size : including moving grid at boundaries) require computation time 146 minutes with $\Delta t = 0.5$ s (Courant number 0.7).

7. Results and Discussion

Results from numerical simulation are shown on figure.2. The initial surface roller for the first wave is detected at the end of plan bathymetry slope of 1: 53 as seen on figure 2.a. The shapes of the first wave have little change (as seen on figure 3.b) when they travel over plan bathymetry slope 1: 150 and 1: 53 . After that, completely wave break and rush to the land and hit the vertical wall.

Behind the first wave, the second wave was coming with different shape from the first wave. The perfect sinusoidal shape of wave is maintained until the surface roller initiated in the middle of plan bathymetry slope 1: 53. The height of second wave was higher than the first wave. It is very interesting because in nature when tsunami attacked the coastal area, the second and third wave usually more higher than the first one.

In figure 2.c, the interaction between wave incoming and rundown process make an oscillation at the top of wave where the surface roller exist. In this stage the surface rollers arenot longer maintain and the equation to calculate the surface roller are stoped. In this area the calculation of wave dynamic (rundown interaction with wave incoming) must be explained well due to their impact to maximum runup of second, third and advance waves. It is very difficult and an artificial diffusion term are not much help due to high nonliniearity of turbulence phenomena which is contribute to that oscillation. More study must be done.

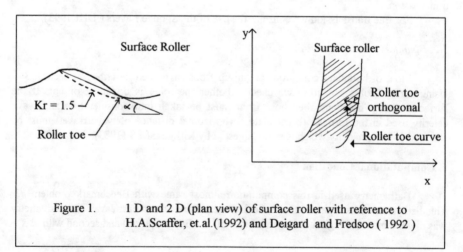

Figure 1.   1 D and 2 D (plan view) of surface roller with reference to
H.A.Scaffer, et.al.(1992) and Deigard and Fredsoe (1992)

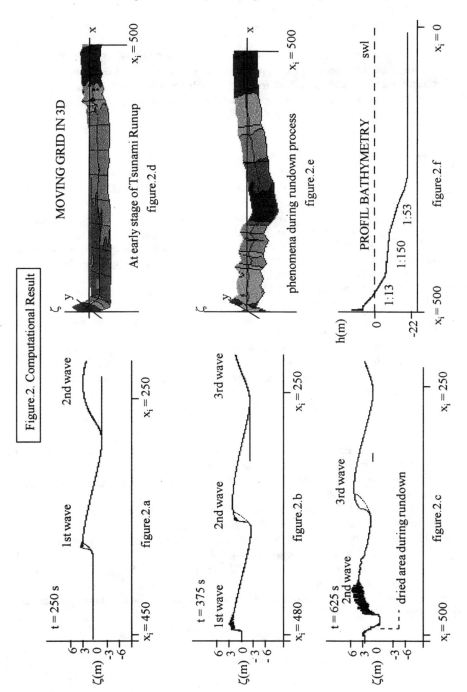

Figure.2. Computational Result

MOVING GRID IN 3D

At early stage of Tsunami Runup
figure.2.d

phenomena during rundown process
figure.2.e

PROFIL BATHYMETRY

figure.2.f

figure.2.a

figure.2.b

figure.2.c

8. Acknowledgments

The authors wishes thanks to Prof. Harry Yeh and National Science Foundation USA for their kndness to give oppurtunity attending the Workshop, Profs. Costas Synolakis, P.L.F.Liu and Peter Raad for their valuable comment and discussion.
This research work was sponsored by Tsunami Research Center, Coastal Engineering Laboratory, LPTP BPP Teknologi.

9.References

1. Bundgard, H.I. and Warren, I.R., *in Science of Tsunami Hazards. Int...Journal of The Tsunami Society*, (DHI ref.53/91).
2. J.Fredsoe and R. Deigaard, *Mechanics of Coastal Sediment Transport, Advanced Series in Ocean Engineering-Vol.3* (World Scientific, Singapore. 1992)
3. H.G. Ramming, and Z. Kowalik, *Numerical Modelling of marine hydrodynamics-applications to dynamic physical process*, Elsevier Oceanography series.26, (Elsevier Scientific Pub.Co.Ltd , Netherland, 19) 360 p.
4. H.A. Schaffer, R.Deigasrd and P. Madsen., *in Proc.23rd.Int.Conf. Coastal Engineering* , Venice, Italy, 1992 (ASCE, New York, 1993)
5. I. Altas, and J.W Stephenson, *in Journal of Computational Physics 94.* (Academic Press, London, 1991).pp.201-224.
6. Q., Chang, G. Wang, and B. Guo, *in Journal of Computational Physics 93* (Academic Press , London, 1991)..pp.360-375.
7. C. Goto, and N.Shuto, *in Tsunami Their science and Engineering*, eds., Iida and T.Iwasaki (Terra Science Pub.Co, Tokyo,1983), pp.439-451.
8. Kirkgoz, M.S.. *in Tsunami Their science and Engineering*, eds., Iida and T.Iwasaki (Terra Science Pub.Co, Tokyo,1983), pp.467-478
9. F. Imamura, *Tsunami Numerical simulation (Numerical code of Tunami-N1 and N2).* School of Civil Engineering, Asian Institute of Technology and Disaster Control Research Center., Tohoku University.Japan (1995).
10. H. Yeh, *in Tsunami'93.* Proc.IUGG/IOC International Tsunami Symposium, Wakayama , Japan. (1993)
11. L P.L.F iu, S.B. Yoon, S.N.Seo, Y.S .Cho., *in Tsunami'93.*Proc.IUGG/IOC international Tsunami Symposium, Wakayama , Japan. (1993)
12. H.Matsutomi., *in Tsunami Their science and Engineering*, eds., Iida and T.Iwasaki (Terra Science Pub.Co, Tokyo,1983), pp.439-451.
13. _____, *Reference Manual MIKE21 Hydrodynamic Modul ver 1.4*, Danish Hydraulic Institute, 1993.
14. _____, *Reference Manual MIKE21 Boussinesq Module*, Danish Hydraulic Institute, 1993

Commentary

Commentary

Other than the discussion on the suitability of the different models, one of the most stimulating discussions was on whether or not a solitary wave is an appropriate initial condition for tsunami runup models and what are the alternatives. We discussed Hammack and Segur[1] comparisons of different lower-order and higher-order theories who found that a typical physical tsunami generated by tectonic motions is so shallow and linear that the sorting distance for a solitary wave to emerge from the initial wave formation is too long for any oceans on the earth. For example, a typical ocean bottom displacement of 3 m at 2000 m deep over 50 km long distance yields the sorting distance of approximately 45,000 km which is greater than the earth's circumference. In this regard, no well-evolved solitary wave can be formed in any ocean basin. Even if a tsunami generated at the edge of continental slope, it still cannot form a single solitary wave because the wave starts shoaling immediately onto the continental shelf (approximately 200 m deep); in this case the wave may be transformed into an undular bore, whose profile resembles that of a series of solitary waves riding on a long wave, similar to the undular bore formation was observed and recorded in video footage during the 1983 Nihonkai Chubu tsunamis in Japan[2]. Highly nonlinear tsunamis could be generated in smaller regions by submarine landslides or volcanic eruptions, and then the sorting distance for a solitary-wave emergence is shorter, in the order of 100 km. While emerging solitary waves might be possible, in reality, such tsunami generation and propagation would be strongly multi-directional with quickly attenuating wave energy, and even at 100 km from the source such waves may have transformed enough through breaking so that they may be insignificant for tsunami hazard. Considerations of many scenario events indicate that a solitary wave formation is more often than not the exception for the tsunami runup models. In fact, the wave formations of the leading depression wave followed by an elevation wave (often termed the N-waves, Tadepalli and Synolakis[3]) or an undular bore may be more common and could be more critical for the determinations of tsunami runup height.

During the discussions at the workshop, it was clear that most numerical model predictions were made for benchmark problem three (see the Appendix), which prima facie was the simplest case since only one wave propagation direction involved. However, the waves generated in the laboratory experiments exhibit rather strong frequency dispersion in the offshore propagation region and pronounced nonlinear effects and wave-breaking during shoaling. While the models based on the non-dispersive, fully-nonlinear shallow-water wave equations are believed capable of modeling wave-breaking, not in detail but as a flow-property discontinuity, most results exhibited major flaws for this simulation: the simulated wave broke immediately in the offshore region, forming the characteristic saw-tooth shape wave, clearly due to the non-dispersive character of the equations. Overall, the maximum runup and maximum wave height were modeled adequately, yet the numerical results displayed significant discrepancies in comparisons of detailed profile shapes. Models based on the weakly-nonlinear and weakly-dispersive Boussinesq equations performed very well, in spite of the fact that wave breaking is not a weakly but a strongly-nonlinear phenomenon. Perhaps, the nonlinearity that is important for local flows at the wave front may not affect significantly the overall wave patterns. A typical example for the comparison among the predictions of the shallow-water theory and the Boussinesq theory and the laboratory measurements is presented in the paper by Watson, Barnes and Peregrine (p. 291). Before the wave breaks, models based on potential-flow theory such as the Boundary- Element Method (BEM) gave unequivocally the best and almost perfect results in the review article by Grilli (p. 116), although no predictions could be produced after breaking, where of course potential theory may be inappropriate.

No attempts were made to implement potential-flow type models to the other benchmark problems, which are essentially three-dimensional. In spite of our hopes at the Catalina workshop in 1990, the extension of the Boundary-Integral-Equation Method to three-dimensional problems is still formidable, possibly due to the requirement of inverting large, fully-populated matrices. On the other hand, an attempt was made to model these three-dimensional problems with Marker-And-Cell (MAC) type free-surface treatment methods by Masamura and Fujima (p. 321). Although fully three-dimensional extension of the Euler models requires extensive computational resources, Masamura and Fujima's results show some success, and the extension of this method to the three-dimensional problems and further improvements are worth warranted mainly due to the advancement of computational capabilities. Furthermore, in comparison with past results, tremendous improvements of the MAC type method were demonstrated for the simulations of breaking and broken waves.

Field applications of any three-dimensional models are still difficult, even for non-breaking waves and none were presented. It is now questionable if they are even desirable or necessary. During the workshop consensus emerged that tsunami evolution from generation to nearshore region before wave breaking can be modelled adequately with the models based on the Boussinesq equations. Within this context, we emphasize that the Boussinesq model reduces to the linear shallow-water model when the dispersion and nonlinear terms are negligible, and to the linearized Boussinesq model when only the nonlinearity is negligible. It is all a question of the scales of the initial wave, and any field-oriented model should be flexible enough to take advantage of the simplifications that may be possible due to the scale of specific events. While the Boussinesq model reduces to the fully nonlinear shallow-water model when the dispersion term is neglected, the classic Boussinesq model cannot replace the shallow-water model. As an example, we note how wave breaking cannot be modelled with the Boussinesq model, because of the resulting large surface curvatures which induce large frequency dispersion, while the actual strong wave-breaking in nature induces large energy dissipation rather than the frequency dispersion that the Boussinesq models produce. Ironically, it is energy dissipation associated with wave breaking which is best modeled with the fully nonlinear non-dispersive shallow-water models. To address this problem and to move forward long before fully 3-D problems become solvable, a 'composite' model was proposed in the workshop, a Boussinesq model in the sense described earlier to calculate wave evolution from the source to the breaker zone, followed by a fully-nonlinear nondispersive shallow-water model for runup computations. Such composite models would generalize the practice of coupling linear models offshore with nonlinear models, which has proven quite powerful in predicting laboratory data, as demonstrated by Watson, Barnes, and Peregrine (p. 291).

For real field applications of depth-integrated models, detailed local bathymetry and topography features have to be modelled to yield accurate runup predictions, and the grid-mesh size must be selected as fine as computational resources allow. Recent field surveys have shown repeatedly that local bathymetric features can significantly affect the maximum runup heights of nearshore tsunamis, yet they may not be at all important for a very long tsunami that would cause gradual rise and fall of tsunami-runup water levels similar to a fast tide. While the maximum runup heights may be predicted adequately with the depth-integrated models, it is yet unclear if the models can predict detailed runup flow patterns and velocity magnitudes, both critical factors for tsunami-induced inundation. For velocity calculations, three-dimensional calculations may be necessary, but equally clearly more detailed work is necessary for comparing the velocity predictions of the existing models with data.

At least one paper presented calculations using hydrodynamic data to determine the initial ground deformation (Tanioka and Satake, p. 249). Such models have been popularized by geophysicists primarily to supplement seismic data. Most existing models rely on inverting tidal records and on determining the Green function from source to the tidal gage location; some models have been reported which invert runup measurements; all models use linear theory. At its face, this practice was questioned on the basis that tidal records are measured inside harbors and may be contaminated with long-period motions such as seiches, which even though tsunami induced, they do not represent the incoming waves; models which invert runup measurements may be more credible, but there is the question of the resolution of the available topographic data or measurement localization. On the other hand, it could be argued that if only the beginning of the records are inverted, then harbor resonance effects may not yet have manifested themselves, and the predictions would identify at least the leading wave. Clearly, inversion models need to be supplemented by transfer functions to identify the tsunami part of the tidal records whenever the latter are obtained inside harbors. Equally clearly, however, the existing inversion results are promising and do warrant comprehensive studies to quantify the effects of harbor resonance or seiching.

Perhaps the best application of the limitations of different models can be found in the analytical solutions of linear and nonlinear theories, which are not affected by either numerical dissipation or numerical dispersion; in that regard analytical solutions are benchmarks of their own. Comparing numerical solutions with analytical results for more generalized initial waveforms such as solitary waves or N-waves is more relevant for evaluating a numerical model's capabilities than the now ubiquitous comparisons of model results with periodic wave solutions; this latter practice is simpler than comparing with solitary wave solutions, but it cannot identify frequency-dependent phase-related numerical artifacts. Therefore, models should be validated with initial waveforms with spectra similar to field-inferred tsunamis, as appropriate for the specific problem, for example with solitary waves, cnoidal waves or N-waves.

Three analytical solutions for wave field around a conical island were presented: one for the trapped waves around the island with a continuously forcing wave field by Meyer (p. 3), other two for benchmark problem two. One of the analytical models approximates the sloping boundary with piecewise horizontal topographies (Kanoglu and Synolakis, p. 214), while the other analytical solution is based on a series expansion of fundamental solutions (Fujima, p. 221). Both analytical solutions produced excellent results for a incident solitary wave with a small amplitude in the same range of accuracy as the numerical models, although the analytical solutions presented do not yet reproduce the enhanced runup on the lee of the island. Further, analytical solutions despite their limitations do have the promise of identifying the dependence of tsunami runup on source parameters such as the dip and slip angles or the deformation area simpler than numerical models.

Overall, in terms of tsunami runup modeling, significant advances have been made since the Catalina Workshop in 1990, both due to the advancement of computational capabilities, but also because of the generation of a large 2-D and 3-D laboratory data set and the fortuitous field measurements in 1992-1995, all of which have contributed to model calibration. We were quite gratified to find that tsunami modeling efforts have become more directed towards their implementation for real tsunami predictions and hindcastings than ever before. It should be also noted that, whereas five years ago large differences between computed runup results and field measurements might have been attributed to both errors in the seismic estimates of the source motion and to the hydrodynamic calculations, now the hydrodynamic calculations are beyond suspicion, at least for non-breaking waves. It was equally clear that reduction and even elimination of numerical dispersion and numerical dissipation effects would, if not already, soon become reality. At the same time, we recognized additional and important problems arising from modeling improvements, such as determination of highly accurate initial wave conditions, modeling the three dimensional flow effects, and turbulence. There is no question that actual tsunami runup motions are turbulent; the runup flow patterns, impacts, scouring effects, and sediment transport are all affected by turbulence in the runup motions. Strangely, no turbulence model was presented nor even discussed during the workshop, and the question of the quantification of the effects of turbulence yet persist.

There is also no doubt that the benchmark-problem exercises used in the workshop proved extremely useful in identifying absolute and comparative modeling capabilities. On the other hand, some participants cautioned and criticized that the benchmark problems selected in the workshop may not have covered all the tsunami characteristics and especially far-field tsunamis. Such tsunamis might even be more linear and nondispersive than the waves used in the benchmark problems one, two or three.

Harry Yeh
Philip Liu
Costas Synolakis

References

1. Hammack, J.L. and Segur, H. 1978. Modelling criteria for long water waves. J. Fluid Mech., 84, 359-373.

2. NHK 1990. A great tsunami eyewitnessed: a record of the Nihonkai Chubu Earthquake. NHK Service Center, Tokyo, Japan.

3. Tadepalli, S. and Synolakis, C.E. 1994. The run-up of N-waves on sloping beaches. Proc. R. Soc. Lond. A 445, 99-112.

Appendices —
Benchmark Problems

BENCHMARK PROBLEM 1
WAVE-PACKET PROPAGATION ALONG A UNIFORMLY SLOPING BEACH

HARRY YEH AND KUO-TUNG CHANG

Model Description

The wave basin used for the experiments is shown in Fig. 1. The dimensions of the basin are 13.4 m long, 5.5 m wide, and 0.9 m deep, and the slope of the beach, β, is 15°. The central section of the beach is constructed of a 2.4 m long, 3 m wide, and 1.27 cm. thick tempered glass plate. As shown in Fig. 1, the glass plate has a water-tight seal to provide a dry observation section under the basin. Outside of the observation section (the glass beach section), the beach modules are constructed with 1.27 cm thick Extren sheets supported by an Extren frame. Extren is a proprietary combination of fiberglass reinforcements and thermosetting polyester; it is relatively stiff, resistant to corrosion, and is hydrophilic (wettable). The Extren surface is just as smooth or smoother than that of a top-grade plywood sheet.

Using a wave paddle, a packet of waves was generated directly at one end of the beach. The wave paddle is wedge-shaped, hinged offshore and swings in the longshore direction about the axis perpendicular to the beach surface. The paddle motion creates the longshore velocity that varies linearly from zero value at the hinge. Wave paddle motion was controlled to push the water mass with a single positive stroke from the initial position, which was perpendicular to the undisturbed shoreline. To avoid excessive high-frequency noise, the paddle trajectory was set to that of a hyperbolic tangent function of time, which achieves a gradual start and cessation of paddle motion, viz.

$$\frac{\theta}{\theta_{max}} = \frac{tanh\left(\frac{6.6t}{t_{max}} - 3.3\right) + tanh\,3.3}{2\,tanh\,3.3}$$

where θ is the angle of the wave paddle from the initial position of 0° (i.e. the position perpendicular to the undisturbed shoreline), $\theta_{max.}$ is the stroke angle, $t_{max.}$ is the stroke time, and t is time. The paddle hinge is located at 82.0 cm offshore along the beach surface from the quiescent shoreline. The water depth for all cases is 68.7 cm in the offshore region where the basin bottom is horizontal. Defining the coordinates that the x-axis points in the longshore direction from the initial wave-paddle position, the y-axis points horizontally offshore, and the z-axis points upward, the following cases are considered:

Required Analysis

Participants were asked to provide the following information for each of the four cases listed below:

> Case A: $\theta_{max.}$ = 6.50°, $t_{max.}$ = 1.5 sec.
>
> Case B: $\theta_{max.}$ = 12.87°, $t_{max.}$ = 1.5 sec.
>
> Case C: $\theta_{max.}$ = 14.69°, $t_{max.}$ = 3.0 sec.
>
> Case D: $\theta_{max.}$ = 9.95°, $t_{max.}$ = 2.2 sec.

Data predicted by the participants at the workshop were temporal water-surface variations, $\eta(x, y, t)$, at (x = 6.89 m, y = 0.1 m), and (x = 5.5 m, y = 0.4 m), and the temporal velocity variations (both longshore and cross-shore components) (u, v) at (x = 6.89 m, y = 0.1 m). The corresponding laboratory data had not been revealed to the participants prior to the workshop: those are shown in Figs. 2, 3, and 4.

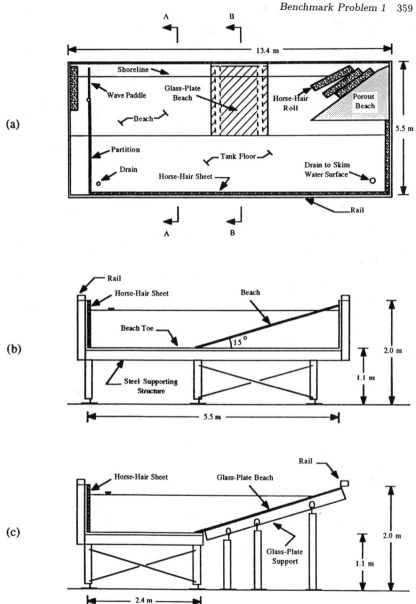

Figure 1 - Schematic drawing of the wave tank. a) plan view, b) elevation view,
Section A-A, c) Section B-B.

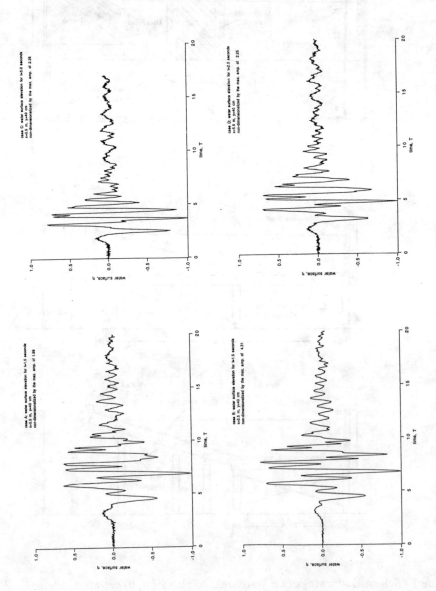

Figure 2 - Temporal wave profiles for Cases A, B, C, and D. The water elevation is normalized by the maximum value, and the time is normalized by the wave-paddle motion period.

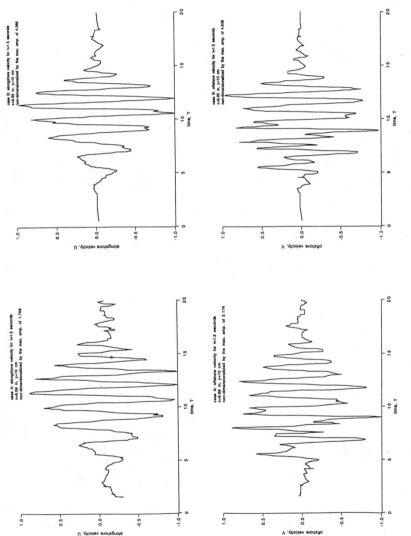

Figure 3 - Temporal profiles of the longshore and offshore velocity components for Cases A and B. The water elevation is normalized by the maximum value, and the time is normalized by the wave-paddle motion period.

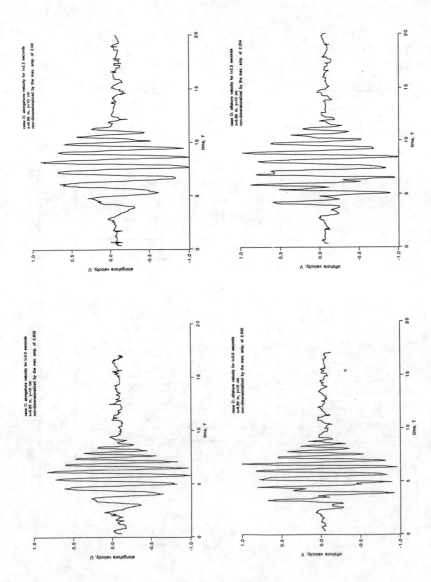

Figure 4 - Temporal profiles of the longshore and offshore velocity components for Cases C and D. The water elevation is normalized by the maximum value, and the time is normalized by the wave-paddle motion period.

Benchmark Problem 2
Runup of Solitary Waves on a Circular Island

Michael J. Briggs[1], Costas E. Synolakis[2], Gordon S. Harkins[1], Debra R. Green[1]

[1]USAE Waterways Experiment Station, Coastal Engineering Research Center, Vicksburg, MS
[2]Department of Civil and Aerospace Engineering, University of Southern California, Los Angeles, CA

Model Description

A physical model of a conical island was constructed in the center of a 30-m-wide by 25-m-long flat-bottom basin at the U.S. Army Engineer Waterways Experiment Station (Figure 1). The island had the shape of a truncated, right circular cone with diameters of 7.2 m at the toe and 2.2 m at the crest. The vertical height of the island was approximately 62.5 cm, with 1V on 4H beach face (i.e. β=14 deg). The surface of the island and basin were constructed with smooth concrete. The basin sides and rear were lined with wave absorber to minimize wave reflections and cross-basin seiche. The water depth was set at 32 cm in the basin.

Coordinate System

The X-axis (X) of the right-hand, global coordinate system was perpendicular to the wavemaker and the Y-axis (Y) was parallel to the wavemaker. The origin was located at the end of the wave-maker, in line with the front surface of all paddles at their rest position. The center of the island was located at X=12.96 m and Y=13.80 m. A local coordinate system was located at the center of the island. Angles increase counterclockwise (polar convention) from the x-axis (x) pointing in the 0-deg direction.

Instrumentation

Twenty-seven capacitance wave gages were used to measure surface wave elevations (Figure 1).

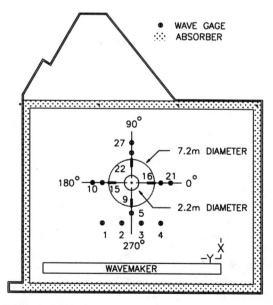

Figure 1: Schematic of Island and Wave Gages in Test Basin.

All gages were calibrated each day to 0.91 mm or better accuracy. The first four gages were located parallel to the wavemaker to measure incident wave conditions. Prior to each run, these gages were moved seaward from the toe of the island a distance equivalent to half a wavelength (i.e. L/2) of the wave to be generated. This procedure ensured that the tsunami wave was always measured at the same relative stage of evolution.

A measurement grid of six concentric circles covered the island to a distance 2.5 m beyond the toe. Measurement points were located at the intersection of these concentric circles and the 90-deg radial lines. The spacing between grid points was a function of the water depth. The shallowest gage was located in an 8-cm depth and the deepest gage was located over the toe. Two gages were evenly spaced between these points along each 90-deg transect. Two additional gages were spaced in the deep water portion at distances of 1.0 m and 2.5 m from the toe (except for the 270-deg transect).

Solitary Waves

A directional spectral wave generator (DSWG) was used to generate solitary waves. The electronically controlled DSWG is 27.4 m long and consists of 60 paddles, 46 cm wide and 76 cm high. The full length of the DSWG was used to generate three solitary wave cases with actual paddle motions as shown in Figure 2. Target wave heights H_{tgt}= 0.05, 0.10, and 0.20 were simulated for Cases A, B, and C, respectively. Measured wave heights H_{meas} were approximately 90 percent of these target values. The decrease in measured wave height from the target was due to losses in the mechanical generation of the solitary waves because of gaps between the floor and the wavemaker. All waves were non-breaking until final stages of transformation near the shoreline (where gentle spilling occurred) except for the H_{tgt}=0.20 waves, which broke nearshore.

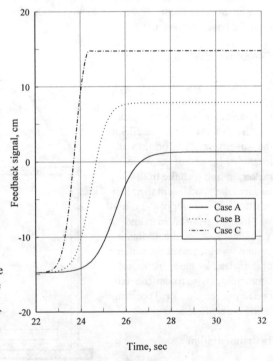

Figure 2: Wave Paddle Trajectories for Cases A, B, and C.

Maximum Runup Height

Maximum vertical runup was measured at twenty locations around the perimeter of the island. Sixteen locations were evenly spaced every 22.5 deg around the perimeter. Four radial transects with uneven spacing were located on the back side of the island (i.e. 90 deg) to improve the resolution in this critical area. At the conclusion of each run, maximum runup along each transect was manually located. A surveyor's rod and transit were then used to measure vertical runup at each transect.

Required Analysis

Participants were asked to provide the following information for each of the three cases listed below:

Case A: Target H_{tgt}=0.05, Measured H_{meas}=0.045
Case B: Target H_{tgt}=0.10, Measured H_{meas}=0.096
Case C: Target H_{tgt}=0.20, Measured H_{meas}=0.181

a) *Time-histories of free-surface displacements* at the following gages:

Gage 6: Front of island over toe (X=9.36m, Y=13.80m, local water depth h=31.7cm)
Gage 9: Front of island at shallowest point (X=10.36m, Y=13.80m, h=8.2cm)
Gage 16: Side of island at shallowest point (X=12.96m, Y=11.22m, h=7.9cm)
Gage 22: Back of island at shallowest point (X=15.56m, Y=13.80m, h=8.3cm)

Participants at the workshop requested information on the four incident gages for each case. They were aligned parallel to the wavemaker and moved a distance equivalent to half a wavelength from the toe for each case. The coordinates are listed below.

Gage 1A: Incident gage 1 for Case A (X=5.76m, Y=16.05m, h=32.0cm)
Gage 2A: Incident gage 2 for Case A (X=5.76m, Y=14.55m, h=32.0cm)
Gage 3A: Incident gage 3 for Case A (X=5.76m, Y=13.05m, h=32.0cm)
Gage 4A: Incident gage 4 for Case A (X=5.76m, Y=11.55m, h=32.0cm)
Gage 1B: Incident gage 1 for Case B (X=6.82m, Y=16.05m, h=32.0cm)
Gage 2B: Incident gage 2 for Case B (X=6.82m, Y=14.55m, h=32.0cm)
Gage 3B: Incident gage 3 for Case B (X=6.82m, Y=13.05m, h=32.0cm)
Gage 4B: Incident gage 4 for Case B (X=6.82m, Y=11.55m, h=32.0cm)
Gage 1C: Incident gage 1 for Case C (X=7.56m, Y=16.05m, h=32.0cm)
Gage 2C: Incident gage 2 for Case C (X=7.56m, Y=14.55m, h=32.0cm)
Gage 3C: Incident gage 3 for Case C (X=7.56m, Y=13.05m, h=32.0cm)
Gage 4C: Incident gage 4 for Case C (X=7.56m, Y=11.55m, h=32.0cm)

Plots are shown for each of these eight gages for each case in Figures 3-5.

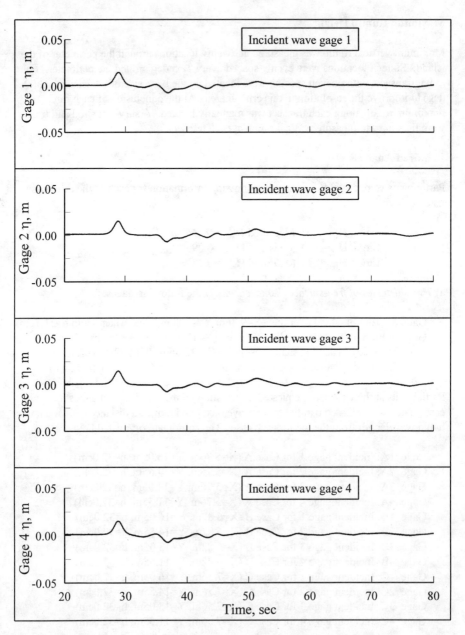

Figure 3: Time-Histories of Free Surface Displacement for Case A, Gages 1-4.

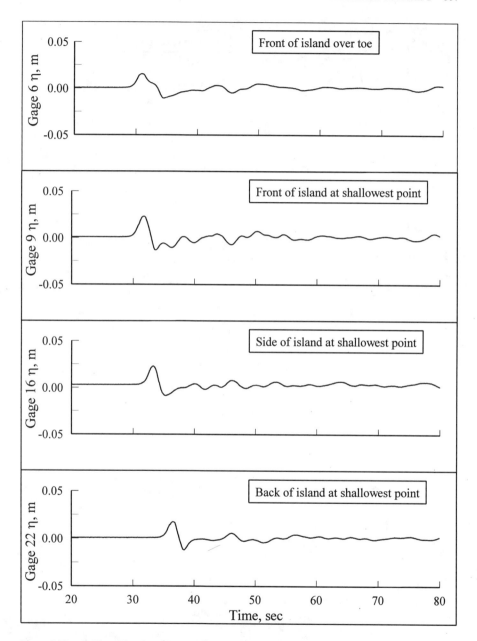

Figure 3 (Cont.): Time-Histories of Free Surface Displacement for Case A, Gages 6, 9, 16, and 22.

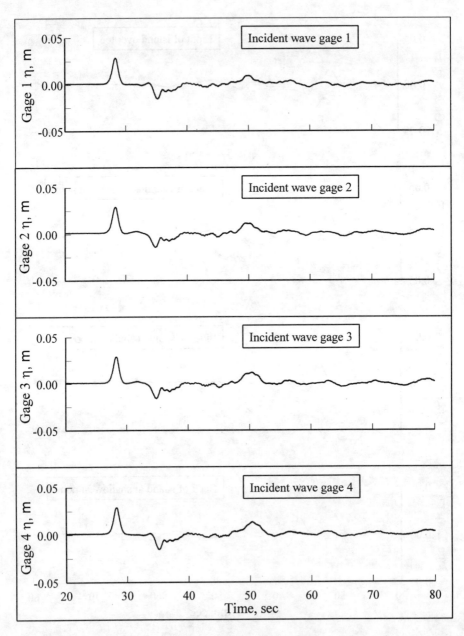

Figure 4: Time-Histories of Free Surface Displacement for Case B, Gages 1-4.

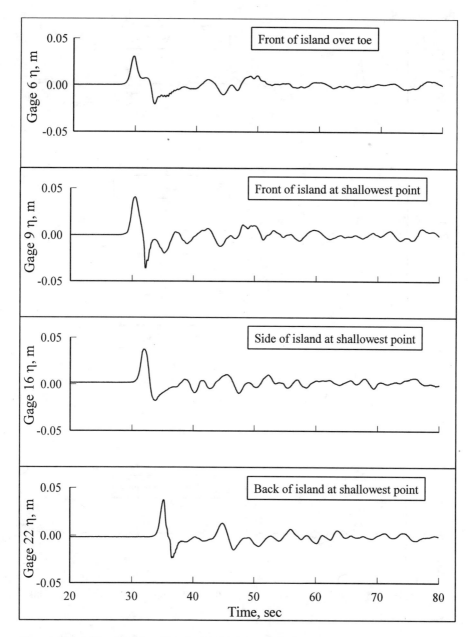

Figure 4 (Cont.): Time-Histories of Free Surface Displacement for Case B, Gages 6, 9, 16, and 22.

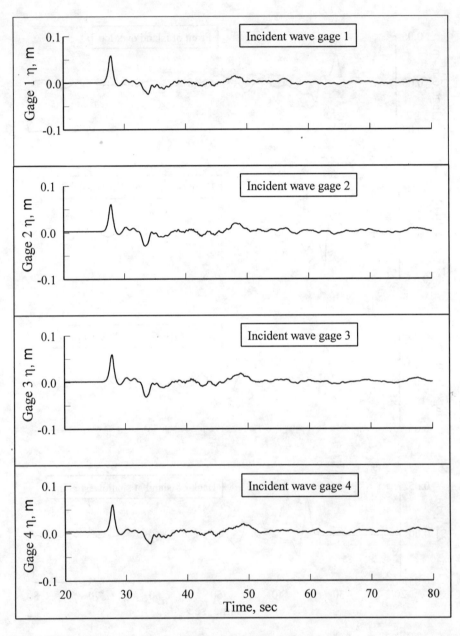

Figure 5: Time-Histories of Free Surface Displacement for Case C, Gages 1-4.

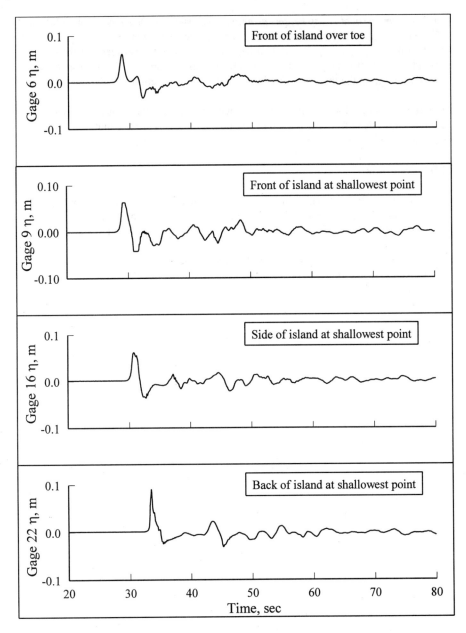

Figure 5 (Cont.): Time-Histories of Free Surface Displacement for Case C, Gages 6, 9, 16, and 22.

b) *Maximum vertical runup heights around the perimeter of the island.* Polar plots or X/Y plots (Figure 7) of runup versus angle around the island may be used. A minimum of sixteen evenly-spaced points at 22.5-deg spacing around the perimeter of the island should be used. It is suggested that angles be increased counterclockwise from the center of the island from the local x-axis pointing in the 0-deg direction (see Figure 1).

Figure 6 is a polar plot of maximum horizontal runup for the three cases. Waves approach the island from the bottom or 270 deg. Figure 7 is an X/Y plot of maximum vertical runup for the three cases.

Measured Data

An FTP site exists for downloading ASCII files containing the measured data from the experiments. The README.TXT file describes the formatting of each file. Access using the "IP Address" listed below is sometimes easier than using the "Site Name.".

Site Name: dswg1.cerc.wes.army.mil/public/tsunami
IP Address: 134.164.156.69

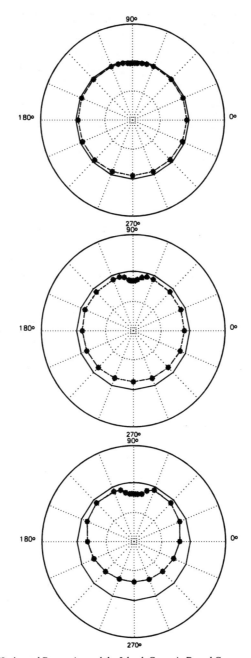

Figure 6: Maximum Horizontal Runup Around the Island, Cases A, B, and C.

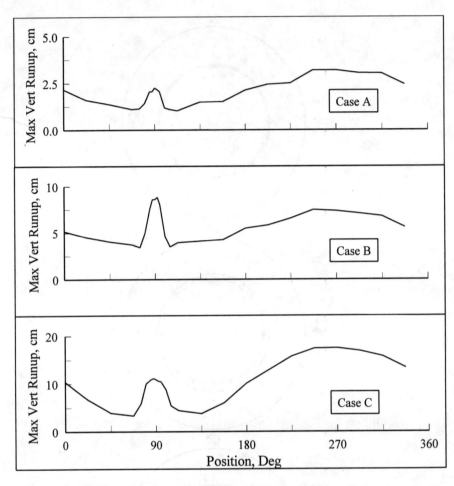

Figure 7: Maximum Vertical Runup Around the Island, Cases A, B, and C.

Benchmark Problem 3
Runup of Solitary Waves on a Vertical Wall

Michael J. Briggs[1], Costas E. Synolakis[2], Utku Kanoglu[2], Debra R. Green[1]

[1]USAE Waterways Experiment Station, Coastal Engineering Research Center, Vicksburg, MS
[2]Department of Civil and Aerospace Engineering, University of Southern California, Los Angeles, CA

Model Description

Two-dimensional (2D) flume experiments of solitary wave runup on a vertical wall were conducted at the U.S. Army Engineer Waterways Experiment Station, Vicksburg, MS. Figure 1 is a schematic of the 23.2-m-long by 45-cm-wide glass-walled flume. The compound-slope, fixed-bed bathymetry consisted of three different slopes (1:53, 1:150, and 1:13) and a flat section in the deep end that simulated the bottom profile at Revere Beach, Massachusetts. The vertical wall was located at the landward end of the 1:13 slope. The water depth in the flat section of the flume measured 21.8 cm.

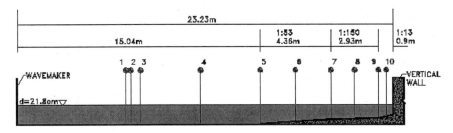

Figure 1: Schematic of Wave Flume and Gage Placement.

Instrumentation

Ten capacitance wave gages were used to measure surface wave elevations along the centerline of the flume ($Y=0$). All gages were calibrated each day to 0.91 mm or better accuracy. The origin of the x-axis was at the wavemaker. The first four gages were located in the constant depth region to measure incident wave conditions. Gages 1 to 3 remained at fixed positions for all tests. Prior to each run, gage 4 was moved seaward from the toe of the 1:53 slope a distance equivalent to half-a-wavelength (i.e. L/2) of the wave to be generated. This procedure ensured that the tsunami wave was always measured at the same relative stage of evolution. Gages 5, 7, and 9 were located over the toes and the remaining gages were spaced approximately midway up each slope of the compound-slope beach profile.

Solitary Waves

A computer-controlled hydraulic wavemaker was used to generate solitary waves. Actual paddle motions to generate three solitary wave cases are shown in Figure 2. Target wave heights H_{tgt} of 0.05, 0.30, and 0.70 were simulated for Cases A, B, and C, respectively. Measured wave heights H_{meas} were usually smaller than target values. Wave breaking occurred for Cases B and C only. For Case B the wave broke at or near the wall. For Case C the wave broke between gages 7 and 8 (i.e. X=19.40 and 20.86 m).

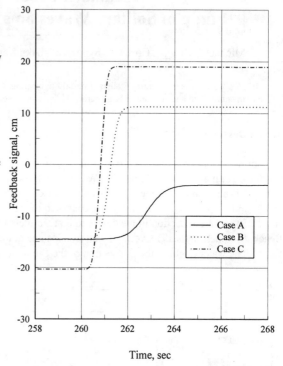

Figure 2: Wave Paddle Trajectories for Cases A, B, and C.

Maximum Runup Height

When the waves reached the vertical wall, a plume of water would shoot upward. The highest point of this excursion was visually noted through the glass walls of the flume and manually recorded after each run. Cases less than H_{tgt}=0.30 continued to shoal to the shallowest gage without wave breaking. Cases larger than this value broke in the nearshore at a normalized local water depth h=h/d between 0.45 and 0.60 (i.e. in front of the toe between the 1:13 and 1:150 slopes) before re-forming and shoaling to the vertical wall. The largest runup at each depth was recorded for the H_{tgt}=0.30 cases, which did not experience wave breaking in the nearshore prior to interacting with the wall.

Required Analysis

Participants were asked to provide the following information for each of the three cases:

Case A: Target H_{tgt}=0.05, Measured H_{meas}=0.039
Case B: Target H_{tgt}=0.30, Measured H_{meas}=0.264
Case C: Target H_{tgt}=0.70, Measured H_{meas}=0.696

a) *Time-histories of free-surface displacements* at the following gages:

  Gage 7: Toe between 1:53 and 1:150 slopes (X=19.40m, Y=0.0m, h=13.6cm)
  Gage 9: Toe between 1:150 and 1:13 slopes (X=22.33m, Y=0.0m, h=11.6cm)

Participants at the workshop requested information on the seven additional gages listed below. The incident gages were moved a distance equivalent to half a wavelength from the toe of the first slope (i.e. 1:53) for each case.

  Gage 4A: Incident gage for Case A (X=12.64m, Y=0.0m, h=21.8cm)
  Gage 4B: Incident gage for Case B (X=14.06m, Y=0.0m, h=21.8cm)
  Gage 4C: Incident gage for Case C (X=14.40m, Y=0.0m, h=21.8cm)
  Gage 5:   Toe between flat and 1:53 slopes (X=15.04m, Y=0.0m, h=21.8cm)
  Gage 6:   Midway on 1:53 slope (X=17.22m, Y=0.0m, h=17.7cm)
  Gage 8:   Midway on 1:150 slope (X=20.86m, Y=0.0m, h=12.6cm)
  Gage 10: Midway on 1:13 slope (X=22.80m, Y=0.0m, h= 8.0cm)

Plots are shown for each of these seven gages for each case in Figures 3-5.

b) *Maximum vertical runup heights at the vertical wall.* Table 1 lists the maximum vertical runup heights R and normalized values H/d at the vertical wall for each case.

Table 1				
Maximum Vertical Runup on the Vertical Wall				
Case	**Target H/d**	**Actual H/d**	**Runup R, cm**	**R/d**
A	0.05	0.039	2.74	0.13
B	0.30	0.264	45.72	2.10
C	0.70	0.696	27.43	1.26

Measured Data

An FTP site exists for downloading ASCII files containing the measured data from the experiments. The README.TXT file describes the formatting of each file. Access using the "IP Address is sometimes easier.

  Site Name: dswg1.cerc.wes.army.mil/public/tsunami
  IP Address: 134.164.156.69

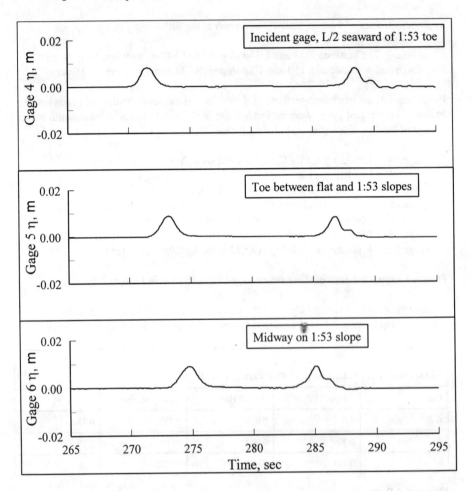

Figure 3: Time-Histories of Free Surface Displacement for Case A, Gages 4-6.

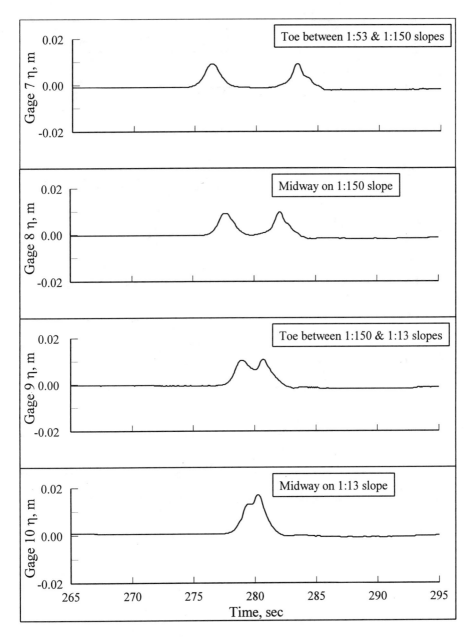

Figure 3 (Cont.): Time-Histories of Free Surface Displacement for Case A, Gages 7-10.

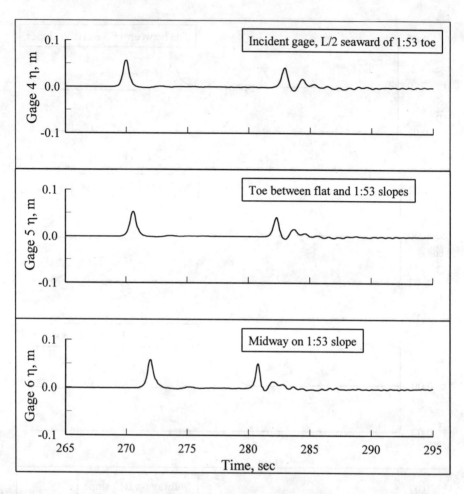

Figure 4: Time-Histories of Free Surface Displacement for Case B, Gages 4-6.

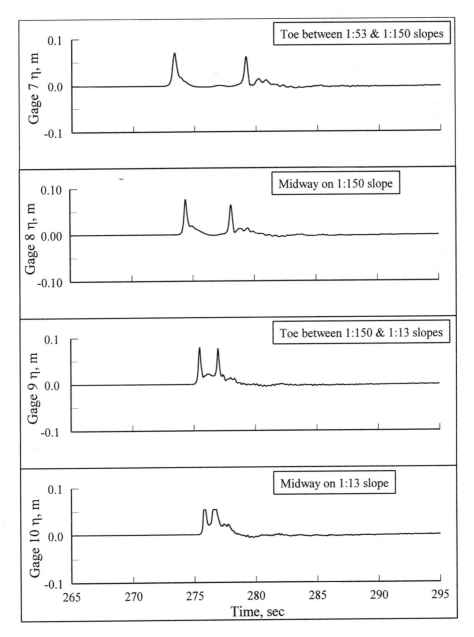

Figure 4 (Cont.): Time-Histories of Free Surface Displacement for Case B, Gages 7-10.

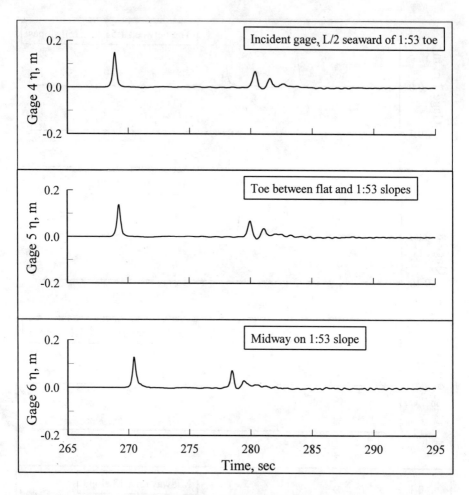

Figure 5: Time-Histories of Free Surface Displacement for Case C, Gages 4-6.

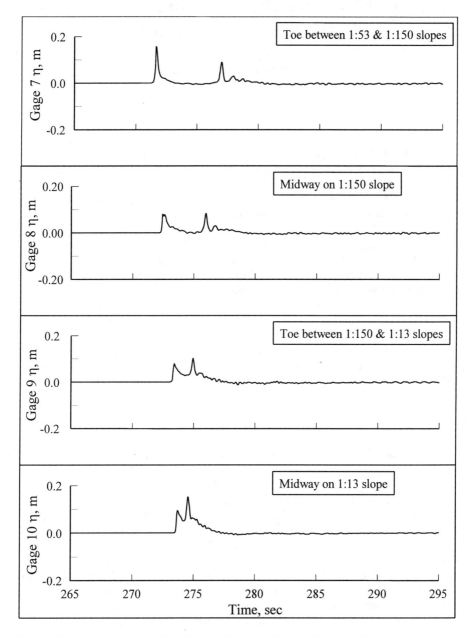

Figure 5 (Cont.): Time-Histories of Free Surface Displacement for Case C, Gages 7-10.

Benchmark Problem 4
The 1993 Okushiri Tsunami — Data,
Conditions and Phenomena

Tomoyuki Takahashi
DCRC, Faculty of Engineering
Tohoku University, Sendai 980-77
Japan

1. Introduction

The 1993 Hokkaido Nansei-oki earthquake tsunami was generated by the earthquake of Ms=7.8, at 22:17, July 12, 1993. The hypocenter determined by JMA was located at the depth of 37 km, at 42.76°N and 139.32°E, off the southwestern coast of Hokkaido. Fault parameters obtained by different institutes are shown in Table 1. All of them were determined from seismic data. The aftershock distribution is shown in Figure 1 [Kodaira et al., 1993].

Around 22:22 when the Sapporo Meteorological Observatory issued the warning "Major tsunami" to the Japan Sea coast of Hokkaido, the west coast of the island of Okushiri was hit by the first wave of the tsunami. In addition, towns of Setana and Taisei on the western coast of Hokkaido which are farther than Okushiri Island from the epicenter were also hit by the small first wave around 5 minutes after the earthquake. It is difficult that this fact is satisfied by any of the initial tsunami profiles predicted based on the aftershock distribution.

Table 2 summarizes the arrival time surveyed by Tohoku University team and others [Shuto and Matsutomi, 1995]. Figure 2 shows the distribution of runup distribution on the island of Okushiri measured by 6 independent teams [Hokkaido Tsunami Survey Group, 1993].

2. Bathymetric and Topographical Maps

The Japan Maritime Safety Agency provides charts of several scales, which any one can purchase in the market. The newest edition of the chart No.6658, on the scale of 1/200,000 published on 24th, March, 1995 covers the area from 40°56'N to 42°36'N and from 138°40'E to 140°10'E. The island of Okushiri is on another chart No.32, on the scale of 1/75,000, from 41°59'30"N to 42°17'N and from 139°18'18"E to 139°43'30"E. The data are also available in digital forms. This chart No.32 also includes the Aonae Harbor and the Okushiri Harbor on the scale of 1/5,000. The chart No.63256 on the scale of 1/50,000 covers from 42°10'N to 42°50'N and from 139°15'E to 139°30'E. If one needs more detailed bathymetry maps, the Hokkaido Development Bureau has charts of the limited areas in and around harbors under its control. The scale of these detailed maps may be larger than 1/5,000.

The Japan Geographical Survey Institute provides maps. Maps on 1/50,000 can cover all the land. Maps on 1/25,000 can also be found for almost all parts of Japan. Maps on 1/10,000 are now being provided. Local governments have detailed maps on larger scale than 1/3,000, for the city and regional planning. In the case of the Okushiri Town, a map of the largest scale provided by the town office is on 1/10,000.

The Disaster Control Resaerch Center (DCRC) digitized the data for the tsunami simulation. The data are stored in the database named "Hokusai" (Hokusai was a painter in the Edo period, famous for his print of huge waves with small Mt. Fuji). Table 3 shows how to obtain the digitized data and contents of Hokusai. Table 4 is for the topographical data. Figure 3 shows the areas covered by the DCRC digitized data.

3. Initial Tsunami Profiles

Figure 4 shows tsunami initial profiles computed with the Manshinha- Smylie method from fault parameters in Table 1. All the results except for the Kikuchi model assume one fault. Kikuchi assumes 3 fault planes.

Okushiri Island is located in the area directly influenced by the fault motion. Subsidence are measured as shown in Figure 5 [Kumaki et al]. Some of the data (marked by Δ) were measured with GPS and are believed to be accurate. Any of the initial tsunami profile should reproduce this vertical displacement of ground, because the initial tsunami profile is considered to be the same as the vertical displacement of sea bottom.

There is no way available to compare the horizontal displacement of ground between the measured and computed. In the case of the 1994 Shikotan Tsunami, the horizontal displacement was larger than the vertical displacement. Therefore, this comparison will be useful as a check of the ground motion.

Hashimoto et al. [1994] obtained a fault model to satisfy both seismic data and subsidence. Their fault is composed of three fault planes as given in Table 5. The corresponding tsunami profile is shown in Figure 6.

DCRC attempted to construct the initial tsunami profile which satisfied seismic, geographic and tsunami runup data. At present, the model DCRC-17a is considered to be the best. The model is also composed of three fault planes. The fault parameters are given in Table 6. The initial profile is shown in Figure 7. In Hokusai, the computed initial profile is given in a digital form. See Table 7.

4. Tsunami Data

No tide records were obtained on Okushiri Island. On the western shore of Hokkaido, two tide stations, Iwanai on the north and Esashi on the south, recorded the tsunami as shown in Figure 8. For comparison of the measured and computed result, hydraulic filtering of the gauges is necessary. Corrections will be made with the method developed by Okada [1985]:

$$\frac{dh}{dt} = W\mathrm{sgn}(H - h)(2g|H - h|)^{\frac{1}{2}} \tag{1}$$

where h and H are the water level inside and outside the tide well, and W is coefficient which is 0.0017 for Iwanai and 0.0060 for Esashi.

It is not recommended to reproduce the "to-be" tide records from the obtained records, because there is no way to reproduce a component lost in the tide records. The correction mentioned above only reproduces the components which are reduced by the hydraulic filtering but still remain in the records. On the other hand it is recommended to filter the computed tsunami outside the gauges to reproduce the time history of water elevation in the tide wells for comparison between the measured and computed.

The original tide records are in analog form. They are digitized and stored in Hokusai: Table 8 is the list of digitized tide records.

The tsunami runup data are shown in Figure 2. For more detailed data, DCRC published their measured runup heights in Tsunami Engineering Technical Report No.11, Part 2, pp.1-120. In Okushiri Island, runups were measured at 125 points.

5. Special Characteristics of This Tsunami to be Simulated or to be Solved

5.1. Early arrival at the west coast of Okushiri Island

About 5 minutes after the earthquake, the first wave hit the west coast including Aonae. This early arrival should be simulated by an assumed initial profile (see Table 2).

5.2. Different incident direction of the first and second destructive wave at Aonae

It is evident that the first wave which destroyed the Aonae came from the west. Ten or more minutes later, the second wave came from the east and partially destroyed some sections of Aonae. The second wave induced a fire which devastated the whole Aonae except for a few houses. It was at a dark night, however, this difference is clearly noticed by residents. The ten minute difference between the first and second wave as well as the different incident direction should be explained by numerical simulation.

5.3. High runup height at Hamatsumae

Hamatsumae is located to the east of Aonae, sheltered against the direct attack of the tsunami by the Aonae Point. Its topography does not suggest any tsunami amplification mechanism such as the V-shaped valley. Figure 9 shows the detailed map of runup distribution at Hamatsumae.

This high runup height is not produced by splash and other local, abnormal causes. When this high runup was found, Tohoku University survey group increased the density of measurement area around this point, and ensured that this high runup was not an isolated phenomenon.

There are several explanations for this high runup:

(1) Refraction on the Okushiri Spar concentrated the tsunami to Hamatsumae. In this case, accuracy of refraction simulation is important; i.e., the accuracy of topography in the simulation and the incident direction of the tsunami.

(2) Effect of soliton fission as discussed by Noguchi during the workshop, the tsunami showed soliton fission at its front and concentration toward Hamatsumae.

(3) Collision of tsunamis coming from east and west. The entrapment around the island could be an important mechanism.

5.4. The highest runup height in a valley north of Monai

The highest runup of 31.7 m was measured at the bottom of a small valley which opens to a small pocket beach. Figure 10 shows the map and the measured runups. The pocket beach 250 m long is bounded by rocky shallows on both, north and south ends. This is a narrow pebble beach backed by high coastal cliffs. A small valley is notched. Its entrance width is only 50 m.

In order to simulate the highest runup, we need a very detailed map, which DCRC specially purchased from an air survey company. The ground height was read on the aerophotos at 5 m horizontal interval, and stored in digital form. Figure 11 is the map drawn by using these data.

5.5. Cause of destruction of breakwaters at Okushiri Harbor

Okushiri Harbor is located at the middle of the eastern coast of Okushiri Island. The tsunami could come round the Inaho Point at the north and the Aonae Point at the south.

Caissons on the north breakwaters in Okushiri Harbor were scattered inward, showing the force of tsunamis which came from the north. A part of parapets on the south sea walls were broken and turned southward. All of these facts indicates the tsunami which came from the north fault was stronger than that from the south fault. Is this true ? Or, the tsunami from the north only arrived earlier ?

5.6. Early arrival at the western coast of Hokkaido

As summarized in Table 2, the first tsunami, although it was not large, arrived at the western coast of Hokkaido about five minutes after the earthquake. This arrival time was abnormally early. There might be other tsunami generation mechanism than the fault motion at the area covered by aftershocks.

1. Fujima, K. and C. Goto (1994): CHARACTERISTICS OF LONG WAVES TRAPPED BY CONICAL ISLANDS, J. Hydraulic, Coastal and Environmental

   Engineering, JSCE, No.497, II-28, pp101-110 [J].
2. Hashimoto, M. et al. (1994):Crustal Movement Caused by the Hokkaido Nansei-Oki Earthquake and Its Fault Model, Kaiyo Monthly, Special No.7, pp.55-61 [J].
3. Hokkaido Tsunami Survey Group (1993): Tsunami Devastates Japanese Coastal Region, EOS, AGU, Vol.74, No.37, pp.417-432.
4. Kodaira, S. et al. (1993): Geometry of the aftershocks of the July 12, 1993, Hokkaido-Nansei-oki Earthquake, Programme and Abstracts, Seismological Soc. Japan, p.15 [J].
5. Kumaki, Y. et al. (1993): Vertical Seismic Crustal Movement of the 1993 Hokkaido Nanseioki Earthquake, Programme and Abstracts, Seism. Soc. Japan, p.63 [J].
6. Okada, M. (1985): Response of Some Tide-wells in Japan to Tsunamis, Proc. ITS,85, pp.208-213.
7. Shuto, N. and H. Matsutomi (1995): Field Survey of the 1993 Hokkaido Nansei-Oki Earthquake Tsunami, PAGEOPH, Edited by Satake and Imamura (in

   press).
8. Tsukuda, H. et al. (1993): Depression of the Okushiri island caused by the 1993 Hokkaido-Nansei-Oki earthquake, Programme and Abstracts, Seism. Soc. Japan, p.61 [J].
9. Tsutsumi, A. et al. (1993): Subsidence of the Okushiri Island caused by the Southwest-off Hokkaido Earthquake, Programme and Abstracts, Seism. Soc. Japan, p.62 [J].
10. Yabuki, T. et al. (1993): Vertical Crustal Movements in the Okushiri Island Associated with the 1993 Off South West Hokkaido Earthquake, Programme and Abstracts, Seism. Soc. Japan, p.60 [J].

Table 1 Fault Parameters Determined Based Upon Seismic Data.

	M_w	$M_0(\times 10^{27})$	depth	Strike	Dip	Slip	Latitude	Longitude
Harvard	7.8	5.6	15km	$1°$	$24°$	$84°$	$42.63°$	$139.24°$
				$187°$	$66°$	$93°$		
USGS	7.3	0.93	18km	$221°$	$37°$	$123°$	$43.158°$	$139.336°$
				$1°$	$60°$	$67°$		
Univ. of	7.7	4.2	10km	$9°$	$35°$	$97°$	$42.22°$	$138.49°$
Tokyo				$181°$	$55°$	$85°$		
Hokkaido	7.8	6.3	30km	$209°$	$43°$	$103°$	$42.789°$	$139.137°$
Univ.				$12°$	$49°$	$102°$		
JMA	7.8	6.3	37km	$206°$	$52°$	$105°$	$42.76°$	$139.32°$
				$3°$	$41°$	$72°$		
Kikuchi	7.8	1.94	8km	$198°$	$34°$	$80°$	$42.96°$	$139.49°$
		0.55	8km	$187°$	$65°$	$111°$	$42.30°$	$139.43°$
		2.29	8km	$192°$	$73°$	$102°$	$41.97°$	$139.43°$
		(Total) 4.78						

Table 2   Tsunami Arrival Time by Shuto and Matsutomi.

Hokkaido	Arrival Time	Remarks
SHAKOTAN TOWN		
Shakotan Point	20 min.	The highest first wave began with an ebb.
Kamui Point	5 – 6 min.	Ditto.
Numamae Point	25 – 35 min.	Ditto.
KAMOENAI VILLAGE		
Kawashiro Point	10 min.	
Sannai	10 min.	
Ryujin Point	5 –10 min.	
IWANAI TOWN		
Iwanai	15 min.	
Iwanai Tide Gauge	22:37	Began with a flood.
SHIMAMAKI VILLAGE		
Enoshima	5 min.	
SETANA TOWN		
Sukki	about 3 min.	The first wave was the highest.
Shimauta	less than 5 min.	Ditto. Began with an ebb. From the northwest.
Setana	5 min.	
Futoro	less than 5 min.	The first wave from the northwest. Begin with an ebb. The second or third was the highest.
TAISEI TOWN		
Ota	5 min.*	
Miyano	5 min.	The highest second wave at 22:27 or 22:28.
Hirahama	5 min.	Began with an ebb from the west. The second wave was the highest.
Esashi Tide Gauge	22:28	Began with an ebb.

Okushiri Island	Arrival Time	Remarks
WEST COAST		
Hoyaishikawa P.S.	22:23*	
Monai	22:21	
SOUTH COAST		
Aonae 5th	4 –5 min.	From the west.
Aonae 1st–4th	22:37, 22:38*	From the east.
Hamatsumae	22:22	

Data with * are from UJNR (1993).

Table 3   Method of Access To "Hokusai"

You can gain access to our database by ftp.

> ftp tsunami2.civil.tohoku.ac.jp
> or
> ftp 130.34.82.31

To log in, input 'anonymous' and your e-mail address.

Our database composition is as follows.

1) This directory includes bathymetry and topography data.

pub/Bathymetry/Hokkaido_SW/	D379-687-450m
	D154-124-150m
	D94-85-150m
	D61-49-150m
	D112-94-50m
	D40-202-50m
	Monai-5m
	Monai-20m

2) This directory includes initial profile data of the tsunami.

Initial_profile/1993_Hokkaido_SW/	I379-687-450m-17a
	I154-124-150m-17a
	I94-85-150m-17a
	I61-49-150m-17a
	I112-94-50m-17a
	I40-202-50m-17a

3) This directory includes observed runup height data of the tsunami.

Observation/1993_Hokkaido_SW/	Okushiri
	Hokkaido
	Honshu

4) This directory includes digitized tide records of the tsunami.

Tide_record/1993_Hokkaido_SW/	Rumoi etc.

You use cd to change directory and get command to get files.

> cd pub/Bathymetry/Hokkaido_SW
> get D379-687-450m

Table 4   List of Bathymetry Data in "Hokusai"

(directory : pub/Bathymetry/Hokkaido_SW)

REGION A

File Name	:	D379-687-450m
Space Increment	:	450m
Dimension	:	(379 , 687)
Limits	:	(138°30'E , 40°31'N) - (140°33'E , 43°18'N)
Format	:	10F7.1
Memory	:	1,900KB

REGION B1

	:	Half south region of Okushiri Island.
File Name	:	D154-124-150m
Space Increment	:	150m
Dimension	:	(154 , 124)
Limits	:	(139°23.22'E , 42°0.16'N) - (139°40'E , 42°10'N)
Format	:	10F7.1
Memory	:	140KB

REGION B2

	:	North-east region of Okushiri Island.
File Name	:	D94-85-150m
Space Increment	:	150m
Dimension	:	(94 , 85)
Limits	:	(139°30'E , 42°10'N) - (139°40'E , 42°17'N)
Format	:	10F7.1
Memory	:	60KB

REGION B3

	:	North-west region of Okushiri Island.
File Name	:	D61-49-150m
Space Increment	:	150m
Dimension	:	(61 , 49)
Limits	:	(139°23.22'E , 42°10'N) - (139°30'E , 42°14'N)
Format	:	10F7.1
Memory	:	20KB

REGION C1

	:	Aonae region of Okushiri Island.
File Name	:	D112-94-50m
Space Increment	:	50m
Dimension	:	(112 , 94)
Limits	:	(139°26'E , 42°2.5'N) - (139°30'E , 42°5'N)
Format	:	10F7.1
Memory	:	80KB

REGION C23

	:	Monai region of Okushiri Island.
File Name	:	D40-202-50m
Space Increment	:	50m
Dimension	:	(40 , 202)
Limits	:	(139°24.5'E , 42°4'N) - (139°26'E , 42°9.5'N)
Format	:	10F7.1
Memory	:	60KB

Table 5   Fault Parameters Determined by Hashimoto et al.

Sub Fault	North	Central	South
Width (Km)	30	30	25
Length (Km)	70	40	25
Strike (Deg)	195	170	130
Dip Angle (Deg)	20	30	55
Slip Angle (Deg)	90	80	50
Depth (Km)	10	5	0
Dislocation (m)	4	4	8
$M_0(\times 10^{27}$dyn $\cdot$ cm) (total:7.7)	3.4	1.9	2.4

Table 6   Fault Parameters of DCRC-17a

Sub Fault	North	Central	South
Width (Km)	25	25	25
Length (Km)	90	30	24.5
Strike (Deg)	188	175	163
Dip Angle (Deg)	35	60	60
Slip Angle (Deg)	80	105	105
Depth (Km)	10	5	5
Dislocation (m)	5.71	2.50	12.00
Origen($^\circ$)	139.40E 42.13N	139.25E 42.34N	139.30E 42.10N
$M_0(\times 10^{27}$dyn $\cdot$ cm) (total:6.62)	3.85	0.56	2.21

Table 7 Digitized Initial Profile of DCRC-17a Stored in "Hokusai"

(directory : pub/Initial_profile/1993_Hokkaido_SW)

REGION A
File Name	: I379-687-450m-17a
Space Increment	: 450m
Dimension	: (379, 687)
Limits	: (138°30'E, 40°31'N) - (140°33'E, 43°18'N)
Format	: 10F7.2
Memory	: 1,900KB

REGION B1 : Half south region of Okushiri Island.
File Name	: I154-124-150m-17a
Space Increment	: 150m
Dimension	: (154, 124)
Limits	: (139°23.22'E, 42°0.16'N) - (139°40'E, 42°10'N)
Format	: 10F7.2
Memory	: 140KB

REGION B2 : North-east region of Okushiri Island.
File Name	: I94-85-150m-17a
Space Increment	: 150m
Dimension	: (94, 85)
Limits	: (139°30'E, 42°10'N) - (139°40'E, 42°17'N)
Format	: 10F7.2
Memory	: 60KB

REGION B3 : North-west region of Okushiri Island.
File Name	: I61-49-150m-17a
Space Increment	: 150m
Dimension	: (61, 49)
Limits	: (139°23.22'E, 42°10'N) - (139°30'E, 42°14'N)
Format	: 10F7.2
Memory	: 20KB

REGION C1 : Aonae region of Okushiri Island.
File Name	: D112-94-50m-17a
Space Increment	: 50m
Dimension	: (112, 94)
Limits	: (139°26'E, 42°2.5'N) - (139°30'E, 42°5'N)
Format	: 10F7.2
Memory	: 80KB

REGION C23 : Monai region of Okushiri Island.
File Name	: I40-202-50m-17a
Space Increment	: 50m
Dimension	: (40, 202)
Limits	: (139°24.5'E, 42°4'N) - (139°26'E, 42°9.5'N)
Format	: 10F7.2
Memory	: 60KB

Table 8 Digitized Tide Records Stored in "Hokusai"

(directory : pub/Tide_records/Hokkaido_SW)

Hokkaido Prefecture
Rumoi
Otaru
Iwanai
Esashi
Ishikari
Yoshioka
Hakodate
Tomakomai
Urakawa
Muroran

Aomori Prefecture
Tappi
Aomori
Ohminato
Fukaura
Noshiro

Akita Prefecture
Funakawa
Akita

Yamagata Prefecture
Sakata

Niigata Prefecture
Kurishima
Iwabune
Niigata
Kashiwazaki
Teradomari
Ryoutsu

Toyama Prefecture
Toyama
Toyama port

Fukui Prefecture
Fukui

Kyoto Prefecture
Maizuru

Osaka Prefecture
Sakai

Fukuoka Prefecture
Hakata

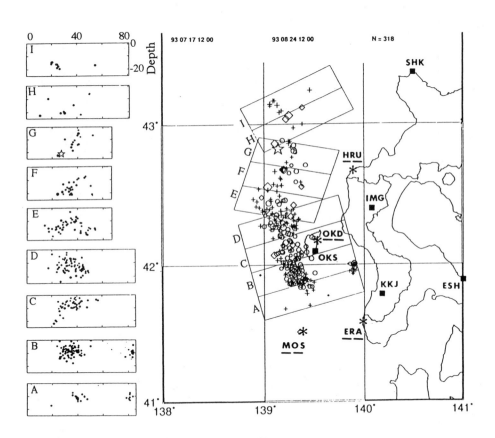

Figure 1   Aftershock Distribution by Kodaira et al.

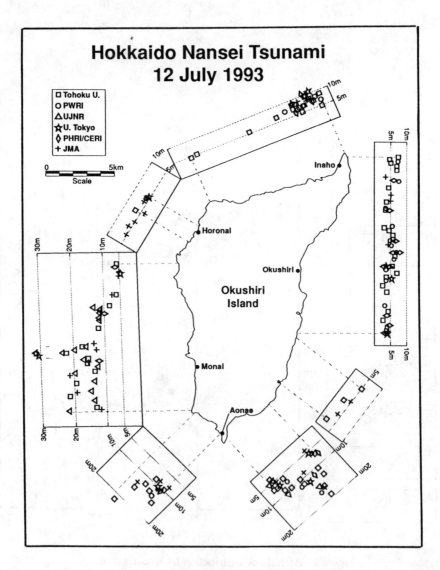

Figure 2   Runup Distribution in Okushiri Island (Hokkaido Tsunami Survey Group).

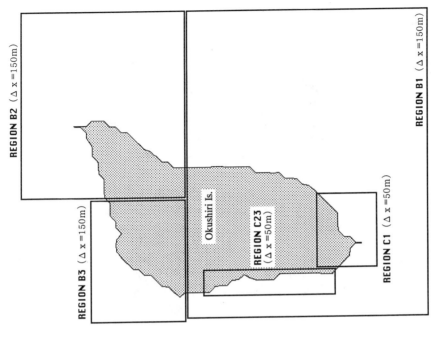

Figure 3(b) Areas of Bathymetry Data Digitized by DCRC and Stored in "Hokusai"

Figure 3(a) Areas of Bathymetry Data Digitized by DCRC and Stored in "Hokusai"

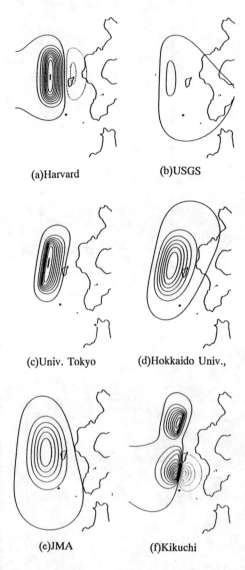

(a)Harvard   (b)USGS

(c)Univ. Tokyo   (d)Hokkaido Univ.,

(e)JMA   (f)Kikuchi

Figure 4   Initial Profiles with Fault Parameters Determined Based Upon Seismic Data.

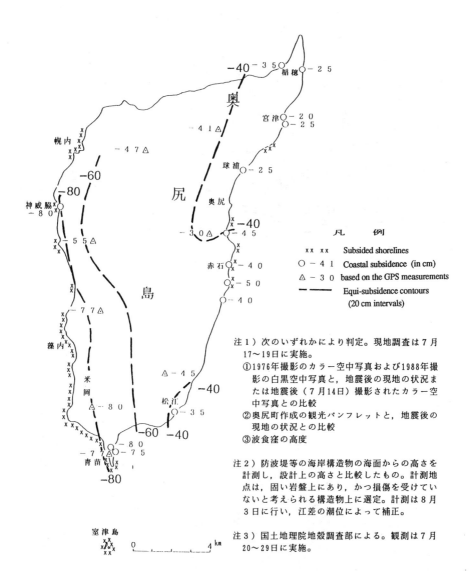

注1）次のいずれかにより判定。現地調査は7月
17〜19日に実施。
①1976年撮影のカラー空中写真および1988年撮
影の白黒空中写真と，地震後の現地の状況ま
たは地震後（7月14日）撮影されたカラー空
中写真との比較
②奥尻町作成の観光パンフレットと，地震後の
現地の状況との比較
③波食窪の高度

注2）防波堤等の海岸構造物の海面からの高さを
計測し，設計上の高さと比較したもの。計測地
点は，固い岩盤上にあり，かつ損傷を受けてい
ないと考えられる構造物上に選定。計測は8月
3日に行い，江差の潮位によって補正。

注3）国土地理院地殻調査部による。観測は7月
20〜29日に実施。

Figure 5   Subsidence Measured by Kumaki et al.

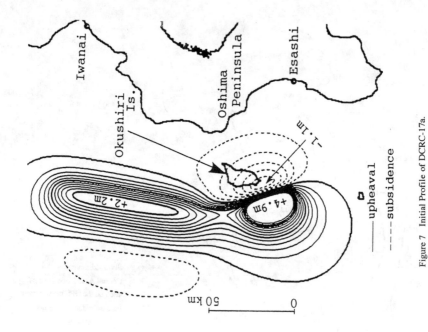

Figure 7   Initial Profile of DCRC-17a.

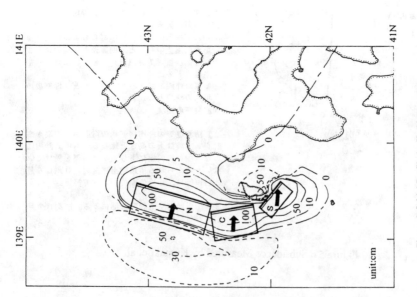

Figure 6   Initial Profile by Hashimoto et al.

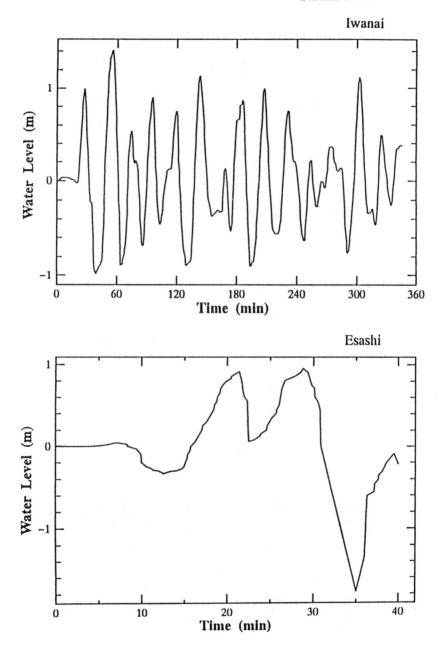

Figure 8   Tide Records at Iwanai and Esashi.

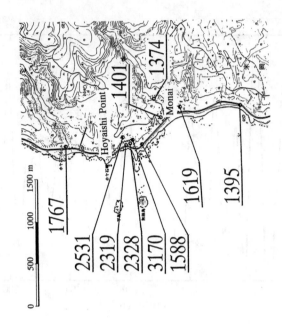

Figure 10   Detailed Map of Runup Distribution North of Monai.

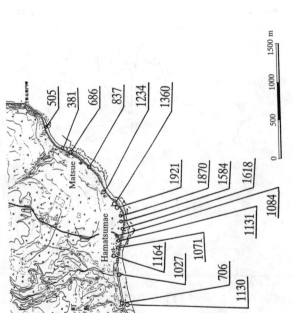

Figure 9   Detailed Map of Runup Distribution at Hamatsumae.

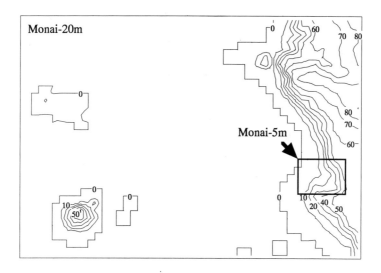

Figure 11(a) Detailed Map of the Small Notch
Where the Maximum Runup was Measured.

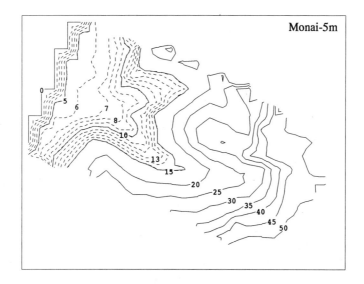

Figure 11(b) Detailed Map of the Small Notch
Where the Maximum Runup was Measured.